RADIATION EFFECTS ON EMBEDDED SYSTEMS

Radiation Effects on Embedded Systems

edited by

RAOUL VELAZCO
TIMA Laboratory, Grenoble, France

PASCAL FOUILLAT
Université Bordeaux 1, France

and

RICARDO REIS
Universidade Federal do Rio Grande do Sul, Porto Alegre, Brazil

Springer

A C.I.P. Catalogue record for this book is available from the Library of Congress.

ISBN-10 1-4020-5645-1 (HB)
ISBN-13 978-1-4020-5645-1 (HB)
ISBN-10 1-4020-5646-X (e-book)
ISBN-13 978-1-4020-5646-8 (e-book)

Published by Springer,
P.O. Box 17, 3300 AA Dordrecht, The Netherlands.

www.springer.com

Printed on acid-free paper

CONTENTS

PREFACE

Since the loss of Telestar satellite in 1962, as a consequence of a high altitude nuclear test, it is known that natural or man-made radiation may perturb the operation of electronic equipments. Today, space vehicles highly rely on electronics and consequently space radiation effects must be considered at the design phase to guarantee the high-reliability and safety requirements of such projects. Even if space-qualified devices exist, so-called "hardened" with respect to radiation effects at the design and/or the manufacturing level, their high cost and lower performances compared with equivalent commercially available devices (Commercial Off The Shelf, COTS) make mass production components, not designed on purpose for space use, quasi mandatory for space applications. This constitutes a major challenge for the feasibility and success of future missions: on one hand it must be understood as far as possible both the nature and variability of space environment, on the other hand the consequences of this environment on electronic devices has to be evaluated taking into account the constant evolution of the technology and the large scope of device's types.

Moreover, with the constant progress achieved in the manufacturing technologies, nano-electronics devices have features (transistor's dimensions, operating frequencies) which make them potentially sensitive to the particles present in the Earth's atmosphere, even at ground level. The very large presence of integrated circuits in all kind of applications result in a non negligible probability of transient errors, gathered under the acronym 'SEE' (Single Event Effects) resulting from the ionization generated by secondary particles issued from the nuclear reactions induced by the impact of atmospheric neutrons with atoms present in the silicon. This makes mandatory, for those applications requiring high reliability/safety, of considering this problem at the early application's design phase to include the needed hardware/software fault tolerance mechanisms.

The estimation of the sensitivity to radiation of a device or architecture is a mandatory step to get an insight about the impact of radiation on the level of reliability/safety. Such a step requires both radiation ground testing, so-called accelerated tests, performed in radiation facilities such as particle accelerators or laser equipment and software tools modeling the applications environment. Such estimation, particularly when they deal with the Single Event Upset (SEU) phenomena, which results in the perturbation of the content of memory cells, may also require the realization of fault injection campaigns in which the consequences of radiation effects are taken into account at a level for which suitable description of the tackled device are available. The realization and exploitation of the results issued from such off-beam fault injection sessions constitute a source of precious information concerning both the estimation of the sensitivity to SEEs of the studied application and the identification of potential "week points" of a design or architecture supposed to be hardened.

This book aims at providing the reader with the major guidelines for coping with radiation effects on components supposed to be included in today's application devoted to operate in space, but also in the atmosphere at high altitude or at ground level. It contains a set of chapters based on the tutorials presented at the International School on Effects of Radiation on Embedded Systems for Space Applications (SERESSA) that was held in Manaus, Brazil, from 20 to 25 November 2005. The book is organized in twelve chapters, the first three dealing respectively with the analysis and modeling of the Space Radiation Environment, the basic mechanisms related with the effects of radiation on electronic devices, a description of a set of known cases of in-flight anomalies on electronic devices due to cumulated or transient effects of radiation.

The following three chapters are devoted to the evaluation of multi-level faults effects and the effects of radiation on analog and mixed-signal circuits that are translated into Single Event Transients (SETs) and Single Event Upsets (SEUs) and the fundamentals of the pulsed laser technique for Single-Event Upset simulation.

The following three chapters deal with mitigation techniques of radiation effects: Hardening By Design" (HBD) methodologies used in CMOS technologies to protect the circuit from radiation effects, the radiation effects on FPGAs and related techniques allowing to mitigate them, and research and development of automatic tools for design hardening.

Finally, the last three chapters are devoted to the description facilities for SEE and dose testing, test methodologies and tools to handle error rate prediction of digital architectures and the possibilities of a pulsed laser system for studying radiation-induced single-event transients in integrated circuits. Three case studies are presented to illustrate the benefits of the spatial and temporal resolution of this technique.

Raoul VELAZCO, Pascal FOUILLAT, Ricardo REIS

April 2007

Radiation Space Environment

Jean-Claude Boudenot

THALES Research & Technology, RD 128, 91767 Palaiseau Cedex (France)
(jean-claude.boudenot@thalesgroup.com)

Abstract. Radiation belts, solar flares and cosmic rays are at the origin of the space radioactive environment. Electrons and protons of the radiation belts as well as protons from the Sun coronal mass ejection lead to total dose effects on electronic devices. Cosmic rays and heavy ions solar flares are responsible for heavy ions effects. In the first part of this paper, we will review these radiation effects. In the second part of this paper, we will address the effects of the atomic oxygen which can affect LEO satellites, the effects of solar U.V, as well as the micrometeoroid effects and finally the man made space debris.

1 Space radiation effects

1.1 Space radiation environment: Van Allen belts, solar flares, solar wind, cosmic rays

The Earth and its immediate environment are protected by the atmosphere, which acts as a semi-permeable "screen", to let throughout light and heat, while stopping radiation and UVs. Because so such natural protection is available in space, human beings and electronics devices must be able to cope with the resulting set of constraints.

Our analysis is limited here to the main components of radiative phenomena encountered in space, by classifying them into four categories by origin: Radiation belts, solar flares, solar wind and cosmic rays. Though somewhat arbitrary (the four types of phenomena are interrelated), this breakdown is best suited to subsequent study of the resulting effects. Note that the particles of interest here are essentially electrons, protons and heavy ions of various origins, with differing energies. Electrons and protons are responsible for total dose; heavy ions and protons cause a number of specific effects that are grouped under the heading "SEE" (Single Event Effect).

Radiation belts. Radiation belts contain trapped electrons and protons. This trapped radiation includes two electron belts. The inner belt contains electrons whose energy is less than 5 MeV. The outer belt contains electrons whose energy may reach 7 MeV, furthermore in the case of outer belt, the electron flux is both more variable and more intense than that of the inner belt. A third belt of electrons was observed after a

R. Velazco et al. (eds.), Radiation Effects on Embedded Systems, 1–9.

magnetic storm on March 24, 1991. This belt is located halfway between the first and second belts and its trapped electron energies are significant, reaching as much as 30 MeV.

Trapped radiation also includes an inner proton belt structure. A second such belt, containing high-energy protons (>100 MeV), appeared after the same magnetic storm mentioned above. A violent magnetic storm can therefore generate new radiation belts whose life expectancy, estimated at more than two years is not precisely known.

Like electrons and protons, heavy ions may also be trapped in the magnetosphere. An abnormal situation known as ACR or Anomalous Cosmic Ray has long been identified. In 1991, Grigorou postulated that ion with energies of less than 50 MeV per nucleon could be trapped. This result was confirmed in 1993 by measurements made by the SAMPEX satellite. Ion belts are primarily made up of light ions (He, C, N, O, Ne, etc.) with low energies. These ions are ionised once only, are very sensitive to solar modulation and exhibit low penetration. They have no impact on electronics, but contribute to the radiation exposure sustained by astronauts: their high biological efficiency factor has significant impact on equivalent dose.

Solar flares. The 11-year sunspot cycle can be roughly subdivided into four low activity years and seven of high activity, punctuated by ordinary events and those of an exceptional nature. There are two types of events to be considered in relation to the radiation environments being considered here. The first is represented by *coronal mass ejection's*, which last several days and emit mainly high-energy protons (up to several hundred MeV). The "Benchmark" here is the proton eruption that took place in August 1972. This single event provided 84% of the solar protons with energies of more than 30 MeV that were inventoried at the time of the 20th solar cycle. The second types of event fall under the transient or "Impulsive Events" category involving large emissions of heavy ions. These include *solar flares* with high-energy ion emissions (several tens of MeV to several hundred GeV per nucleon) that are often ionised only once and vary in composition from one event to another. The references here are the heavy ion flares of September 1977 and October 24, 1989.

Solar wind. The high temperature of the Sun corona (# 2 millions K) inputs sufficient energy to allow electrons to escape the gravitational pull of the sun. The effect of the electron ejection's is a charge imbalance resulting in the ejection of protons and heavier ions from the corona. The ejected gas is so hot that the particles are homogenised into a dilute plasma. The energy density of the plasma exceeds that of its magnetic field so the solar magnetic field is "frozen" into the plasma. This electrically neutral plasma streams radially outward from the sun at a velocity of approximately 300 to 900 km/s with a temperature on the order of 10^4 to 10^6 K. The energies of the particles range from approximately 0,5 to 2 keV per nucleon. The average density of the solar wind is 1 to 30 particles / cm^3. The composition of solar wind is approximately : 95% p^+ ; 4% He^{++} ; < 1% other heavy ions ; and the number of electron needed to make solar wind neutral.

Major perturbations in the geomagnetic field can occur with changes in the solar wind density (e.g. solar flares), the solar wind velocity (e.g. coronal mass ejection's), and the orientation of the embedded solar magnetic field. The coronal mass ejection's and solar flares cause disturbances of the solar wind, and it is the interaction between theses disturbances and the earth's magnetosphere that causes perturbations called magnetic storms and substorms. The correlation of the number of storm with the level of solar activity is very strong, and the major magnetic storms are closely associated with coronal mass ejection's. During the period of high solar activity, fluctuations in the sun's magnetic field are observed, and these fluctuations result in a compression of the earth's magnetic field lines (geomagnetic storms). When the compression of the earth's magnetic field lines occurs, plasma on the night side of the earth is pushed towards the earth surface. As this plasma pushes closer to the earth, the electrons and ions are deflected by the earth magnetic field and, as a result, spacecraft orbiting between local midnight and 6 a.m will see an abundance of energetic electrons. Consequently, during geomagnetic storms the greatest concern is for spacecraft operating between midnight and 6 a.m.

While at GEO the plasma is quite hot (# 2 keV for electrons and 10 keV from ions on the average) and with a low density (# 10 to 100 cm^{-3}) in normal condition, and hot (# 10 keV for electrons and 14 keV from ions on the average) but with a lower density (# 1 cm^{-3}) during magnetic storms, the plasma founded at LEO, is colder and incapable of inducing significant charging. However, because energetic particles may move along the magnetic field lines, spacecraft in low altitude polar orbits may encounter the more energetic plasma that is seen to originate at higher altitudes. In situ observations confirm that auroral electrons can be accelerated to several kilovolts, producing a plasma environment capable of more severe charging. This energetic plasma is confined to an annular region near the poles, in the region where the magnetic field lines enter the lower altitudes. Because a spacecraft will only pass through this region periodically during the course of its orbit, charging in the auroral regions is typically of very short duration. Severe charging is more likely when the ambient plasma density is lower because the presence of the low energy ambient plasma acts as a source of neutralising current.

Solar wind: effects and mitigation. Differences in the emission and absorption characteristics of materials, differences in sunlight exposure, and localised effects resulting in unequal electron populations produce voltage differences between insulated satellite surfaces (this phenomenon is called differential surface charging). Furthermore, electrons having sufficient energy to pass through the thermal blanket result in internal charging of surfaces and assemblies. Typical insulated objects include cable jackets, ungrounded thermal wrap, thermal paint, component encapsulants, etc. Higher energy electrons penetrating subsystem chassis assemblies may deposit charge onto circuit board and wire insulators, connectors, capacitors, etc. In this process, termed deep dielectric charging, high-energy electrons penetrate circuit elements and devices, leading to trapped charge build-up within the dielectric material.

Mitigation's included the use of filters to prevent the propagation of discharge event created signal transients, the use of surface coatings (paint, etc.) and materials which provide dissipation of deposited charge. Shielding of electronic assemblies to reduce the flux of high-energy electrons impinging on sensitive devices and materials is also recommended to address charging effects.

Cosmic rays. Cosmic rays are highly energetic heavy ion nuclei (without the surrounding neutron cloud). In actual fact, these heavy ions only represent 1% of the nucleonic component of cosmic radiation, which otherwise contains 83% protons, 13% helium nuclei and 3% electrons. The origin of this radiation has not been truly identified; however, we know that part of it (i.e. the most energetic ions) comes from outside the Milky Way Galaxy and the rest from inside it. Ion energies are very high (the most energetic ion ever detected had an energy of 3.10^{20} eV, i.e. nearly 50 J !) and the corresponding acceleration mechanisms are not always well understood. Cosmic radiation is nearly isotropic when it arrives in the vicinity of the magnetosphere. However, because the radiation couples to the Earth's magnetic field, its isotropy is not preserved. While its composition is nearly identical to that of matter found in the local galaxy, it does seem to be affected by interaction with interstellar matter. At energies of less than 1 GeV per nucleon, particle flux depends on solar activity.

1.2 Dose effects: origins, effects on electronic devices, order of magnitude

The total doses sustained in space environments are almost exclusively attributable to trapped particles contained in radiation belts and to protons emitted by solar flares.

Total Ionising Dose Evaluation. To evaluate the Total Ionising Dose (TID) on a component, one have to use the "dose profile curve" that indicates the dose received through a shield of varying thickness (most often a hollow aluminium sphere). This curve is often used as a specification, since it depends only on the mission of interest. Then, according to the exact location of the component considered in the satellite, and the various shields that protect it from space (satellite insulation, printed circuit card, housing component, etc.) the TID is computed. One can use for that two methods. The first is an analytical method based on "sectoral analysis", i.e. weighting of the dose profile curve. The second makes use of a Monte-Carlo approach.

Influence of the orbit. At Low Earth Orbits (the altitude of LEO satellite is in the range of 300-5000 km), the average space distribution of particles is inhomogeneous: the outer electron radiation belt is close to the Earth at high latitudes (polar horns) and the region centred on the south Atlantic has a high level of trapped particles (electrons and protons). This means that:

- A satellite placed in very low equatorial orbit (300 km) sustains little radiation,
- A satellite in low orbit at an inclination of less than 45° is subjected to the SAA,
- A satellite in low polar orbit at an inclination greater than 55° (e.g. heliosynchronous satellite) is subjected both to the SAA and to the impact of the polar horns,
- A satellite placed at altitude over 1400 km (e.g. constellation satellite) is heavily impacted by dose effects. The proton belt makes a further contribution that sometimes leads to a total dose greater than that received in geostationary orbit.

At GEostationary Orbit (GEO at 36000 km of altitude) and Medium Earth Orbit (the altitude of MEO satellite is in the range of 5000-36000 km) the main source of dose radiation is due to the outer electron belt. As an example, for a 18 years GEO, TID is 100 krad behind 5 mm of aluminium, and 10 krad behind 10 mm. In this kind of orbit it is possible to use some local shielding, like shielded box, because the efficiency of shielding with respect to electron is quite good. (As a comparison a TID received by a satellite placed at 2000 km of altitude is, for 5 years and behind 10 mm aluminium shielding, in the range of 300 krad)

Dose effects: order of magnitude. The TID that electronic devices can withstand depends on the technology. Standard CMOS COTS is in the range of 1 to 10's of krad, while CMOS rad hard devices are in the range of 100 krad to 1 Mrad. Standard bipolar is better than standard CMOS and can withstand a TID in the range of 10's to 100 krad. AsGa is intrinsically TID hardened and can withstand a TID of 1 Mrad or even more. But one have to be very careful on these orders of magnitude, because there are a lot of dependence factors on TID effects, like: 1) Dose rate (there is a low dose rate enhancing for bipolar technologies contrary to high dose rate enhancing for MOS technologies); 2) Bias during and after irradiation; 3) Time after irradiation (annealing or rebound) ; 4) Lot to lot dependence ; and so on.

1.3 Displacement effects: origin, effects on electronic devices, order of magnitude

Low earth orbits higher than 1400 km are also impacted by the effects of atomic displacement due to trapped proton. This effect, which is a familiar in military situations (due to neutrons) was, up until now, more or less neglected in space applications. Because new orbits are increasingly located in the proton belt, the space industry has now to take into account proton-induced displacements and their inclusion in radiation analysis. The displacement effect is quantitatively measured by its Non Ionising Energy Loss (or NIEL) per opposition to ionising loss measured by the dose deposition (note that proton, which is both charged and massive particle, has the ability to induce both dose and displacement effect).

The order of magnitude of displacement effects are as follow: 10^{11} n(eq 1 MeV).cm^{-2} for CCD & optolinks ; 10^{12} n(eq 1 MeV).cm^{-2} for bipolar ; 10^{14} n(eq 1 MeV).cm^{-2} for MOS ; and 10^{15} n(eq 1 MeV).cm^{-2} for AsGa.

As an example the level of displacement damage is between 10^{12} n/cm^2 to 310^{12}/cm^2 for a LEO orbit which altitude is comprised between 1400 km and 2000 km. This level is almost independent of the thickness of the shielding. So for this kind of orbit the problem of displacement effects has to be taken into account mostly for high analogic devices such as CCD or optolinks.

1.4 Heavy ion effects: origin, effects on electronic devices, order of magnitude

When a heavy ion pass through matter it goes straight a line. The more it is heavy, the more the ionisation along ii pass is important. In fact we measure the effect of heavy ions by its Linear Energy Transfer, that is the ionising energy it loss per unit length (roughly the LET increases with the Z of the ion). It is possible to compute the LET from various ions from various energy, by doing so, we have to notice that the maximum LET it is in the order of magnitude of 100 MeV cm^2 mg^{-1}. When an ion pass trough the active volume of an electronic device it deposes a charge along it's trace and this charge is collected per the electric field of the device. The associated "iono-current" can induced several effects such as :

- SEU (Single Event Upset) which is a transient effect, affecting mainly memories,
- SEL (Single Event Latch-up), which can destroy the component, affecting mainly CMOS structure
- SEB (Single Event Burnout), which has destructive impact; affecting mainly power MOSFET

- SEGR (Single Event Gate Rupture), which is also potentially destructive, affecting mainly submicronic structure,
- SHE (Single Hard Error), yet another destructive effect.

Two parameters are needed to quantify the vulnerability of an electronic device to heavy ion. The first one is the threshold LET, the second one is the cross section. If the LET is greater than a threshold, the energy deposition can triggered the effect. Furthermore the device offer a cross section which represents the probability for an ion to hit a sensitive volume of the component. The higher the cross section is, the more sensitive the device is. From technology point of view all kind of technology may be sensitive to SEE, but the larger the active volume the more sensitive the device. This is the reason why the bipolar technology is more sensitive than the MOS bulk, and the MOS bulk is more sensitive than the MOS SOI technology. Increasing use of electronics in onboard systems and enhanced circuit integration have revolutionised things to such an extent that, today, allowance for heavy ions is of vital importance in selecting components.

The environment itself is characterised by a LET spectra : that is to say the number of ion which LET is greater than a given LET. GEO orbit corresponds to the maximal constraint because it doesn't benefit of magnetosphere shielding. As the altitude of the orbit decreases, and as the inclination of the orbit decreases to, the magnetosphere shielding is more and more effective and the flux of cosmic rays decreases. Due to this shielding effect, one can have several orders of magnitude differences between orbit (e.g. GEO and low altitude, low inclination LEO).

1.5 Proton Effects: Origin (direct & indirect), effects on electronic devices, order of magnitude

The first observation of single event upset induced per proton was made 10 years ago, in 1990. As for heavy ion we can distinguish non destructive effects, like SEU, and destructive effects, like SEL or SEB. Furthermore we have to distinguish indirect effects, due to the interaction between incident proton and a nuclei of the component (spaliation reaction), and direct effect due to ionisation induce per the proton inside a sensitive volume of the device.

Based on the critical energies currently encountered in electronic components, it is clear that direct upsets are only exceptionally produced by protons. By contrast, nuclear reaction of these particles with silicon is possible and can lead to recoil of the heavy residual nucleus (together with emission of lighter fragments) or to formation of two ions of similar masses, by fragmentation of the silicon nucleus. These "secondary" ions then cause the indirect SEE.

The three main sources of indirect heavy ion events are:

- Proton-emitting solar flares, for geostationary and low polar orbits,
- Trapped protons for medium orbits (MEOs),
- The SAA for low earth orbits.

As for heavy ions, the magnetosphere offers a natural screen against protons. Its degree of "screen effect" depends on the type of orbit and the date of the mission. This effect is weak for geostationary and highly inclined low orbits (polar areas) but very strong for low orbits with small inclinations. Moreover, proton flux, as heavy ion flux, is weakest during periods of maximum solar activity, since the concomitant increase in interplanetary magnetic fields accelerates scattering of ions before the latter reach the magnetosphere. Note that the relative significance of heavy ion-induced direct effects and indirect effects depends on the type of orbit and, for a given orbit, on the type of component considered.

2 Other effects

2.1 Atomic oxygen: origin and effects

The Atomic oxygen is the main specie of the atmosphere from 200 km; the associated density is in the range of 10^9 – 10^{10} atoms/cm^3 at 200 km, and in the range of 10^5 atoms/cm^3 at 800 km. The density varies with solar activity (e.g. 10^4 (minimum) to 10^8 (maximum) atoms/cm^3 at 800 km). The oxidation power of atomic oxygen is enhanced by it relative velocity of 8 km/s and it's temperature (800 K equivalent to 5 eV energy).

The effects of atomic oxygen are various:

- Material erosion. The reaction efficiency depend on the material :
  - very low for Al or Au
  - high for Kapton (# 3 10^{-24} cm^3 / atom) : 500 μm for ISSA for 30 years
  - very high for silver (# 10.5 10^{-24} cm^3 / atom)
- Electrical interconnects oxidation
- Mirror reflectivity decrease
- Thermal parameters (α_s ; ε) evolution
- Luminescence

Mitigation provisions for atomic oxygen environment effects include the choice of altitude which reduces the exposure, choice of materials which are resistant to chemical degradation and exhibiting high sputtering threshold, the use of coatings to protect surfaces, orienting sensitive surfaces and devices away from the ram direction, and reducing aerodynamic drag by reducing the space vehicle cross section.

2.2 Sun U.V: origin and effects

The Sun spectra is a black body with temperature of 5600 K. In absence of atmosphere there is no U.V filter (ozone layer filters $\lambda < 0.3$ μm) and so spacecraft has to withstands U.V ray. Due to high energy of U.V (e.g. 9.2 eV for 0.13 μm ; 3.2 eV for 0.39 μm) chemical bonds may be broken. As an example the energy needed to break some classical chemical link is list here:

- C≡C (0.14 μm) ; C≡N (0.13 μm) ; C≡O (0.16 μm)
- C=C (0.20 μm) ; C=N (0.19 μm) ; C=O (0.16 μm)
- C-C (0.36 μm) ; C-N (0.39 μm) ; C-O (0.33 μm)

The associated effects included:
- Fibber degradation
- Optical darkening (creation of "colour centres")
- Thermal parameters (α_s ; ε) evolution ($\Delta\alpha_s = 0.01$ for many material during typical spacecraft lifetime)
- Weakening.

2.3 Micrometeoroids: origin and effects

The origin of micrometeoroids is twofold 1) a continuous background (omnidirectional and sporadic) ; 2) a meteor shower (directional and periodic) like Perseid and Leonid meteor shower and many others. Typical characteristics of such a micrometeoroids are a mass in the range of 10^{-10} g to 1 g, a density comprised between 0.5 to 2 $g.cm^{-3}$ and a velocity in the range of 10 to 70 km/s (with an average of 17 km/s). The micrometeoroids effects included:

- Erosion of surface materials
- Changes in thermal control properties : $\Delta\alpha_s$; $\Delta\varepsilon$
- Contamination of sensitive surfaces
- Tank perforation hazard
- These effects are very dependant on diameter :
    - 0.1 mm $\Rightarrow$ erosion
    - 1 mm $\Rightarrow$ serious damage
    - 3 mm particle moving at 10 km/s carries the kinetic energy of a bowling ball moving at 100 km/h
    - 1 cm particle has a kinetic energy of a 180 kg safe

2.4 Orbital debris: origin and effects

Due to their origin, orbital debris is present in the most often used orbits, and will impact spacecraft mainly in the ram direction, with velocities which are less than micrometeoroids. The orbit debris population is influenced by the solar cycle through increased aerodynamics drag effects.

- The source of orbital debris are mainly :
    - Non operational spacecraft, boost vehicles, spacecraft explosion
    - Break-ups, collisions
    - Solid rocket fuel particulates, surface erosion particulates

The number of these orbital debris is continuously increasing:

- 30.000 to 100.000 pieces of debris > 1 cm
- 20.000 pieces of debris > 4 cm
- 7000 pieces of debris > 10 cm

Due to mass and velocity (0 to 16 km/s with an average of 11 km/s) the orbital debris has a significant effects as:

- Tank perforation hazard
- Changes in thermal control properties : $\Delta\alpha_s$; $\Delta\varepsilon$
- Contamination of sensitive surfaces
- Erosion of surface material

Mitigation provisions for the micrometeoroid and orbital debris threat is to choose altitude and inclination orbit architecture with minimal presence of orbit debris, locate critical devices and structures away from the ram direction, and the use of a layered "bumper" structure to protect critical elements.

References

Concerning radiation effects there are two major conferences, both with short course and proceeding published by IEEE/NS. These conferences (NSREC = Nuclear and Space Radiation Conference, and RADECS = RADiation Effets sur les Composants et Systèmes), constitute an immense bibliographical data base. In addition some review books exist on this field:

[1] Alan Tribble : The space environment. Implications for spacecraft design., Princeton University Press (1995)
[2] Andrew Holmes-Siedle and Len Adams : Handbook of radiation effects., Oxford science publications (1993)
[3] N.J. Rudie : Principles and techniques of radiation hardening., Western Periodicals Company (1986)
[4] Jean-Claude Boudenot : L'environnement spatial., Coll. Que sais-je? 2d Edn. PUF (1996)
[5] Jean-Claude Boudenot & Gérard Labaune : Compatibilité électromagnétique et nucléaire, Ed. Ellipses (1998)

Radiation Effects in Microelectronics

R. D. Schrimpf
Institute for Space and Defense Electronics
Vanderbilt University
Nashville, TN 37235
USA
ron.schrimpf@vanderbilt.edu

Abstract. Understanding the effects of radiation on electronic devices and circuits is particularly important for space applications because the electronics may be exposed to a variety of energetic particles and photons. The resulting effects may be manifested as long-term parametric degradation or as transient changes in the state of the circuits. This paper presents an overview of these effects, emphasizing those device-level effects that are particularly relevant for space environments. MOS and bipolar technologies are considered. A new simulation method for analyzing single-event effects, based on detailed descriptions of a large number of individual events, is described. This method has the potential to provide more accurate analysis than conventional methods based on simulation of the device response to an average event.

1. Introduction

The combined effects of advances in microelectronic materials and device structures have resulted in more changes to underlying integrated-circuit technologies over the past five years than in the previous forty years. Some of these changes are still in research labs, but many of them now are beginning to appear in mainstream products. These changes have profound implications for radiation hardness. Energy absorption, carrier generation, carrier transport, charge trapping and defect formation and dynamics depend on the specific materials used in the ICs. Sensitivity to the electrostatic effects of radiation-induced trapped charge, lifetime degradation, and device-edge and inter-device leakage depend on the detailed device geometries and doping profiles. Moreover, high-speed circuits exhibit increased vulnerabilities to single-event effects, including multiple-bit upsets that result from aggressive scaling, and large enhancements relative to ion-strike angle that are much greater than for earlier generations of technology. While it was possible to study total dose and single event effects separately in larger devices, the boundary is now increasingly less clear as a single event may induce charging or damage in the entire device. This paper reviews the mechanisms responsible for radiation effects in advanced technologies. Two major categories of effects are considered: long-term and transient.

R. Velazco et al. (eds.), *Radiation Effects on Embedded Systems*, 11–29.

1.1 Long-term effects

Exposure to radiation produces relatively stable, long-term changes in device and circuit characteristics that may result in parametric degradation or functional failure. The total ionizing dose primarily impacts insulating layers, which may trap charge or exhibit interfacial changes. Non-ionizing energy loss results in displacement damage and defects in both insulator and semiconductor regions. In older technologies, these effects were well described by a spatially uniform representation of the cumulative amount of energy deposited. The accuracy of this description relies on the relatively large size of the devices to average the energy deposited by individual particles or photons; in very small devices (less than approximately 130 nm), this approach is no longer valid.

Oxide trapped charge (N_{ot}) refers to radiation-induced charges, typically net positive, that are relatively stable. In ultrathin, high quality gate oxides, effects of oxide-trapped charge are minimal because of the small volume in which charge is generated and the ease with which it can tunnel from the oxide. However, high-κ dielectrics are currently more susceptible to ionizing radiation than thermal oxides of comparable effective thickness [1-3]. In state-of-the-art MOS integrated circuits (ICs), field oxides and isolation structures are usually less radiation-tolerant than the active device regions [4]. Ionizing radiation also results in formation of interface traps (N_{it}) at semiconductor/insulator boundaries that exchange charge with the semiconductor on relatively short time scales. In MOSFETs, interface traps stretch out the subthreshold *I-V* characteristics and reduce the inversion-layer mobility. In BJTs, the current gain decreases with total dose due to increased surface recombination caused by interface-trap formation [5]. Border traps are defects that are similar to oxide traps in microstructure, but electrically behave like slow interface traps [6].

The non-ionizing energy deposited by particle irradiation displaces atoms and creates electrically active defects. These defects reduce carrier lifetimes and mobilities, change carrier densities, and increase non-radiative transitions in optical devices, among other effects. Minority-carrier devices are particularly susceptible to displacement damage.

1.2 Transient effects

While the total-dose hardness of commercial ICs has generally improved in recent years, primarily because of reductions in gate oxide thicknesses and increases in doping densities, reduced device dimensions and accompanying technological changes have resulted in increased sensitivity to transient radiation effects [7]. Transient effects can be caused by individual ionizing particles (single-event effects) or high dose-rate ionizing radiation ($\dot{\gamma}$ radiation).

Single-event effects (SEEs) are a serious problem for electronics operated in space and they are becoming an issue for advanced technologies in avionics, and even at sea level. The charge deposited by a single ionizing particle can produce a wide range of effects, including single-event upset, single-event transients, single-event functional interrupt, single-event latchup, single-event dielectric rupture, and others. In general, the sensitivity of a technology to SEE increases as device dimensions decrease and as circuit speed increases [8]. These effects can be produced by direct ionization or by secondary particles resulting from nuclear reactions or elastic collisions. Recent experimental results from heavy ion and proton irradiations of advanced devices have demonstrated unpredictable SEE responses (e.g., [9]).

In a high dose-rate environment, energy is generated relatively uniformly throughout the IC. The resulting photocurrents produce effects that include rail-span collapse, cell upset, and burnout of metal lines [10]. Depending on system requirements, it may be necessary to operate through a dose-rate event or it may be possible to circumvent it by temporarily removing power.

2. MOS devices

2.1 Threshold-voltage shifts

Historically, the dominant effect of ionizing radiation on MOS devices has been a shift in the threshold voltage caused by the electrostatic effect of trapped charge and charged interface traps. While radiation-induced threshold voltage is a less significant concern in advanced MOS technologies with thin gate oxides for reasons that will be discussed below, the same physical mechanisms affect isolation oxides and may result in large leakage currents.

The net charge trapped in the oxide is usually positive, while the charge state of the interface traps depends on the surface potential and the physical nature of the defect responsible for the interface trap. The electrostatic effects of oxide trapped charge (qN_{ot}) and interface trapped charge (qN_{it}) are described by Poisson's equation. The vertical field in a MOSFET can usually be analyzed accurately by considering Poisson's equation in one dimension:

$$\frac{dE}{dx} = -\frac{d^2V}{dx^2} = \frac{\rho}{\varepsilon} \tag{1}$$

where E is the electric field, V is the electrostatic potential, ρ is the volume charge density, and ε is the dielectric constant.

When a MOSFET is exposed to ionizing radiation, electron-hole pairs are generated throughout the device. The number of carriers generated depends on the amount of energy absorbed by the material; this is described by the total ionizing dose (TID), which is typically specified in units of Grays (1 Joule/kg) or rads (100 ergs/g). Note that 1 Gray = 100 rads. On average for high-energy incident radiation, the amount of energy required to produce one electron-hole pair in Si is 3.6 eV; the corresponding value for SiO_2 is approximately 18 eV.

In the Si, the carriers generated by the radiation are transported by drift and diffusion and some of them recombine. The fraction of carriers that survive initial recombination is a function of the carrier concentration and the electric field [11]. The transporting carriers result in transient currents that may affect circuit operation, but have no long-term effect on device characteristics. However, in the oxide, the electrons are relatively mobile and, under typical operating biases, leave the oxide under the influence of the electric field. The remaining holes move slowly by a process that involves defect sites in the oxide. These defects trap holes for a time that depends on the trap energy level. While the holes initially are distributed throughout the oxide, the density of deep hole traps is typically highest near the Si/SiO_2 interface [11]. When a positive bias is applied to the gate (the on state for *n*-channel MOSFETs), the holes drift toward the interface where a significant fraction of them are stably trapped.

In addition to trapped charge in the oxide, ionizing radiation also creates electronic states within the Si bandgap that are physically located at the Si/SiO_2 interface (interface traps). Interface traps can be either donorlike (positively charged when empty and neutral when occupied by an electron) or acceptorlike (neutral when empty and negatively charged when occupied by an electron). The most common process for interface-trap formation is believed to comprise the following steps [12]: (1) release of hydrogen trapped in the oxide by radiation-generated holes, (2) transport of the hydrogen (typically in the form of protons) to the Si/SiO_2 interface through drift or diffusion, (3) reaction of the protons with hydrogen-passivated defects at the interface, forming H_2 molecules and electrically active defects [13-15], and (4) transport of the H_2 molecules away from the interface. Charge in the interface traps affects device operation through electrostatic effects and the energy levels associated with the traps also increase surface recombination velocity by serving as recombination centers.

The threshold voltage of a MOSFET that has no charge in the gate oxide or at the Si/SiO_2 interface is given by:

$$V_T = \begin{cases} \Phi_{MS} + 2\phi_F + \dfrac{1}{C_{ox}}\sqrt{2\varepsilon_s q N_A \left(2\phi_F\right)} & n-channel \\ \Phi_{MS} + 2\phi_F - \dfrac{1}{C_{ox}}\sqrt{2\varepsilon_s q N_D \left(2\left|\phi_F\right|\right)} & p-channel \end{cases} \tag{2}$$

where Φ_{MS} is the metal-semiconductor workfunction difference, ϕ_F is the bulk potential, C_{ox} is the oxide capacitance per unit area, ε_s is the dielectric constant of Si, q is the electronic charge, N_A is the doping density in *p*-type Si, and N_D is the doping density in *n*-type Si.

Radiation-induced charge in the oxide affects the threshold voltage according to the density of the charge and its physical location:

$$\Delta V_T = -\frac{1}{\varepsilon_{ox}}\int_0^{x_{ox}} x\rho_{ox}\left(x\right)dx \tag{3}$$

where V_T is the threshold voltage, ε_{ox} is the dielectric constant of the oxide, x_{ox} is the oxide thickness, ρ_{ox} is the volume density of charge in the oxide, and x is the position in the oxide. If all the charge is located at the Si/SiO_2 interface, its effect is maximized and the threshold-voltage shift becomes:

$$\Delta V_T = -\frac{Q_{\text{int}}}{C_{ox}} \tag{4}$$

where Q_{int} is the total areal density of charge at the interface and $C_{ox} = \varepsilon_{ox}/x_{ox}$.

Although trapped charge is usually distributed throughout the oxide, the details of the spatial distribution often are not known. It is convenient to represent the electrostatic effects of the charge in the oxide by an equivalent areal density of charge, projected to the interface, which is defined as the oxide trapped charge density, Q_{ot} (the number density of the oxide charge is N_{ot}). The contribution of this charge to the threshold-voltage shift is:

$$\Delta V_{ot} = -\frac{Q_{ot}}{C_{ox}}. \tag{5}$$

The charge in the interface traps depends on the surface potential, so the contribution of the interface traps to voltage shifts depends on the specific value of the surface potential. The surface potential at threshold is defined by the condition in which the minority-carrier concentration at the surface is equal to the majority-carrier concentration in the bulk Si:

$$\phi_s = 2\phi_F. \tag{6}$$

For this specific condition, the charge in the interface traps is defined uniquely and the threshold-voltage shift due to interface trapped charge is given by:

$$\Delta V_{it} = -\frac{Q_{it}}{C_{ox}}, \tag{7}$$

where Q_{it} is the areal density of charge in the interface traps at threshold (the number density of the traps that are charged at threshold is N_{it}). Radiation-induced interface traps tend to be acceptorlike in the upper half of the Si bandgap and donorlike in the lower half of the bandgap. Thus, when the Fermi level is located at midgap, the net charge in the interface traps is approximately zero. For *n*-channel MOSFETs, the Fermi level is above midgap at threshold, so the donorlike traps in the lower half of the bandgap are filled and neutral, while the filled acceptorlike traps above midgap are negatively charged. Similarly, the net charge in the interface traps of *p*-channel MOSFETs at threshold is usually positive.

The total threshold-voltage shift is given by the sum of the shifts due to oxide trapped charge and interface trapped charge:

$$\Delta V_T = \Delta V_{ot} + \Delta V_{it}. \tag{8}$$

For *n*-channel MOSFETs, the shift due to oxide trapped charge tends to be negative while that due to interfaced trapped charge tends to be positive; the net shift can be either positive or negative. However, for *p*-channel MOSFETs, both terms are negative. While it is sometimes possible for *n*-channel MOSFETs to exhibit relatively small radiation-induced threshold shifts even though the individual contribution of oxide trapped charge and interface trapped charge are large, this is not a desirable situation because the result depends on dose rate, temperature, and other factors. In addition, other device parameters, such as subthreshold slope and carrier mobility may degrade even though the threshold voltage does not change significantly.

The amount of charge generated in the oxide by radiation is proportional to the oxide thickness. In addition, the effect of the oxide charge on the threshold voltage is proportional to the distance of the charge from the gate electrode, as described above.

The combined effects of the generation volume and the electrostatic moment arm are described by:

$$\Delta V_T = -\frac{Q_{ot}}{C_{ox}} \propto x_{ox}^2 . \qquad (9)$$

Thus, thin oxides, such as modern gate oxides, are much less sensitive to ionizing radiation than field oxides or older gate oxides. In fact, thin gate oxides are even less sensitive to total ionizing dose than this estimate would suggest because most of the charge within 5 nm of the interface is quickly removed by tunneling.

2.2 Mobility degradation

The inversion-layer mobility may change significantly due to Coulomb scattering from radiation-induced charges. The effectiveness of the charges in scattering the carriers depends on the proximity of the charges to the interface; charged interface traps have a greater effect on mobility than charges located in the oxide bulk or near-interfacial region. The dependence of inversion-layer mobility on interfacial charge in MOSFETs was originally identified in devices that had not been irradiated [16]. The form of the relationship between mobility and interface charge density was adapted for use in describing irradiated MOSFETs [17] and extended to include both interface traps and oxide trapped charge [18]:

$$\mu = \frac{\mu_0}{1+\alpha_{it}N_{it}+\alpha_{ot}N_{ot}} , \qquad (10)$$

where α_{it} and α_{ot} are parameters that quantify the effects of interface traps and oxide trapped charge, respectively, on mobility. Since interface traps have a greater effect on mobility than oxide trapped charge, $\alpha_{it} > \alpha_{ot}$. The mobility is plotted qualitatively vs. total ionizing dose (including effects of both interface traps and oxide trapped charge) in Figure 1.

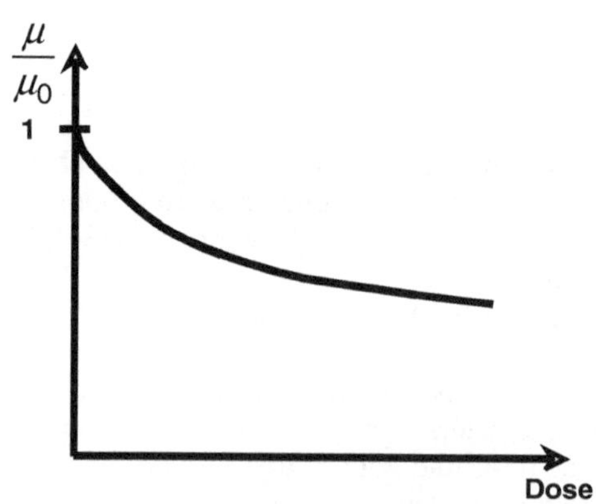

Figure 1. Inversion-layer mobility vs. total ionizing dose for an irradiated MOSFET.

2.3 Subthreshold slope

The drain current of a MOSFET does not vanish abruptly as the gate voltage (V_G) is decreased below the threshold voltage. The region of operation in which $V_G < V_T$ is described as the subthreshold region of operation. The subthreshold current of an *n*-channel MOSFET varies exponentially with the surface potential and is described by [19]:

$$I_D = \frac{1}{2}\mu_n\left(\frac{W}{L}\right)\left(\frac{kT}{q}\right)^2 \frac{\sqrt{2\varepsilon_s q N_A}}{\sqrt{\phi_s - kT/q}} \exp\left[\frac{q\left(\phi_s - 2\phi_F - V_{SB}\right)}{kT}\right]\left(1 - e^{-(qV_{DS}/kT)}\right), \quad (11)$$

where μ_n is the electron mobility, W is the channel width, L is the channel length, k is Boltzmann's constant, T is the absolute temperature, V_{SB} is the source-body voltage, and V_{DS} is the drain-source voltage. Using the approximation that the surface potential is near its strong-inversion value allows estimation of the inverse subthreshold slope, S [19]:

$$S = \left(\frac{\partial \log I_D}{\partial V_{GS}}\right)^{-1} = \left(\frac{kT}{q}\right)\ln(10)\left[1 + \frac{C_D}{C_{ox}} + \frac{C_{it}}{C_{ox}}\right], \quad (12)$$

where C_D is the depletion-layer capacitance per unit area and C_{it} is the capacitance per unit area associated with the interface traps. The ideal value of S (59.6 mV/decade of current) occurs when the oxide capacitance is much larger than either the depletion capacitance or the interface-trap capacitance.

As the interface-trap density increases due to irradiation, the subthreshold slope decreases (S increases) and the MOSFET turns off more slowly. This is an important issue as MOSFET dimensions decrease, since the supply voltage and threshold voltage also decrease. If the threshold voltage becomes too small, significant current will flow even when no gate voltage is applied, as illustrated in Figure 2. This can lead to excessive power dissipation or functional failure if devices cannot be turned off.

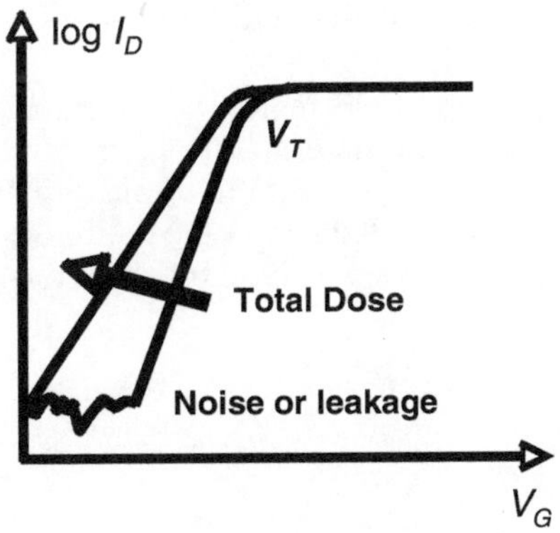

Figure 2. Drain current vs. gate voltage for a MOSFET, illustrating the subthreshold region of operation.

2.4 MOSFET leakage currents

As described in Section 2.1, the effects of radiation-generated charge on MOSFET threshold voltage are relatively small in scaled devices with thin gate oxides. However, the isolation oxides that surround the active devices are relatively thick and the behavior of these oxides dominates the radiation response of most unhardened CMOS integrated circuits. As positive charge is trapped in these oxides, negative charge is induced in the nearby Si. For *p*-type substrates or wells, an inversion layer will form when the positive charge density in the oxide becomes sufficiently high. The inversion layer can short the source and drain of a transistor together at the edge of the active area, as illustrated in Figure 3. The edge of the active area behaves as another transistor in parallel with the main transistor, as illustrated in Figure 4 for the specific case of silicon-on-insulator (SOI) transistors.

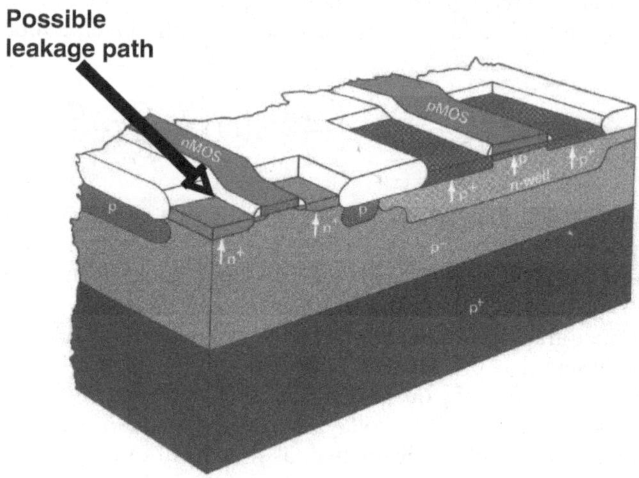

Figure 3. Illustration of possible radiation-induced source-drain leakage path in an *n*-channel MOSFET (after [20]).

It also is possible to invert the field region between adjacent *n*-channel devices, leading to leakage currents between the drain of one transistor and the source of another. This manifests itself as a large increase in power supply current if the field oxide over the *p*-regions becomes inverted across the entire die. The net charge created by ionizing radiation is almost always positive, so field inversion is only considered to be a problem for *n*-channel MOSFETs.

3. Bipolar devices

3.1 Introduction

The principle effect of total ionizing dose and displacement damage on bipolar junction transistors (BJTs) is an increase in the number of defects that participate in

Shockley-Read-Hall recombination (either at the Si/SiO_2 interface or in the Si bulk), resulting in increased base current and decreased current gain [21, 22]. In addition to gain degradation, bipolar integrated circuits also may suffer from device-to-device or collector-to-emitter leakage current, as described above for MOSFETs [23-27]. In circuits in which the absolute value of the current gain is not important as long as it exceeds some minimum value, radiation-induced leakage current may be the dominant failure mechanism.

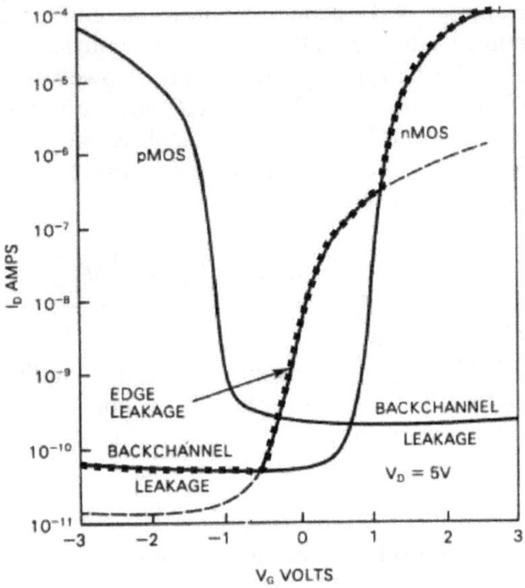

Figure 4. Drain current vs. gate voltage for CMOS SOI transistors. The bold line illustrates the composite *I-V* characteristic that would be observed for an irradiated *n*-channel device (after [20]).

BiCMOS integrated circuits combine bipolar and MOS transistors on a single substrate. In principle, each device type can be used for the applications at which it excels. The bipolar and MOS transistors in BiCMOS technologies are both affected by radiation and thus the unique radiation-induced degradation characteristics of both device types (gain degradation, leakage currents, threshold-voltage shifts, surface-mobility degradation, etc.) must be considered. However, BiCMOS technologies exhibiting very high levels of radiation tolerance have been developed [28, 29].

3.2 Current components

The collector current in an *npn* BJT operating in the forward-active region consists of electrons that are injected from the forward-biased emitter-base junction, diffuse across the neutral base, and are swept through the collector depletion region by the electric field. If all of the electrons injected into the base reach the collector, the collector current density is equal to the diffusion current of electrons in the neutral base:

$$J_C = \frac{qD_{nB}n_{B0}}{W_B}\exp\left(\frac{qV_{BE}}{kT}\right), \tag{13}$$

where J_C is the collector current density, D_{nB} is the electron diffusivity in the base, n_{B0} is the equilibrium electron concentration in the base, W_B is the neutral base width, and V_{BE} is the base-emitter voltage.

The main components of base current in an *npn* BJT are back-injection of holes from the base to the emitter (J_{B1}), recombination in the emitter-base depletion region (J_{B2}), and recombination in the neutral base (J_{B3}). In unirradiated devices, the back-injection component of the base current dominates. However, for devices that have been exposed to ionizing radiation, both J_{B2} and J_{B3} increase, with J_{B2} normally dominating. When the degradation is dominated by displacement damage, J_{B2} and J_{B3} both may increase significantly. The back-injected hole-current density from the base to the emitter, J_{B1}, is a diffusion current:

$$J_{B1} = \frac{qD_{pE}p_{E0}}{L_{pE}}\exp\left(\frac{qV_{BE}}{kT}\right), \tag{14}$$

where D_{pe} is the hole diffusivity in the emitter, p_{E0} is the equilibrium hole concentration in the emitter, and L_{pE} is the diffusion length for holes in the emitter. Analogous equations describe the operation of *pnp* transistors, with the roles of holes and electrons interchanged.

3.3 Recombination in the emitter-base depletion region

The primary effect of ionizing radiation on BJTs is usually an increase in the base current resulting from enhanced recombination in the emitter-base depletion region. The amount by which the base current increases above its pre-irradiation value is called the excess base current (defined as $\Delta I_B = I_B - I_{B0}$, where I_{B0} is the pre-irradiation base current). The recombination-rate increase occurs mainly where the depletion region intersects the Si/SiO_2 interface, due to formation of interface traps that serve as recombination centers (the surface recombination velocity increases with the interface-trap density). When a BJT is exposed to energetic particles, displacement damage occurs in the bulk Si. The resulting defects reduce the minority-carrier lifetime.

The recombination rate is a function of position within the depletion region, exhibiting a strong peak where $n = p$. The ideality factor (defined as n_B in $\exp(qV/n_BkT)$) is 2 for recombination occurring at this peak and 1 for the ideal component of the base current. However, the recombination rate must be integrated throughout the depletion region to obtain the contribution of recombination to the base current. The excess base current due to surface recombination thus has an ideality factor between 1 and 2, which combines the effects of the different spatial locations at which the recombination takes place. However, it is a reasonable approximation in many cases to assume that the radiation-induced excess base current is dominated by the peak recombination rate, which leads to the results [22]:

$$J_{B2,surf} \propto v_{surf} \exp\left(\frac{qV}{2kT}\right) \tag{15}$$

and

$$J_{B2,bulk} \propto \frac{1}{\tau_d} \exp\left(\frac{qV}{2kT}\right), \tag{16}$$

where $J_{B2,surf}$ and $J_{B2,bulk}$ are the base current densities due to recombination in the depletion region where it intersects the Si/SiO_2 interface and in the depletion region in the bulk Si, respectively, v_{surf} is the surface recombination velocity, and τ_d is the minority-carrier lifetime in the depletion region. Since the base current due to recombination in the depletion region increases more slowly with voltage than the back-injected (J_{B1}) component of the base current, its effect is greatest at low emitter-base voltages.

3.4 Recombination in the neutral base

When minority carriers are injected from the emitter into the neutral base, some of them recombine before they reach the collector junction. This neutral-base recombination may take place in either the bulk Si or at the Si/SiO_2 interface (if the Si surface is neutral, as determined by the charge in the oxide or the bias on any electrodes lying over the base region). The fraction of carriers that recombine is small in most unirradiated devices, but it may become significant following irradiation.

If the average lifetime of minority carriers in the neutral base is τ_B, the base current density required to supply the recombination process is:

$$J_{B3} = \frac{Q_B}{\tau_B} = \frac{qW_B n_{B0} \exp\left(\frac{qV_{BE}}{kT}\right)}{2\tau_B}. \tag{17}$$

Unlike recombination in the emitter-base depletion region, which exhibits an ideality factor close to 2, the ideality factor associated with recombination in the neutral base is 1. Recombination in the neutral base is usually the dominant effect of particle-induced displacement damage on BJT characteristics.

3.5 Current gain

The common-emitter current gain (β) is defined as the ratio of the collector current to the base current:

$$\beta = \frac{I_C}{I_B}. \tag{18}$$

When a BJT is irradiated, the base current increases, but the collector current typically remains relatively constant, causing the current gain to decrease [30, 31]. An example Gummel plot (log I_C and I_B vs. V_{BE}) is shown in Figure 5. For this device, the collector current remains virtually unchanged except at very low bias levels, while the base current increases significantly. This causes a large reduction in the current gain, especially at low bias levels where the base current increases most rapidly. The current gain of an irradiated *npn* BJT is plotted vs. V_{BE} in Figure 6.

Increased recombination in the emitter-base depletion region does not reduce the collector current at a given bias level because the number of carriers injected into the base depends only on the doping of the base and the applied bias. If recombination increases in the depletion region, the emitter and base currents increase, but the collector current remains constant. However, when injected carriers recombine in the neutral base, they do not reach the collector junction and the collector current decreases.

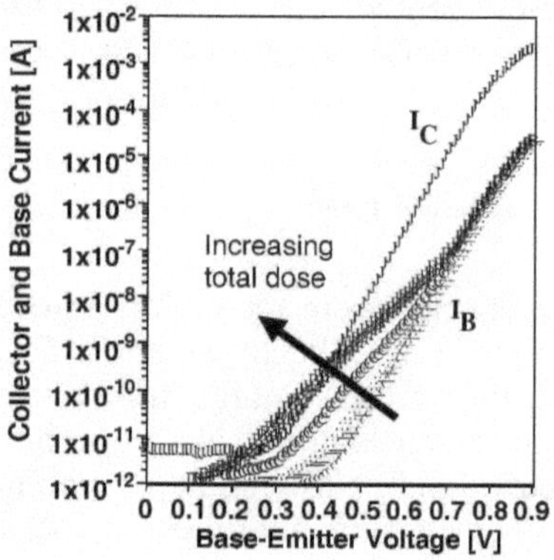

Figure 5. Collector and base current vs. base-emitter voltage for an irradiated *npn* BJT.

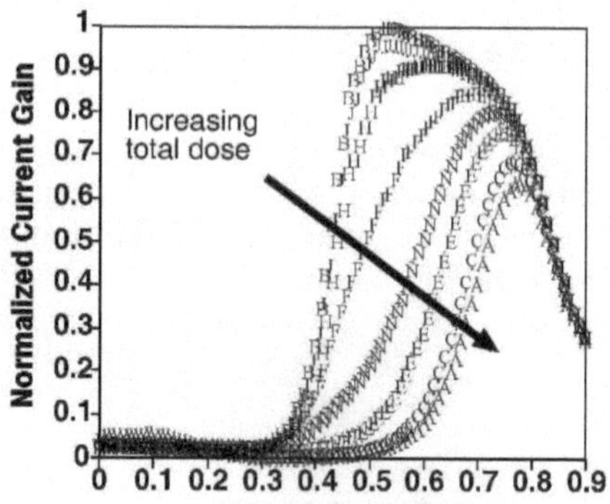

Figure 6. Normalized current gain vs. base-emitter voltage for an *npn* BJT irradiated to various total doses.

The increase in surface recombination velocity is approximately proportional to the density of recombination centers at the silicon/silicon dioxide interface that covers the emitter-base junction. Traps with energies near the middle of the silicon bandgap are the most effective.

Since the net charge introduced into the oxide by ionizing radiation (N_{ox}) is positive, the depletion region spreads on the *p*-side of a *pn* junction. For *npn* transistors, the surface of the relatively lightly doped *p*-type base region becomes depleted. The recombination rate is maximized when the electron and hole populations are equal; this condition occurs at a point within the depletion region.

The increased recombination occurs around the edge of the emitter, so the amount of excess base current caused by surface recombination is proportional to the perimeter of the emitter [32]. Device layouts with large perimeter-to-area ratios increase the sensitivity to ionizing radiation because the excess base current (proportional to emitter perimeter) is large relative to the ideal component of the base current (proportional to emitter area).

Vertical *pnp* transistors are relatively radiation hard compared to vertical *npn* devices because positive radiation-induced trapped charge tends to accumulate the *n*-type base, reducing the width of the emitter-base depletion region. The positive charge depletes the *p*-type emitter, but the effect is typically small because of the high emitter doping.

In lateral *pnp* BJTs, the excess base current scales with the surface recombination velocity, which increases approximately in proportion to the number of interface traps. However, this increase is moderated by the effects of positive charge in the oxide over the base. This positive charge accumulates the surface of the base, suppressing surface recombination. Thus, the effects of radiation-induced positive oxide charge and increased surface recombination velocity oppose each other. However, the combined effects of the radiation-induced increase in surface recombination velocity and the moderation due to the oxide trapped charge almost always result in significant gain degradation in lateral *pnp* transistors.

Lateral *pnp* transistors are affected more significantly by ionizing radiation than substrate *pnp* transistors, since the current flow pattern in the LPNP devices is lateral and directly under the oxide where the recombination centers occur, while the current flow path in the SPNP devices is vertical. In typical bipolar technologies, lateral *pnp* transistors are the most susceptible of all the available BJT types; if high levels of radiation tolerance are required, it may be desirable to avoid use of LPNP transistors. Vertical BJTs of either polarity are usually much more resistant to ionizing radiation.

4. Single-event effects

4.1 Introduction

The subject of transient radiation effects, particularly single-event effects, is very broad and a comprehensive discussion of the topic is beyond the scope of this paper. However, it is interesting to consider physically based models of single-event effects since this level of treatment is comparable to the treatment of parametric effects discussed above. A comprehensive approach to simulating radiation effects requires tools that describe processes at the following levels:

- a quantitative description of the relevant radiation environment (particle flux, energy, etc.),

- energy deposition in the electronic materials resulting from interaction with the impinging radiation,
- conversion of energy into charge or defects,
- radiation-induced charge transport in semiconductor and insulating layers,
- charge trapping,
- device current-voltage characteristics, including the effects of radiation-induced charge and defects,
- compact-model parameter extraction, and
- circuit simulation, including radiation-induced transients and parametric changes.

In this section, we describe an approach for simulating single-event effects based on detailed descriptions of large numbers of individual particle interactions. Spatially and temporally realistic representations of the charge deposited by individual energetic particles are used as input for device simulations [33], which in turn are used to determine the circuit-level response. An overall view of this approach is illustrated in Figure 7 [34].

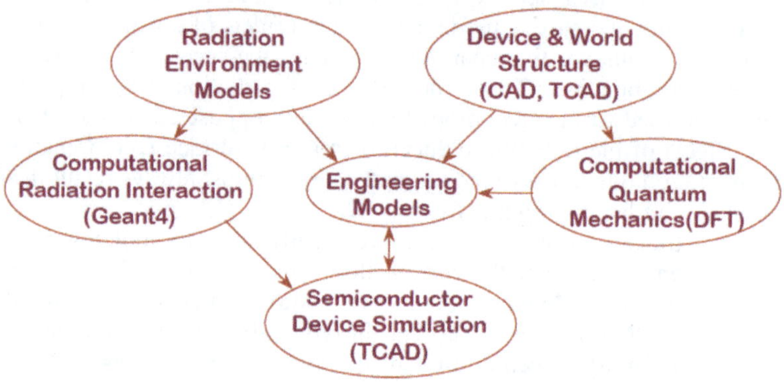

Figure 7. Hierarchical organization of radiation-effects simulation tools.

Until very recently, the interaction of radiation with devices usually has been quantified by the average deposited energy along a particle track (e.g., linear energy transfer (LET) or non-ionizing energy loss (NIEL)). The validity of this approach relies on the large size of the devices to integrate the effects of the radiation, a situation that no longer exists in highly scaled devices, where all events may be thought of as single events. By describing radiation environments using a large number of events initiated by individual primary radiation quanta and studying device response to these individual events, this approach allows us to obtain both average device response and statistical variability. Unlike the radiation-effects simulation methods commonly used, processes that depend on the microstructure of radiation interactions with emerging devices, such as multiple bit upsets, secondary radiation from materials near active devices, microdose, and highly localized displacement damage, can be analyzed quantitatively.

4.2 Simulation approach

The approach described here uses the MRED (Monte Carlo Radiative Energy Deposition) code developed at Vanderbilt University) and commercially available device simulation tools to simulate radiation-deposited energy, the resulting charge transport in devices, and subsequent circuit-level effects. MRED is based on Geant4, which is a comprehensive library of C++ routines designed for Monte Carlo simulation of the interactions of radiation with matter. Fidelity at the individual particle level is required because, for an excited system, the average response is not, in general, the response to the average excitation. This means that for very small devices, all events are single events.

MRED is used to generate very large numbers of individual event descriptions. These events are intrinsically three-dimensional and the resulting particles may include electrons, protons, neutrons, other subatomic particles, larger atomic fragments, and photons. The amount of energy is provided as a function of space and time, along with the fraction of the energy that results in ionization. Custom tools are used to process this information, converting it to a 3-D charge distribution that is automatically meshed for use in a device simulation tool. The device-simulation tool provides terminal currents vs. time, and if the device is embedded in a circuit, the occurrence of an upset can be determined. Analysis of a large number of these events allows determination of higher-level representations of circuit response, such as upset cross-section vs. incident particle characteristics.

4.3 Device-level effects

If a track structure based on the average event is used as the input for device simulations, the result will differ dramatically from that obtained if a realistic event structure is used. Accounting for this variability requires that a large number of individual events be simulated in order to estimate the error rate of the circuit. In addition to the variability in the amount of energy deposited, the spatial non-uniformity further complicates the situation. However, this situation parallels that encountered in an actual space environment or in an accelerator test.

Figure 8 represents the variation in device-level response that can result from four different 100-MeV proton events. The track structure of each event is shown, as well as the drain current-vs.-time response that results. These results were obtained for a 0.18-µm MOSFET subjected to four different events initiated by 100-MeV protons: (1) an "average" event, represented by LET, (2) an event that includes a delta electron, (3) a (p,) reaction, and (4) a spallation reaction.

The event labeled “LET” in Figure 8 represents the response to an “average” event, while the other three panels illustrate particular events selected to demonstrate the range of possible responses. The peak drain current for the other three events ranges from 10^4-10^6 times that for the “LET” event for a delta electron and a spallation reaction, respectively. The delta electron passed directly through a sensitive device region, making it more effective than would otherwise be expected based on the amount of energy it deposited.

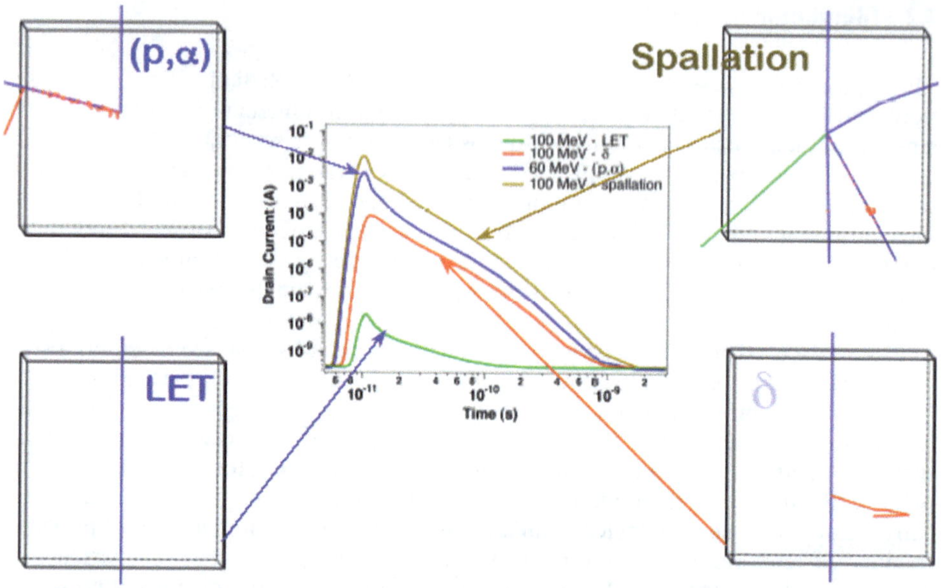

Figure 8. Drain current vs. time for a 0.18 μm MOSFET subjected to four different events initiated by 100-MeV protons: (1) an "average" event, represented by LET, (2) an event that includes a delta electron, (3) a (p,α) reaction, and (4) a spallation reaction. After [34].

5. Summary

Understanding radiation effects in electronics is a complex and continually evolving challenge. With each change in the underlying technology, new effects appear and old models must be revised. In recent years, the change has accelerated with the introduction of new materials and device structures. This paper describes some of the underlying principles that affect the radiation response of semiconductor devices. In particular, the basic mechanisms responsible for degradation of MOSFETs and BJTs are reviewed. The emphasis is on the underlying physical phenomena that will determine the susceptibility of current and emerging technologies to radiation.

A fundamentally new approach to simulating radiation effects in electronics also is described. A large ensemble of individual event descriptions is generated using Geant4-based software. Specific events are selected for device simulations, based on user-defined criteria (e.g., amount of energy deposited). Extreme events, as well as more typical events, can be simulated with much improved fidelity. This simulation approach provides improved capability for dealing with single-event effects in very small devices or other physical situations in which conventional methods break down.

Acknowledgments

Technical contributions from Hugh Barnaby, Dan Fleetwood, Ken Galloway, Lloyd Massengill, Sokrates Pantelides, Robert Reed, Bob Weller, and Marcus Mendenhall are gratefully acknowledged.

References

[1] J. A. Felix, D. M. Fleetwood, R. D. Schrimpf, J. G. Hong, G. Lucovsky, J. R. Schwank, and M. R. Shaneyfelt, "Total-Dose Radiation Response of Hafnium-Silicate Capacitors," *IEEE Trans. Nucl. Sci.*, vol. 49, pp. 3191-3196, 2002.

[2] J. A. Felix, H. D. Xiong, D. M. Fleetwood, E. P. Gusev, R. D. Schrimpf, A. L. Sternberg, and C. D'Emic, "Interface trapping properties of $Al_2O_3/SiO_xN_y/Si(100)$ nMOSFETS after exposure to ionizing radiation," *Microelectron. Engineering*, vol. 72, pp. 50-54, 2004.

[3] J. A. Felix, M. R. Shaneyfelt, D. M. Fleetwood, T. L. Meisenheimer, J. R. Schwank, R. D. Schrimpf, P. E. Dodd, E. P. Gusev, and C. D'Emic, "Radiation-induced charge trapping in thin $Al_2O_3/SiO_xN_y/Si[100]$ gate dielectric stacks," *IEEE Trans. Nucl. Sci.*, vol. 50, pp. 1910-1918, 2003.

[4] M. Turowski, A. Raman, and R. D. Schrimpf, "Nonuniform total-dose-induced charge distribution in shallow-trench isolation oxides," *IEEE Transactions on Nuclear Science*, vol. 51, pp. 3166-3171, 2004.

[5] R. D. Schrimpf, "Gain Degradation and Enhanced Low-Dose-Rate Sensitivity in Bipolar Junction Transistors," *Int. J. High Speed Electronics and Systems*, vol. 14, pp. 503-517, 2004.

[6] D. M. Fleetwood, "'Border Traps' in MOS Devices," *IEEE Trans. Nucl. Sci.*, vol. 39, pp. 269-271, 1992.

[7] P. E. Dodd and L. W. Massengill, "Basic mechanisms and modeling of single-event upset in digital microelectronics," *IEEE Trans. Nucl. Sci.*, vol. 50, pp. 583-602, 2003.

[8] P. E. Dodd, M. R. Shaneyfelt, J. A. Felix, and J. R. Schwank, "Production and propagation of single-event transients in high-speed digital logic ICs," *IEEE Trans. Nucl. Sci.*, vol. 51, pp. 3278-3284, 2004.

[9] R. A. Reed, P. W. Marshall, H. S. Kim, P. J. McNulty, B. Fodness, T. M. Jordan, R. Reedy, C. Tabbert, M. S. T. Liu, W. Heikkila, S. Buchner, R. Ladbury, and K. A. LaBel, "Evidence for angular effects in proton-induced single-event upsets," *IEEE Trans. Nucl. Sci.*, vol. 49, pp. 3038-3044, 2002.

[10] L. W. Massengill, S. E. Diehl, and J. S. Browning, "Dose-Rate Upset Patterns in a 16K Cmos Sram," *IEEE Trans. Nucl. Sci.*, vol. 33, pp. 1541-1545, 1986.

[11] G. F. Derbenwick and B. L. Gregory, "Process Optimization of Radiation Hardened CMOS Circuits," *IEEE Trans. Nucl. Sci.*, vol. 22, pp. 2151-2158, 1975.

[12] N. S. Saks and D. B. Brown, "Interface Trap Formation via the Two-Stage H^+ Process," *IEEE Trans. Nucl. Sci.*, vol. NS-36, pp. 1848-1857, 1989.

[13] S. T. Pantelides, S. N. Rashkeev, R. Buczko, D. M. Fleetwood, and R. D. Schrimpf, "Reactions of Hydrogen with $Si\text{-}SiO_2$ Interfaces," *IEEE Trans. Nucl. Sci.*, vol. 47, pp. 2262-2268, 2000.

[14] S. N. Rashkeev, D. M. Fleetwood, R. D. Schrimpf, and S. T. Pantelides, "Proton-Induced Defect Generation at the $Si\text{-}SiO_2$ Interface," *IEEE Trans. Nucl. Sci.*, vol. 48, pp. 2086-2092, 2001.

[15] S. N. Rashkeev, D. M. Fleetwood, R. D. Schrimpf, and S. T. Pantelides, "Defect generation by hydrogen at the Si-SiO_2 interface," *Phys. Rev. Lett.*, vol. 87, pp. 165506.1-165506.4, 2001.
[16] S. C. Sun and J. D. Plummer, "Electron Mobility in Inversion and Accumulation Layers on Thermally Oxidized Silicon Surfaces," *IEEE Trans. Electron Devices*, vol. ED-27, pp. 1497-1508, 1980.
[17] K. F. Galloway, M. Gaitan, and T. J. Russell, "A Simple Model for Separating Interface and Oxide Charge Effects in MOS Device Characteristics," *IEEE Trans. Nucl. Sci.*, vol. NS-31, pp. 1497-1501, 1984.
[18] D. Zupac, K. F. Galloway, P. Khosropour, S. R. Anderson, R. D. Schrimpf, and P. Calvel, "Separation of Effects of Oxide-Trapped Charge and Interface-Trapped Charge on Mobility in Irradiated Power MOSFETs," *IEEE Trans. Nucl. Sci.*, vol. 40, pp. 1307-1315, 1993.
[19] C. Y. Chang and S. M. Sze, ULSI Devices. New York, NY: Wiley-Interscience, 2000.
[20] J. Y. Chen, CMOS Devices and Technology for VLSI. Saddle River, NJ: Prentice Hall, 1990.
[21] R. D. Schrimpf, "Recent Advances in Understanding Total-Dose Effects in Bipolar Transistors," *IEEE Trans. Nucl. Sci.*, vol. 43, pp. 787-796, 1996.
[22] S. L. Kosier, A. Wei, R. D. Schrimpf, D. M. Fleetwood, M. DeLaus, R. L. Pease, and W. E. Combs, "Physically Based Comparison of Hot-Carrier-Induced and Ionizing-Radiation-Induced Degradation in BJTs," *IEEE Trans. Electron Devices*, vol. 42, pp. 436-444, 1995.
[23] R. L. Pease, R. M. Turfler, D. Platteter, D. Emily, and R. Blice, "Total Dose Effects in Recessed Oxide Digital Bipolar Microcircuits," *IEEE Trans. Nucl. Sci.*, vol. NS-30, pp. 4216-4223, 1983.
[24] J. L. Titus and D. G. Platteter, "Wafer Mapping of Total Dose Failure Thresholds in a Bipolar Recessed Field Oxide Technology," *IEEE Trans. Nucl. Sci.*, vol. 34, pp. 1751-1756, 1987.
[25] E. W. Enlow, R. L. Pease, W. E. Combs, and D. G. Platteter, "Total dose induced hole trapping in trench oxides," *IEEE Trans. Nucl. Sci.*, vol. 36, pp. 2415-2422, 1989.
[26] J. P. Raymond, R. A. Gardner, and G. E. LaMar, "Characterization of radiation effects on trench-isolated bipolar analog microcircuit technology," *IEEE Trans. Nucl. Sci.*, vol. 39, pp. 405-412, 1992.
[27] W. C. Jenkins, "Dose-rate-independent total dose failure in 54F10 bipolar logic circuits," *IEEE Trans. Nucl. Sci.*, vol. 39, pp. 1899-1902, 1992.
[28] M. Dentan, E. Delagnes, N. Fourches, M. Rouger, M. C. Habrard, L. Blanquart, P. Delpierre, R. Potheau, R. Truche, J. P. Blanc, E. Delevoye, J. Gautier, J. L. Pelloie, d. Pontcharra, O. J.; Flament, J. L. Leray, J. L. Martin, J. Montaron, and O. Musseau, "Study of a CMOS-JFET-bipolar radiation hard analog-digital technology suitable for high energy physics electronics," *IEEE Trans. Nucl. Sci.*, vol. 40, pp. 1555-1560, 1993.
[29] M. Dentan, P. Abbon, E. Delagnes, N. Fourches, D. Lachartre, F. Lugiez, B. Paul, M. Rouger, R. Truche, J. P. Blanc, C. Leroux, E. Delevoye-Orsier, J. L. Pelloie, J. de Pontcharra, O. Flament, J. M. Guebhard, J. L. Leray, J. Montaron, and O. Musseau, "DMILL, a mixed analog-digital radiation-hard BICMOS technology for high energy physics electronics," *IEEE Trans. Nucl. Sci.*, vol. 43, pp. 1763-1767, 1996.
[30] A. Wei, S. L. Kosier, R. D. Schrimpf, W. E. Combs, and M. DeLaus, "Excess Collector Current Due to an Oxide-Trapped-Charge-Induced Emitter in Irradiated NPN BJTs," *IEEE Trans. Electron Devices*, vol. 42, pp. 923-927, 1995.
[31] H. J. Barnaby, R. D. Schrimpf, D. M. Fleetwood, and S. L. Kosier, "The Effects of Emitter-Tied Field Plates on Lateral PNP Ionizing Radiation Response," in *IEEE BCTM Proc.*, pp. 35-38, 1998.

[32] R. N. Nowlin, R. D. Schrimpf, E. W. Enlow, W. E. Combs, and R. L. Pease, "Mechanisms of Ionizing-Radiation-Induced Gain Degradation in Modern Bipolar Devices," in *Proc. 1991 IEEE Bipolar Circuits and Tech. Mtg.*, pp. 174-177, 1991.

[33] A. S. Kobayashi, A. L. Sternberg, L. W. Massengill, R. D. Schrimpf, and R. A. Weller, "Spatial and temporal characteristics of energy deposition by protons and alpha particles in silicon," *IEEE Trans. Nucl. Sci.*, vol. 51, pp. 3312-3317, 2004.

[34] R. A. Weller, A. L. Sternberg, L. W. Massengill, R. D. Schrimpf, and D. M. Fleetwood, "Evaluating Average and Atypical Response in Radiation Effects Simulations," *IEEE Trans. Nucl. Sci.*, vol. 50, pp. 2265-2271, 2003.

In-flight Anomalies on Electronic Devices

Robert Ecoffet

CNES
18 Avenue Edouard Belin
31401 Toulouse, France
robert.ecoffet@cnes.fr

Abstract. Spacecraft anomalies due to radiation effects on electronic devices have been known since the very beginning of the space era. This chapter will describe a sample of known cases of cumulated or transient effects. A brief overview of environment monitors and technology experiments will also be made.

1. Introduction

Radiation-induced spacecraft anomalies have been known since the very first days of space exploration. The measurement of radiation in space has been one of the very first concerns of the newly born space community.

The first US artificial satellite, Explorer I, designed and built by the Jet Propulsion Laboratory, and launched on January 31, 1958, carried a Geiger counter proposed by J.A. Van Allen. The counter suddenly stopped counting cosmic rays when the spacecraft reached some altitude. It was later found that the counter was in fact saturated by an extremely high particle count rate. This was the day of the discovery of the Van Allen belts, whose existence was confirmed by Explorer III, which was launched on 26 March 1958. The discovery of the Van Allen Belts by the Explorer satellites was one of the outstanding discoveries of the International Geophysical Year 1958. The evidence of the existence of trapped particles in Earth's radiation belts can be considered, in this respect, as the very first scientific output of the Space Age.

The knowledge of the radiation space environment grew together with the development of space techniques. Both areas have the same age, both are still evolving. Better knowledge of the environment helped in building better hardware which in turn allowed to develop better instruments. Contrary to other disciplines involved in spacecraft design, such as celestial mechanics or electromagnetism, for instance, which pre-existed before the space era, the knowledge of the space environment was a consequence of space exploration itself.

A few years after the Explorer-I launch, on 10 July 1962, was launched Telstar, designed and build by the Bell Telephone Laboratories with AT&T funds and supported by NASA. Telstar definitively opened the era of satellite telecommunication with a live television picture transmission from the USA to France on that same 10 July 1962. It was the first satellite equipped with transponders, and thus amplification, and thus the first active telecommunication satellite. On 9 July 1962, the day before Telstar launch, the USA proceeded to a high altitude nuclear test. The

R. Velazco et al. (eds.), Radiation Effects on Embedded Systems, 31–68.

extremely high radiation levels induced by electrons injected in the radiation belts caused degradations of some electronic components, (diodes in the command decoder) and finally, the loss of the satellite on 21 February 1963. This was the first spacecraft loss due to radiation effects.

Although these failures were created by man-made conditions, they inaugurated, from the very first days of space exploration, a long series of radiation induced spacecraft anomalies.

Then came a new class of effects, starting from first observations in 1978 when Intel Corporation discovered that anomalous upsets occurred at the ground level on dynamic random access memories (DRAMs). These effects, called “single event effects” because they were triggered by one particle alone, were attributed to alpha particles produced by fission reactions of trace amounts of unstable elements in the components’ package. Rapidly, it was established that ions, protons and neutrons could also produce single event effects, and it soon became one of the major causes of component dysfunction in space.

Since these early days, radiation induced spacecraft anomalies have been regularly observed.

2. Overview of radiation effects

1.1 The space environment

The natural radiation environment in space has been detailed in the previous chapters. Let us just roughly remind the main components constituting this environment .

Table 1. Main sources of the natural space radiation environment.

Radiation belts	Electrons	eV ~10 MeV
	Protons	keV- 500 MeV
Solar flares	Protons	keV- 500 MeV
	Ions	1 to a few 10 MeV/n
Galactic cosmic rays	Protons and ions	Max flux at about 300 MeV/n

1.2 The main classes of effects on components

The effects of space radiation on electronic components will be described in further details in the following chapter of this book, let us nevertheless introduce the main classes of effects in order to have elements for analysing in-flight anomaly cases.

The basic effect of radiation-matter interaction is to bring energy deposition into the target object. Depending on physical processes involved in the target of interest, this energy will be transformed in a variety of effects.

Energetic charged particles have two ways of depositing energy within the matter constituting the electronic components and the spacecraft structure. They can interact with the electrons of the target atoms, pull them out of their orbits around the nucleus, and release them in the surrounding medium. These electronic interactions constitute ionisation effects. In solid state matter, the result is the creation of electron-hole pairs whose behaviour will be governed, after the interaction, by the electrical properties of the device of interest.

A second way of producing ionisation comes from a secondary effect due to the sudden loss of velocity when particles enter dense matter. When breaking suddenly, particles convert some of their kinetic energy into photons, this is the Bremstrahlung, or "braking radiation". These high energy X or gamma ray photons can in their turn induce ionisation along their path. They also penetrate deeply into matter and can ionise strongly shielded targets. Secondary radiation is also a prime concern for detector noise.

Charged particles can also interact, although with a much lower probability, with the nucleus itself of the target atom, either through electromagnetic or nuclear interaction. A fraction of the energy of the incoming particle is transferred to the target nucleus, which is excited and can be displaced. This effect is called displacement damage. The target nucleus may be fragmented in some cases of nuclear interaction. The consequence is a local modification the lattice structure and the creation of crystalline defects and interstitial-vacancies complexes. The accumulation of these defects, or clusters of defects, modify the crystalline and electronic properties of the crystal. The effects range from changes in optical properties (transparency, colour, optical absorption) and mechanical properties (Young module, oscillation frequencies), to changes in the electronic structure of the device. In semiconductors, these defects contribute to the creation of additional trapping levels in the forbidden band and these levels can modify in their turn the response of the device to electron-hole pair creation through ionisation processes.

When energies are high enough, the nuclear interaction can result in one or many recoil atoms with sufficient remaining energy for travelling some noticeable distance in the device, and being able in their turn to ionise the matter along their track. This secondary radiation can be considered as an ion source inside the device and can add an extra contribution to all effects due to primary cosmic or solar heavy ions.

Depending on the local or diffuse nature of energy deposition, and depending also on the orders of magnitude of the energies involved, we come to two main classes of effects from a system point of view.

The first class is a degradation of system lifetime through the accumulation of ionising dose and displacement damage over time. These effects induce a gradual modification of the electrical properties of the component, finally leading to component failure.

The second class is composed of functional flaws, and sometimes destructive effects, they correspond to sudden and localised energy depositions. These effects are related to system dependability and performance, and are treated as a probabilistic and risk estimation problem.

Table 2 shows the correspondence between physical and system effects.

Table 2. Main classes of radiation effects on electronic components.

Small and homogenous $\Delta E_{ionisation}$ over a long time	Most numerous particles (e-, p+) Secondary Bremstrahlung photons	Direct or secondary ionisation	Total cumulated ionising dose effects System lifetime
Sudden high $\Delta E_{ionisation}$ at the wrong place and time	Energetic heavy particles (p+, ions)	Direct ionisation	Single event effects Functional flaws Destructive effects
Accumulation of $\Delta E_{nuclear}$	Numerous heavy (p+) or very high energy light particles (e-)	Displacement damage	Crystalline effects Enhancement of dose effects System lifetime
Sudden high $\Delta E_{nuclear}$ at the wrong place and time	Energetic heavy particles (p+, ions)	Ionisation by recoil atoms	Single event effects Functional flaws Destructive effects

If we now compare the radiation sources in space with the expected effects, we come to a correspondence summarised in the chart below.

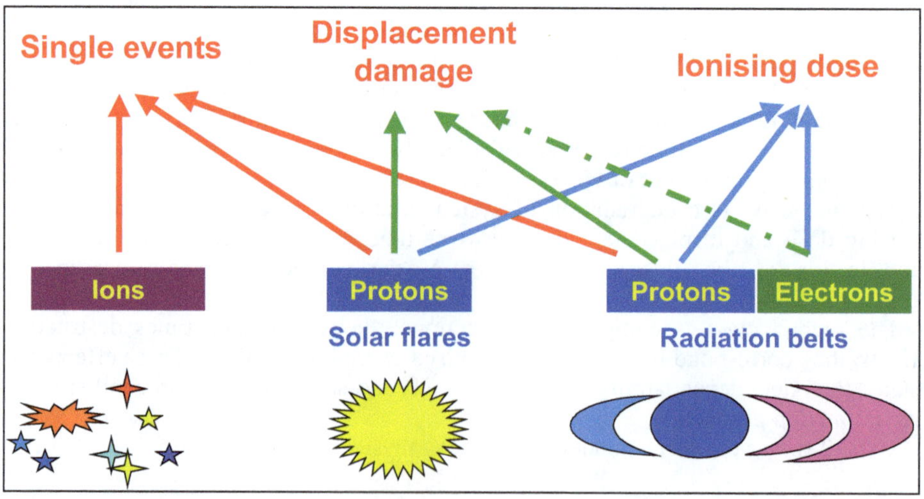

Figure 1. Correspondence between radiation sources and effects on components.

The Earth's radiation belts, which are composed of high energy electrons and protons, can contribute to all classes of effects, so do high energy protons from solar flares.

The galactic cosmic rays are composed of rare high energy ions, capable of inducing single event effects, but not numerous enough to contribute to accumulated degradation of electronic components (but they do contribute to degradation of living matter, which is much more sensitive).

2. In-flight anomalies and the space environment

2.1 Sources of data

Sources of spacecraft anomaly data are few, and it is difficult to have a general overview of the extent of the problem. A large part of the satellite fleet is composed today of commercial and defence spacecraft, and information about possible anomalies may be classified for many reasons. Space agencies such as NASA or ESA are generally much less reluctant to give information, but even in this case, it is not obvious that all information is known by all and centralised somewhere.

There are anyway some excellent public sources of spacecraft anomaly information, such as : a NASA reference publication written by MSFC [1] in August 1996, an Aerospace Corp. study [2] from September 2000, and the outstanding internet reference Satellite Digest News [3]. For statistical comparisons trying to identify the weight of different factors in spacecraft anomalies, there is always a possible bias due to the sample of anomalies used for the comparison. For example, an anomaly set consisting mainly of problems encountered on geosynchronous satellites (i.e. particular environment conditions) may lead to a general classification of anomaly causes that would not be the same if the set used was mainly composed of LEO satellite anomalies.

It should thus be stressed that every spacecraft is a new case, and that survivability efforts in the design phase should be done with respect to an analysis of the particular environment of the spacecraft compared with the expected mission performance, and should not rely on general purpose statistical figures.

2.2 Statistical comparisons

We will limit here to the results of the NASA and Aerospace studies. In "Spacecraft system failures and anomalies attributed to the natural space environment", NASA reference publication 1390 [1], MSFC compares more than 100 anomaly cases attributed to the environment of the satellite proposes the following classification :

- plasma (charging)
- radiation (TID, SEE, DDD)
- meteroids and space debris
- neutral thermosphere (satellite drag, effects on surface materials)
- solar (effects of solar events, may fall in "plasma", "radiation" or "geomagnetic" categories)

- thermal (thermal conditions in vacuum – may be considered as a design problem)
- geomagnetic (effects on magneto-torquers for example).

The chart in figure was made using this classification. For classifying anomalies by their physical cause, we have break out the "solar" category into "plasma", "radiation" or "geomagnetic" categories.

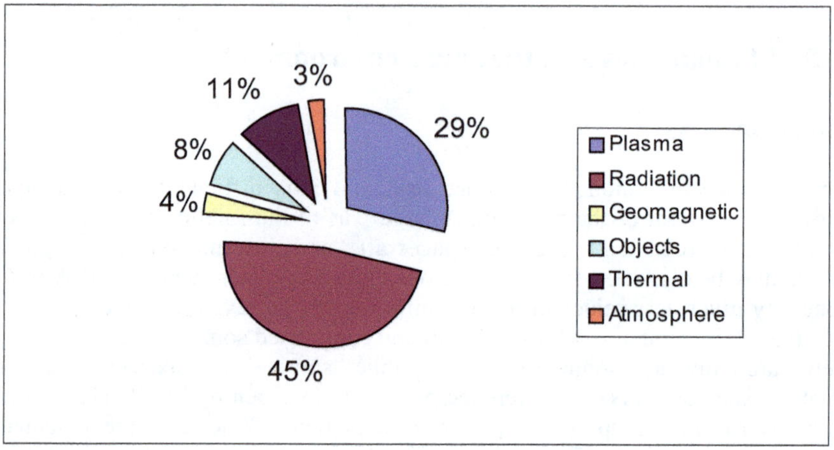

Figure 2. Spacecraft anomalies due to the space environment, from [1].

From this study, the two main environment related causes of spacecraft anomalies are radiation and plasma effects. If now we break out the radiation part into the main radiation effects, we come to the distribution in figure. The "upset" category should be understood in a wide sense, and may include SET effects.

The set used in the Aerospace study [2] includes more recent anomaly cases and has an overlap with the set used by MSFC. This set is composed of 326 anomaly cases, a large number of which is related to GEO satellites, thus emphasising the ESD issue. Table 3 from [2] gives the classification used and the repartition of anomalies.

Although the results may differ from one study to another, a general conclusion can be drawn out :

The two major sources of environment related spacecraft anomalies are, statistically, plasma and radiation effects, i.e. effects related to the charged particles from the space environment.

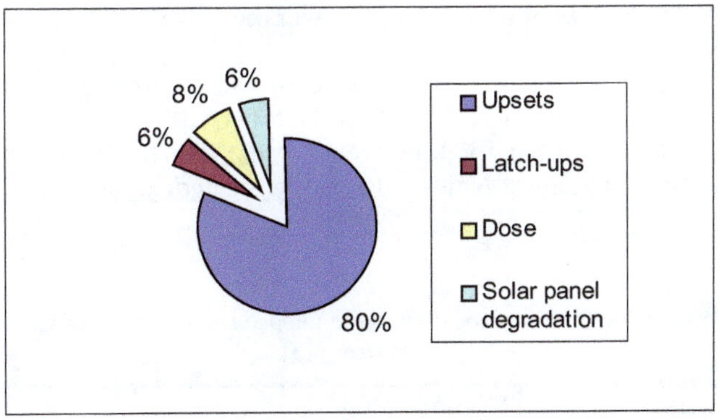

Figure 3. Repartition of radiation spacecraft anomalies, from [1].

Table 3. Repartition of causes for 326 spacecraft environment related anomalies, from [2].

Diagnosis	Number of Forms
ESD - Internal Charging	74
ESD - Surface Charging	59
ESD - Uncategorized	28
Surface Charging	1
Total ESD & Charging	162
SEU - Cosmic Ray	15
SEU - Solar Particle Event	9
SEU - South Atlantic Anomaly	20
SEU - Uncategorized	41
Total SEU	85
Solar Array - Solar Proton Event	9
Total Radiation Dose	3
Materials Damage	3
South Atlantic Anomaly	1
Total Radiation Damage	16
Micrometeorid/Debris Impact	10
Solar Proton Event - Uncategorized	9
Magnetic Field Variability	5
Plasma Effects	4
Atomic Oxygen Erosion	1
Atmospheric Drag	1
Sunlight	1
IR background	1
Ionospheric Scintillation	1
Energetic Electrons	1
Other	2
Total Miscellaneous	36

2.3 Example of the effects of a major space weather event

The growing concern for space weather effects had a striking confirmation when major solar flares occurred at the end of October 2003. At least 33 spacecraft anomalies have been reported for this event. A list can be found on the excellent internet address Satellite Digest News, http://www.sat-index.com and is reproduced in Table 4 :

Table 4. Spacecraft anomalies associated to the intense solar activity of October / November 2003, from [3].

Date	Satellite	Event
23 Oct	Genesis	Enters safe mode. Operations resumed on 3 Nov
24 Oct	**Midori**	**Safe mode, power dropped, telemetry lost - total loss**
24 Oct	Stardust	Enters safe mode because of read errors. Recovered
24 Oct	Chandra	Observations halted because of high radiation levels. Restarted 25 Oct
24 Oct	GOES 9, 10	High bit error rates
24 Oct	GOES 12	Magnetic torquers disabled
25 Oct	RHESSI	Spontaneous reset of CPU
26 Oct - 5 Nov	SMART-1	Several auto-shutdowns of electric propulsion
26 Oct	INTEGRAL	One instrument goes into safe mode because of increased radiation
26 Oct	Chandra	Observations halted again autonomously. Later resumed
27 Oct	NOAA 17	AMSU-A1 lost scanner. Possibly power supply failure
27 Oct	GOES 8	X-ray sensor turned itself off. Could not be recovered
28 Oct	SIRTF	Science experiments turned off for 4 days due to high proton fluxes
28 Oct	Chandra	Observations halted autonomously. Resumed on 1 Nov
28 Oct	DMSP F16	SSIES sensor lost data. Recovered
28 Oct	RHESSI	Spontaneous reset of CPU
28 Oct	Mars Odyssey	MARIE instrument has temperature red alarm and is powered off. Not yet recovered
28 Oct	Microwave Anisotropy Probe	Star tracker reset and backup tracker autonomously turned on. Prime tracker recovered
28 Oct	**Kodama**	**Safe mode, signals noisy, probably lost**
29 Oct	AQUA, Landsat, TERRA, TOMS, TRMM	All instruments turned off or safed. Orbital maintenance of TRMM had to be increased

29 Oct	CHIPS	Computer went offline, contact lost for 18 hrs. Spacecraft went tumbling; recovered successfully later. Offline for a total of 27 hrs
29 Oct	X-ray Timing Explorer	Proportional Counter Assembly (PCA) experiences high voltages. All Sky Monitor autonomously shuts off
29 Oct	RHESSI	Spontaneous reset of CPU
29 Oct	Mars Odyssey	Memory error. Corrected with a cold reboot on 31 Oct
30 Oct	DMSP F16	Microwave sounder lost oscillator; switched to redundant system. Primary sounder recovered 4 Nov
30 Oct	X-ray Timing Explorer	Both instruments recovered, but PCA shuts down again. Recovery delayed
28-30 Oct	(Inmarsat)	Two Inmarsat satellites experienced speed increases in momentum wheels requiring firing of thrusters, and one had outage when its CPU tripped out
28-30 Oct	FedSat	Despite triaxial stabilisation, satellite starts wobbling; has Single Event Upset
28-30 Oct	ICESat	GPS resets. Turning on of the UARS/HALOE instrument delayed
28-30 Oct	SOHO	CDS instrument commanded into safe mode for 3 days
28-30 Oct	ACE	Plasma observations lost
28-30 Oct	WIND	Plasma observations lost
28-30 Oct	GOES	Electron sensors saturated
28-30 Oct	MER 1, MER 2	Enter Sun idle mode after excessive star tracker events. Stable, waiting for recovery
28-30 Oct	GALEX	Two ultraviolet experiments turned off because of high voltages caused by excess charge
28-30 Oct	POLAR	Despun platform went out of lock three times. Recovered automatically each time
28-30 Oct	Cluster	Some of the four spacecraft had processor resets. Recovered.
2 Nov	Chandra	Observations halted again autonomously due to radiation
3 Nov	DMSP F16	SSIES sensor lost data. Recovered.
6 Nov	POLAR	TIDE instrument reset itself . High voltage supplies were disabled. Recovered

3. Cumulated effects

There are surprisingly few reports on satellite anomalies due to the total dose failure of electronic components. For finding such records, we have to turn to situations when spacecraft were exposed to extreme radiation environments, or to situations when such failures did occur, but after a time largely exceeding the original design lifetime of the on-board systems.

This situation is most probably due to excessive design margins coming from radiation environment models, radiation test procedures, component shielding estimates, and finally, additional design safety margins introduced in the parts procurement selection.

The most important contribution to this excessive margin is probably situated in shielding calculations. Shielding has a strong impact on the level received, especially on the electron contribution to total dose. The main problem with shielding had been that, for years, complex mechanical structures were difficult to take into account, so the effective shielding thickness were systematically under-estimated. They still are, but, with the incredible evolution of computer power in the years 1990, it became possible to cope with complex shielding structures, and to take quantitatively into account component packages, and equipment and satellite shielding. Now, it is even possible to run representative Monte-Carlo simulations on complete satellite structures in a reasonable time.

Table 5 gives a simple example of shielding effectiveness for instruments located on ESA INTEGRAL gamma ray space telescope [4]. The table compares the dose received at various mm Al thickness in different geometrical hypothesis. The original general procurement specification for this project was 120 krad (3 mm Al sphere). The instruments studied could be simply simulated by cubes with 5 mm Al thick walls. When this cube was placed on the mechanical model of the telescope tube, it turned out that the expected received dose would be in the range of 6 krad. Component procurement was specified at 12 krad using a factor 2 design safety margin. To change the procurement specification from 120 to 12 krad made a big difference, and allowed to mount high performance commercial 16-bit ADCs. With such margins, it is understandable that few total dose anomalies have been reported.

Table 5. Example of the effect of shielding description on the estimated mission levels for AFEE / DFEE equipment on INTEGRAL – doses in rad [4].

Thickness mm Al	Sphere	Cube	Cube on satellite
1	1.291E+6	8.53E+05	3.80E+05
2	3.435E+5	1.77E+05	9.21E+04
3	1.219E+5	6.34E+04	3.02E+04
4	4.978E+4	2.96E+04	1.22E+04
5	2.279E+4	1.61E+04	5.99E+03

Nevertheless, caution should more than ever be taken for future projects. The very possibility of undergoing precise dose calculations leads to consequent margin reductions, but there are still uncertainties in the calculation processes. If we reduce as far as possible the margin due to shielding calculation, then we remain with the uncertainties related with space environment models and radiation tests procedures. This is the reason why such efforts are undertaken at the international level for progressing in these two fields.

3.1 Artificial radiation belts

As quoted in the introduction, exo-atmospheric nuclear experiments were accomplished by the USA and the USSR before this was banned by international treaties in 1967. These experiments had dramatic consequences not only on the ground (the EMP effect was sensible as far as Hawaii after the Starfish test), but also in space through long-term enhancements of the radiation belt fluxes.

The most noticeable US experiment, because of the power of the weapon and the altitude at which it was exploded, was the Starfish Prime experiment which lead to the formation of enhanced flux zones in the inner belt between 400 and 1600 km and beyond. The USSR experiments are not well documented, but it is believed that they contributed to flux enhancements in the outer belt. These modified fluxes finally extended in all the radiation belt regions and were detectable as late as the beginning of the years 1970. In some zones of the inner electron belt, fluxes increased by a factor 100. Electrons from the Starfish blast dominated the inner belt fluxes for five years. These artificial fluxes delayed the issue of representative natural radiation belts models, and Starfish artefacts might still be present in some zones of the EA8 and AP8 models in use today.

Table 6. Historical artificial radiation belts.

Explosion	Location	Date	Yield	Altitude km	Nation
Argus I	South Atlantic	8-27-58	1 kt	~200	US
Argus II	South Atlantic	8-30-58	1 kt	~250	US
Argus III	South Atlantic	9-6-58	1 kt	~500	US
Argus III	South Atlantic	9-6-58	1 kt	~500	US
Starfish	Johnson Island (Pacific)	7-9-62	1 Mt	~400	US
?	Siberia	10-22-62	? 100s of kilotons	?	USSR
?	Siberia	10-28-62	submegaton	?	USSR
?	Siberia	11-1-62	megaton	?	USSR

In the period directly following these tests in 1962, a series of at least 10 satellites successively failed (7 in 7 months), among which TELSTAR, TRANSIT 4B, TRAAC, and ARIEL, and a number of other ones were impaired such as OSO-1 (Orbiting Solar Observatory).

On 9 July 1962, the day before TELSTAR launch, the US proceeded to the Starfish Prime experiment. TELSTAR orbit, 942/5646 km 44.8°, took the satellite through the inner radiation belt and parts of the outer belt. The radiation exposure was very intense, and increased again after Soviet tests in 1962. It is believed that TELSTAR experienced fluxes 100 times higher than those expected. Four months later, some transistors failed, and finally the loss of the spacecraft (end of life 21 February 1963) was attributed to total ionising dose degradation of diodes in the command decoder.

ARIEL, the first international science satellite, designed and built by NASA/GSFC to study the ionosphere and solar radiation, was launched on 26 April 1962 and provided science data until September 1962, after which it degraded rapidly because of the degradation of its solar panels by the Starfish increased radiation. It continued to function erratically and was cut off in November 1964. TRANSIT 4B and TRAAC mission losses were also attributed to solar cell degradation.

3.2 HIPPARCOS

The ESA star mapping mission HIPPARCOS was launched on 8 August 1989 for a 12°W geostationary position, but its apogee motor failed and the satellite ended in an initial GTO orbit 498/35889 km, 6.5°. This caused the satellite to be exposed to much higher radiation fluxes than it was designed for. Furthermore, a major solar event in March 1991 caused flux enhancements in the radiation belts. This event was one of the major observations of the US CRRES satellite and has revived the interest for understanding the dynamic behaviour of Van Allen belts. The event was observed by HIPPARCOS itself through background noise increase in its instruments [5].

The change in HIPPARCOS orbit lead to an exposure to total dose levels 5 to 10 times more severe than those expected for the nominal orbit. HIPPARCOS flew 5 gyroscopes, 3 of which were active at a time. After 3 years of mission, the five gyroscopes (one of which was already degraded) successively failed in a period of 6 months. For four of them, the spin down and final stop were attributed to the total dose degradation of the access times of bipolar PROM which digitally stored the sinusoidal excitation of fields spinning the wheels. The dose received on the PROM was estimated to about 40 krad. Before the final spin downs, problems appeared when trying to re-start the gyroscopes at cold temperature because of degradations of transistors in the DC/DC converter. Noise and erratic data were also due to the degradation of the 262 kHz clock of the wheel motor power supplies. The last gyroscope, kept in cold redundancy until the very last days of the mission, also failed a few months after it was turned on. This time, the failure was attributed to the radiation degradation of an optocoupler in the thermal regulation system whose received dose was estimated to 90 krad. Finally, communications were lost with HIPPARCOS.

During the last period of its life, HIPPARCOS was operated with only two of the three gyroscopes normally required, and schemes were drawn for operating it without gyroscopes at all. The first gyro-less data were being acquired when a communication failure with the on-board computer put an end to the science mission on 24 June 1993. Further attempts to restore operations were unsuccessful and the mission terminated on 15 August 1993, 4 years after launch.

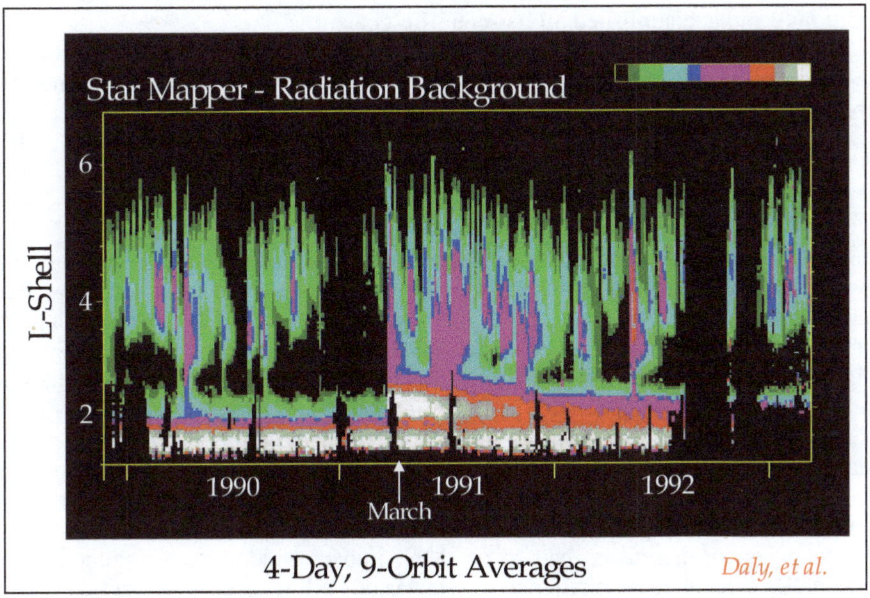

Figure 4. Background noise versus time for Hipparcos Star Mapper, from [5].

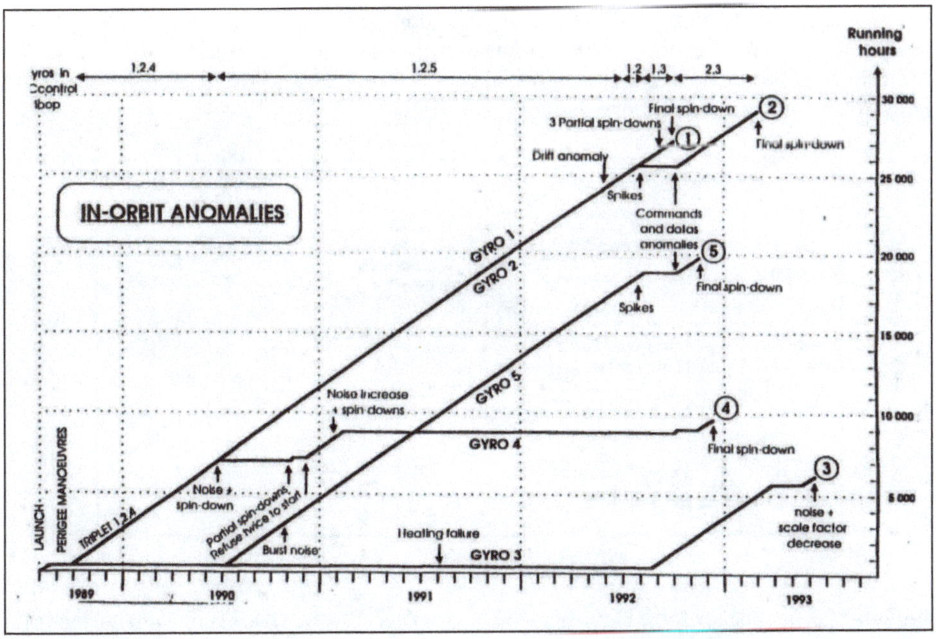

Figure 5. Timeline for Hipparcos gyroscope failures.

In spite of these problems, HIPPARCOS observed 118274 stars with an extreme precision. HIPPARCOS was designed for an operational life of 2.5 years in GEO orbit. It conducted during more than 3.5 years an extremely successful mission in the much more aggressive environment of a GTO orbit (and even in enhanced flux conditions), and accomplished all its scientific goals.

The reason for such a resistance probably lies again in design known and unknown margins. As stated before, the few available cases of total dose anomalies corresponded all to conditions far exceeding the original specifications.

3.3 The GALILEO probe at JUPITER

Extreme environments can also be found around other planets of the Solar System. In this respect, the Jupiter environment is a pandemonium of intense radiation fluxes trapped in the colossal magnetic field of the giant planet. This environment has been first studied by JPL's PIONEER and VOYAGER probes, whose results helped designing the JPL GALILEO spacecraft. Since then, JPL has produced excellent and careful studies of the effects of Jovian radiation belts on GALILEO systems [6][7][8].

GALILEO was almost entirely equipped with rad-hard components and the radiation issue had been as carefully as possible studied by JPL in the design phase. The use of rad-hard components eliminated the SEL risk and most of the SEU problem. For this project, SEE, in the "classical" sense, was not an issue, but background noise in sensors had to be taken into consideration. Cumulated effects (TID, DDD) were a strong challenge : in its various orbits around Jupiter, Galileo was submitted to doses between 10 to 50 krad at each periapsis. A characteristic of these orbits, having a long period, one to three months, was that this exposure was concentrated at periapsis passes, and the components could anneal during the remaining part of the orbit. This characteristic was used by JPL for extending the lifetime of some critical systems. This gave also time for the flight team to work out circumvention solutions before the next pass.

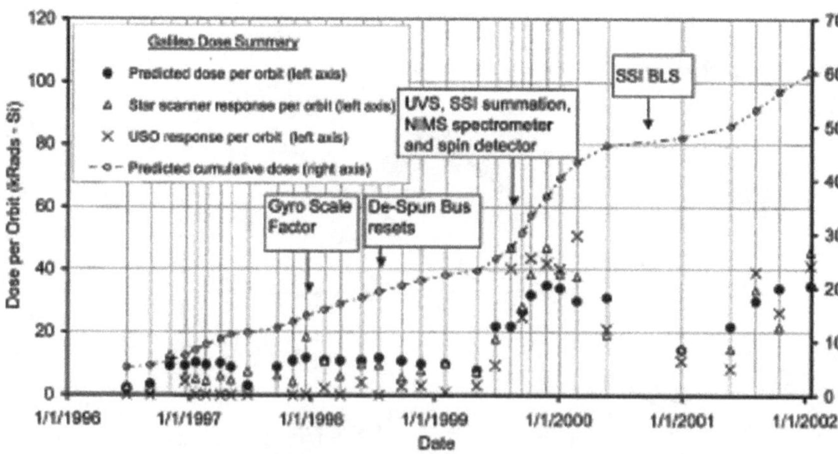

Figure 6. Timeline for some Galileo anomalies, compared with TID accumulation, from [6].

Various effects were observed during the GALILEO mission, we will give some examples from references [6], [7] and [8].

The star scanner was sensitive to high energy electrons (>1.5 MeV), resulting in a radiation noise that could be misidentified for a star. These misidentification lead to erroneous celestial attitude estimates. At the periapsis of one of the most exposed orbits ("E12"), the attitude estimate from the star scanner was in strong disagreement with the inertial attitude estimates from the gyroscopes (themselves affected by radiation, see below). The flight software thought that GALILEO scan platform was not pointed where it has been commanded. The fault-monitors commanded swapping to redundant systems twice in 8 hours. Another problem was related to the averaged radiation signal that increased the signal to noise ratio and caused some stars to fall outside of the expected brightness range, resulting in "missing stars" in the scanner's acquisitions. For passes inside 8 Jupiter radii, only the 10 or 20 brightest stars could be distinguished.

The scale factor of the gyroscopes used for pointing instruments was also affected. A discrepancy between the expected pointing direction and gyro's position estimates grew with mission time, adding its contribution to the hardware swapping in orbit E12. After this orbital pass, the gyros gave unreliable information, and the flight software thought that the spacecraft's antenna was no longer pointed at Earth. The AOCS fired thrusters, maneuvering the spacecraft to a wrong attitude. Six unexpected maneuvers occurred before the flight team succeeded in taking control. Fortunately, the antenna was still pointed sufficiently toward the Earth for receiving telecommands. The gyroscope problem was traced back [7] to leaking currents from a DG181 switch that resets the slew rate integrator to zero after the measurements are performed.

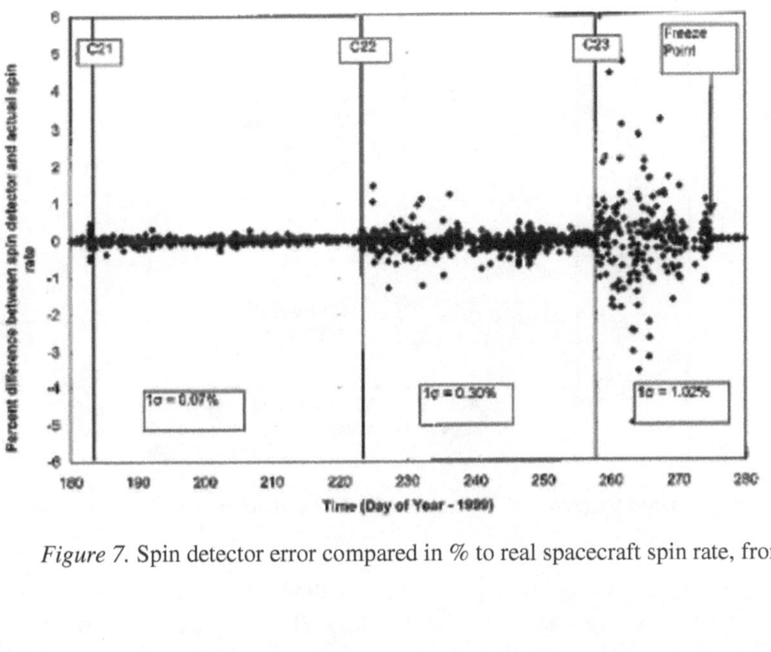

Figure 7. Spin detector error compared in % to real spacecraft spin rate, from [6].

On one of its last orbits, while recording data from the Amalthea moon, data collection stopped abruptly just after this encounter that have brought GALILEO deeper in the radiation belts than ever before. The cause was a sudden switch of the spacecraft to safe mode attributed to proton SETs in the four phase lock loops of the command and data system. The data from Amalthea have been recorded but could not be plaid back to Earth. The problem was traced back to the tape motor drive electronics, and more specifically, to three GaAs OP133 LEDs used in the wheel position encoder. Displacement damage was incriminated as the cause for drops in the LEDs light output. Knowing this, JPL derived a circumvention strategy based on current enhanced annealing properties of the LED. They used a special operating mode allowing current to be applied continuously to the LEDs without trying to move the motor. The recorder began to operate for a few seconds. After each annealing steps, the recorder operating period was longer, and finally JPL succeeded in retrieving all the missing data [8].

GALILEO used also an USO (ultra stable oscillator) for tuning the telemetry 2.29 GHz download frequency. This USO frequency was subject to gradual and permanent drifts due to cumulated radiation . It was also subject to frequency jumps during the passes through the radiation belts, some of which were close to 1 Hz. These drifts were precisely measured because operators of the Deep Space Network had to find the signal by manually sweeping through frequencies. Sometimes, the spacecraft was already transmitting before the carrier wave was locked-up.

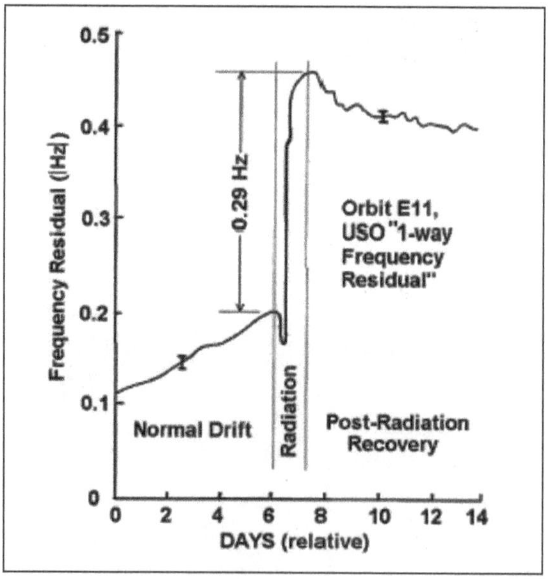

Figure 8. USO frequency shift during orbit E11 radiation belt pass, from [6].

Other effects are described in reference publication [6]. Table 6 from [6] gives a summary of the various systems affected and how JPL found clever circumvention schemes. These schemes could not have been derived without a detailed comprehension of the radiation phenomena.

Table 6. Summary of Galileo in-flight radiation issues, from [6].

SYMPTOM	CAUSE	FIX
Spurious signals at slip rings.	+++,1A	Reprogram software to ignore signals
Camera returns white images.	+++,1A	Drop signal input to sensitive FET.
Infrared spectrometer (NIMS) memory resets.	+++,1B	Scheduled software reloads in radiation.
Instrument (EPD) memory resets.	+++,1C, 4C	Scheduled software reloads in radiation.
Quartz oscillator frequency changes.	+++,2A,3A	Receivers widen bandwidths.
Spin detector signal noise increase.	+++,2A	Reprogrammed to output a constant spin rate determined by other means.
Gyro electronics suffer signal bias.	+++,2B	Frequent characterization tests. Less use of gyros.
Star Scanner sees false stars, blinded.	+++,3A	Use bright stars.
Visible camera (SSI) image noise.	+++,3A	Adjacent pixel averaging.
Polarimeter (PPR) signal noise.	+++,3A,1C	Strip out "impossible" values from data set.
Infrared spectrometer (NIMS) signal noise.	+++,3A,1C	Hand removal of noise from data set.
Dust detector (DDS) signal noise.	+++,3A,1B	Instrument design allows noise/data discrimination.
Voltage controlled oscillator frequency jump.	++,1C, 2C	Pulse current to neutralize ion drift in electronic device.
Particle detector (EPD) sensitivity loss.	++,2B	Park detector behind nearby mass to provide shielding. Loss of channel in one case.
Spectrometer (UVS) grating failure.	++,2B	None - loss of instrument.
Photomultiplier tube (Star Scanner) gain loss.	++,2B	Use bright stars, adjust predicted intensities.
Camera (SSI) image compression failure.	++,2B	None - some forms of on-CCD compression lost.
S-band fr degradation in Io torus.	++,3B	De-weight data for navigation
Magnetometer processor lock-up.	+,1C, 4C	Scheduled power-cycles & memory reloads.
Voltage controlled oscillator frequency drift.	+,2C	Input frequency adjusted to VCO's new base frequency.
Dust detector sensitivity decrease.	+,2C	None
Analog to digital converter shift.	+,2C	None
CMOS Memory cell failures.	+,4C	Reprogram around failed cells.

3.4 Ultra sensitive systems

Some systems are so finely tuned and depend so much on general spacecraft environment stability that dose effects manifestations can be observed even in moderate radiation environments. This is for example the case of the USO (ultrastable oscillator) of the DORIS positioning payload on board JASON-1. JASON-1 is a joint CNES and JPL satellite devoted to altimetry and positioning applied to ocean sciences. The satellite moves on a 1335 km, 66° orbit that makes it pass through moderate flux regions of the proton belt. This orbit is still much more exposed than more conventional LEO Earth observation orbits such as 800 km, 98°. To give an idea of the proportion factor in rough numbers, the dose received on JASON behind a

1 g/cm² spherical shield is about 10 times the dose on an Earth observation satellite. Nevertheless, the order of magnitude of dose exposure at each pass through the SAA is less than 1 rad. Some very sensitive systems such as USOs can be sensitive to dose increments of this order.

The effect of radiation on USO systems is to induce frequency shifts related to the amount of dose received (see also GALILEO effects in the previous paragraph). DORIS sensitivity is such that even slight frequency shifts can have an impact on the final system performance. In the case of the JASON application for DORIS, the satellite passes through the SAA, leading to tiny dose depositions, can clearly be seen in the processed data. The effect of rising or descending passes through the SAA can even be observed. Figure 9 plots the frequency shift amplitudes versus geographical position : the correlation with SAA fluxes is clearly visible.

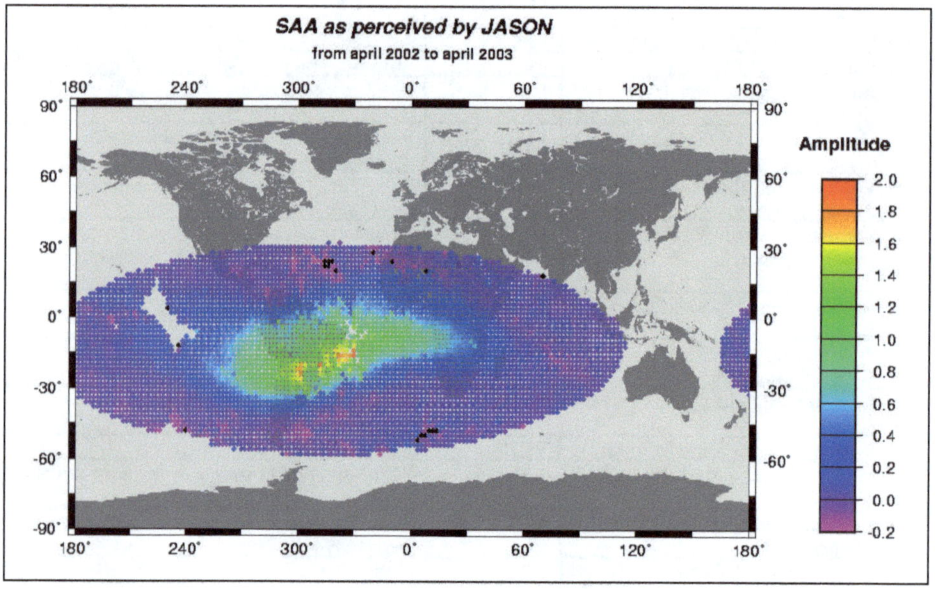

Figure 9. DORIS on JASON frequency shifts, courtesy J.M. Lemoine, GRGS.

Other very sensitive systems are finely calibrated imaging systems, which may record even a slight modification of their pixels' dark current. Figure 10 from [9] shows a cartography of dark current increases for the MISR imager on NASA TERRA satellite when the instrument shutter is closed. The effect of trapped protons is clearly visible.

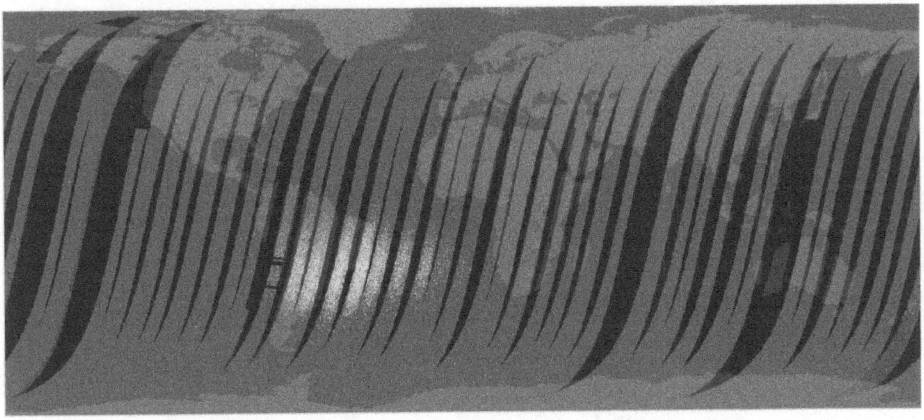

Figure 10. MISR on TERRA dark current increases, from [9], NASA image.

4. Single event effects

4.1 Galactic cosmic rays

Galactic cosmic rays (GCR) are the most efficient particles for inducing single event upsets, but their flux is very weak. Most of today's technologies being sensitive to proton induced upsets as well, GCR anomalies are found in either "old" satellites, or "new" ones situated in a proton free environment.

The CNES SPOT-1,-2 and –3 satellites (820 km, 98.7°) were all equipped with a central processing unit (CPU) whose memory array was made out of 1000 1-kbit HEF4736 static RAMs. These memories, procured in 1986, were only sensitive to CGR. Their LET threshold [10] was 40 MeV.mg-1.cm², so they responded only to heavy ions from the group of Iron.

During the lifetime of these satellites, single event upsets (SEU) were regularly recorded in the CPU memory array []. About half of these SEUs lead to operational problems of various importance, including switching the satellite to safe mode. Later generations of SPOT satellites were equipped with completely different CPUs, not sensitive to these effects.

These SEUs correlate very well with cosmic ray fluxes and their frequency of occurrence follows the expected CGR modulation by the solar cycle. In solar maximum conditions, a fewer SEU rate was observed, whereas solar minimum conditions lead to higher rates [10].

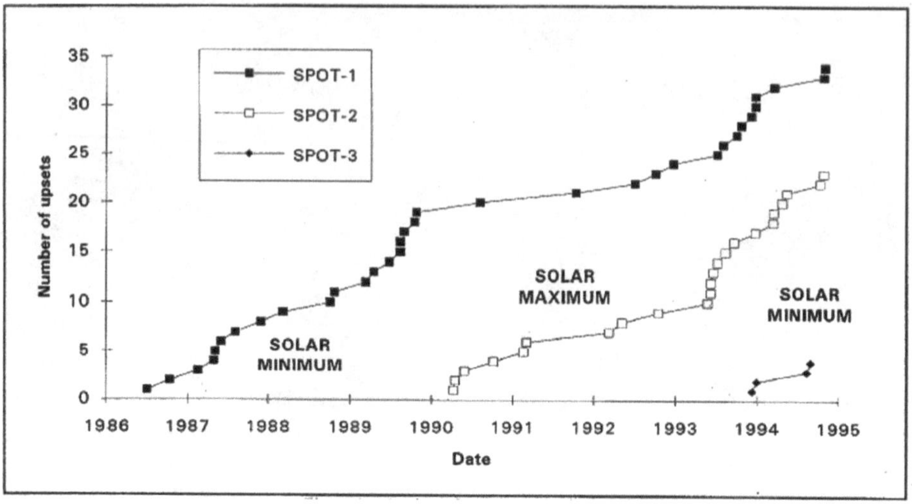

Figure 11. Cumulated number of upset on SPOT-1,-2,-3 OBCs, from [10].

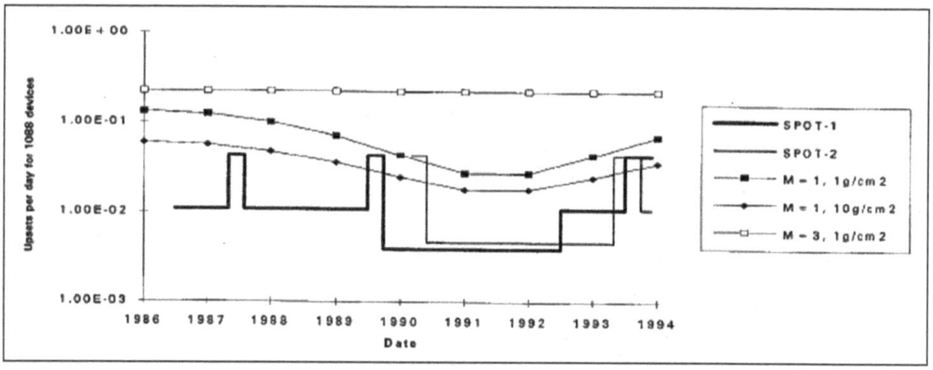

Figure 12. Comparison between measured and estimated upset rates for SPOT-1-2 OBCs, from [10].

As indicated earlier, SEUs are also now associated with particles other than CGR, and associated in-flight anomalies are harder to find. Nevertheless, there is a category of effects which is still related to GCR and is beginning to become a major contribution to operational spacecraft anomalies. This category of effects is made out of single event transients (SETs) in analog electronics. The typical SET effect is an unexpected switch-off or reset, but other manifestations can occur. Such unexpected switch-offs occurred on SOHO, along with various other phenomena, and were studied by ESA-ESTEC [11]. Table 7 from [11] summarises these anomalies. For power supply events SET, PM 139 comparators, UC1707J dual channel power driver, were suspected. ESTEC proceeded to ground test on engineering models and confirmed this cause as the origin of the anomalies.

Table 7. Summary of SOHO unexpected events, from [11].

TABLE I
ESR EVENTS

Date	Unit	Event
04/12-1996	ESR	Attitude Control Unit – PSU reset
19/11-1997	ESR	Attitude Control Unit – self switch-off
03/03-1998	ESR	Centrale Data Mana. Unit – switched
28/11-1999	ESR	Attitude Control Unit – PSU reset
07/01-2000	ESR	Attitude Anomaly Detector – spurious
28/11-2000	ESR	Attitude Control Unit – PSU reset
14/01-2001	ESR	Attitude Control Unit – PSU reset

TABLE II
BDR EVENTS

Date	Unit	Event
12/01-1997	BDR1.2	Switch-off triggered by protection
01/04-1997	BDR1.1	Switch-off triggered by protection
16/05-1998	BDR2.1	Switch-off triggered by protection

TABLE III
VIRGO EVENTS

Date	Unit	Event
09/09-1996	VIRGO	Crashed – self switch-off event
07/05-1997	VIRGO	Latch-up - self switch-off event
20/05-1997	VIRGO	Latch-up – self switch-off event
26/05-1998	VIRGO	Power fail – self switch-off event
12/07-1999	VIRGO	Latch-up in DAS – (1st SEL)
11/02-2000	VIRGO	Latch-up in DAS – (2nd SEL)
30/03-2001	VIRGO	Latch-up in DAS – (3rd SEL)

TABLE IV
LASCO EVENTS

Date	Unit	Event
19/03-1996	LASCO	Voltage anomaly – requiring reboot
10/06-1996	LASCO	Voltage anomaly – requiring reboot
19/12-1996	LASCO	Voltage anomaly – requiring reboot
26/04-1998	LASCO	Hung-up – requiring reboot
28/03-2000	LASCO	PROM off – requiring reboot

SETs are comparatively less sensitive to protons than SEUs are. In contrast with SEUs, SET system effects depend on a variety of bias and use conditions. Generally, these conditions are such that only large transients can have an effect on the system. These large transients being generally due to large energy depositions induced by heavy ions, the SET phenomenon remains closely related to GCR causes.

Many private communications have convinced the authors that SETs are a growing concern for geosynchronous telecommunication satellites. Some cases of operational problems were recorded and attributed to SET effects. These cases were not correlated to the large solar events of the present solar cycle, and seem to be GCR induced. The susceptibility of the word telecommunication fleet in GEO is not surprising simply because these satellites constitute the bigger sample of spacecraft in a GCR exposed orbit. The geosynchronous orbit is far above the Earth trapped proton belt, and, in the absence of solar events, the only contribution to single event effects thus comes from galactic cosmic rays.

4.2 Solar particles (protons, ions)

Few cases exist for anomalies that would have been triggered by solar heavy ions. Nevertheless, there is at least one recently documented case reported by NASA/GSFC [] on the Microwave Anomaly Probe (MAP), and the long term observation of upsets in TDRS-1 AOCS memory. Solar proton effects are very common.

4.2.1 MAP processor reset

MAP was launched on 30 June 2001 and was injected in a stabilised orbit around L2 Earth Lagrangian point. On 5 November 2001, the spacecraft AOCS system switched MAP to a safehold condition caused by a reset of the spacecraft's processor. About 15 hours later, the ground control succeeded in returning the spacecraft to normal operations.

NASA/GSFC attributed the root cause of the processor reset to a single event transient (SET) on a voltage comparator (PM139), which caused a voltage drop leading to the validation of the reset signal [12].

This event is correlated to a large solar storm in the period 3 to 7 November 2001. All the available monitors showed an increase in particle fluxes. In the MAP case, the SETs on the PM139 in the processor reset circuitry can cause a reset only if they correspond to high energy ions whose LET > 2 MeV.mg-1.cm2. The occurrence of the event was thus associated with both solar flare conditions and the presence of high energy ions in the flare composition, so the investigators looked for data evidencing that the flare was an ion flare.

These data were available through the DERA's (now QinetiQ, UK) CREDO instrument on the MPTB (Microelectronics and Photonics TestBed) payload. CREDO measured an ion flux increase in the range 0.1 < LET < 10 MeV.mg-1.cm2.

As a conclusion, all the available data support the assessment that the MAP reset event was caused by a SET induced by solar flare ions.

4.2.2 TDRS-1 AOCS RAM upsets

The Tracking and Data Relay Satellite (TDRS-1) has experienced single event upsets since its launch in April 1983. These upsets take place on very sensitive bipolar 93L422 memories which are used in the RAM of the Attitude and Orbit Control System (AOCS). These upsets were mission critical because they could cause a tumbling of the satellite. Heavy load was put on the ground control teams to keep the spacecraft in the proper attitude.

Considering the extreme sensitivity of the 93L422 bipolar SRAMs, all particle types could induce an upset. The comparison of the upset rate and the environment conditions have been very well documented in [13] and [14] for example. The first anomalies, due to radiation belt protons, occurred during the transfer of the satellite to its GEO position, between April and July 1983. As soon as the spacecraft was in its GEO position, the anomalies followed the cosmic ray modulation of the solar cycle. Then, in 1989, during the maximum of solar cycle 22, a series of large flares occurred. The later one, on 19 October 1989, the day after GALILEO launch, was one of the most powerful solar flare ever recorded. The upset count on TDRS-1 jumped from about 15 to 20 upsets / week, to 249 events in 7 days (19-25 October 1989).

Fortunately, the GALILEO Jupiter probe was still near Earth and its ion counter instruments could measure the ion composition of the flare. The NOAA GOES-7 proton data could also be used for proton flux estimations. Considering the sensitivity of the 93L422 memories, the investigators concluded that about 30% of the upsets were due to solar protons and 70% due to solar ions. This balance depends on component sensitivity and satellite shielding and should not, of course, be extrapolated to other applications.

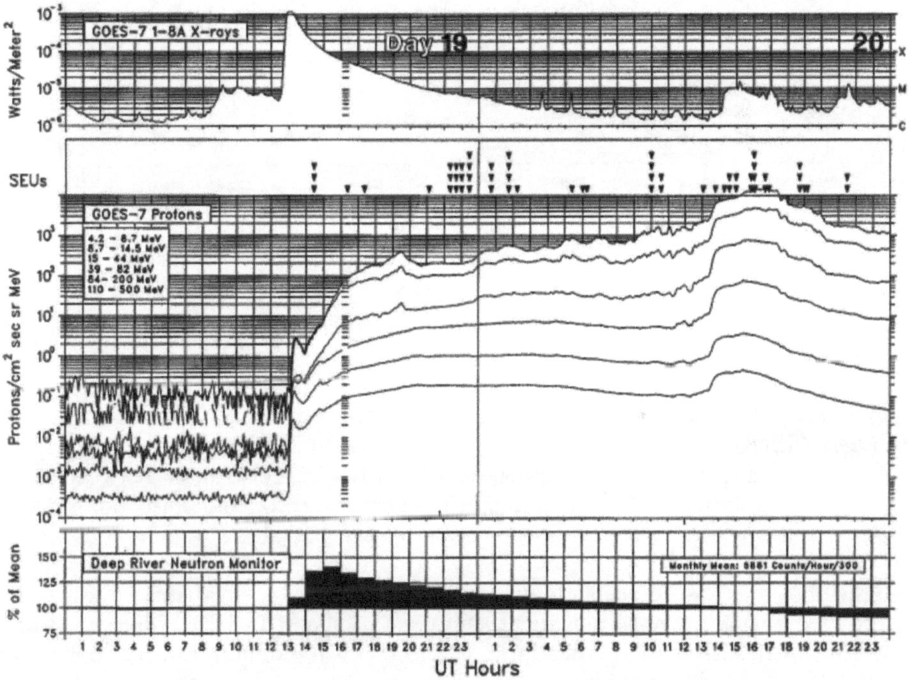

Figure 13. TDRS-1 AOCS RAM upsets compared with GOES flux measurements during October 1989 solar flare, from [13].

4.2.3 SOHO SSR upsets

Another particularly well documented correlation between solar events and radiation issues [11] is the case of single event upsets in the ESA / NASA SOHO spacecraft at the L1 Lagrangian point. Figure shows the evolution of the SEU rate between 1996 and 2001. The background rate is due to galactic cosmic rays. The solar cycle modulation of this background rate is visible in the plot. The sudden increases in the SEU rate corresponds to solar events.

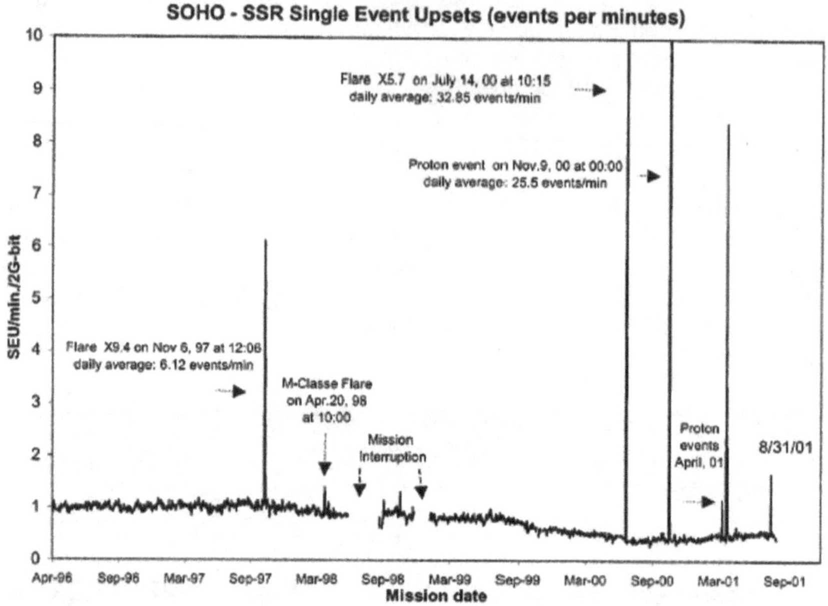

Figure 14. SOHO SSR upsets versus time, from [11].

4.3 Trapped protons

NASA/GSFC has carefully studied [15] the upsets observed on some solid state recorders (SSRs) on SAMPEX (512 x 687, 81.7°) and TOMS/METEOR-3 (1183 x 1205, 82.6°). Figures 15 [15] show very clearly the correlation between upset occurrence and the SAA. Note that due to the error detection and correction routines implemented on these applications, these upsets had little impact on system performance.

SAMPEX was also a technology demonstrator using up to date techniques not before tried in space applications. For example, it used a MIL-STD-1773 optical fibre data link [15]. These data links use a LED as an emitter, a fibre optics for transmission, and a photoreceptor as a receiver. When an upset occurs in a data transmission (mainly due to particle transients on the photoreceptor), a bit is corrupted and the system asks for a retry. Figure 15 shows again a clear correlation between MIL-STD-1773 bus retries and the SAA.

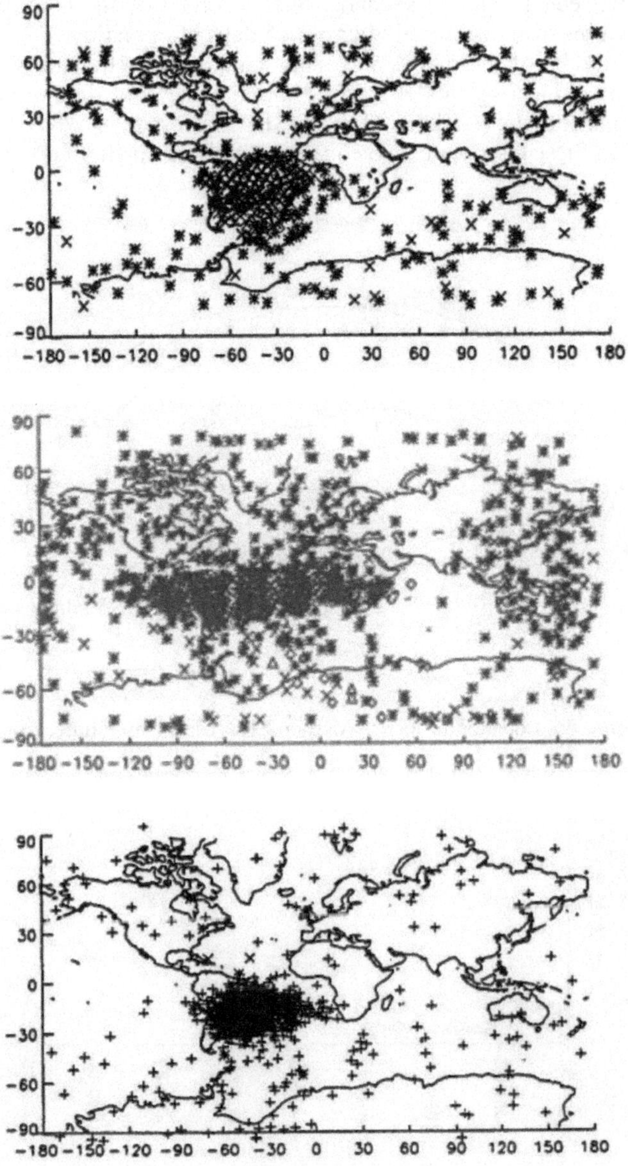

Figure 15. SAMPEX and TOMS / METEOR-3 events, upper chart : SAMPEX SSR upsets, middle chart : TOMS SSR upsets, lower chart : SAMPEX MIL-ST-1773 bus transmission retries, from [15].

Latch-ups have also been attributed to trapped protons. One of the most famous cases, the first to evidence proton induced latch-up, was the failure of the non-ESA PRARE altimeter on-board the ESA Earth observation satellite ERS-1 (774 km, 98.5°). At the moment of the switch-off event, an increase in primary power of about

9 W, lasting between 16 and 32 seconds, was recorded by the housekeeping system of the satellite. This event occurred after only 5 days of operation. The failure location is right in the middle of the SAA zone. The anomaly was studied by ESA [16], and ground tests on engineering models proved that the origin of the failure was due to a latch-up on one of the 64-kbit CMOS SRAMs used in PRARE. Occurrence estimations from these tests were also consistent with the in-flight observation.

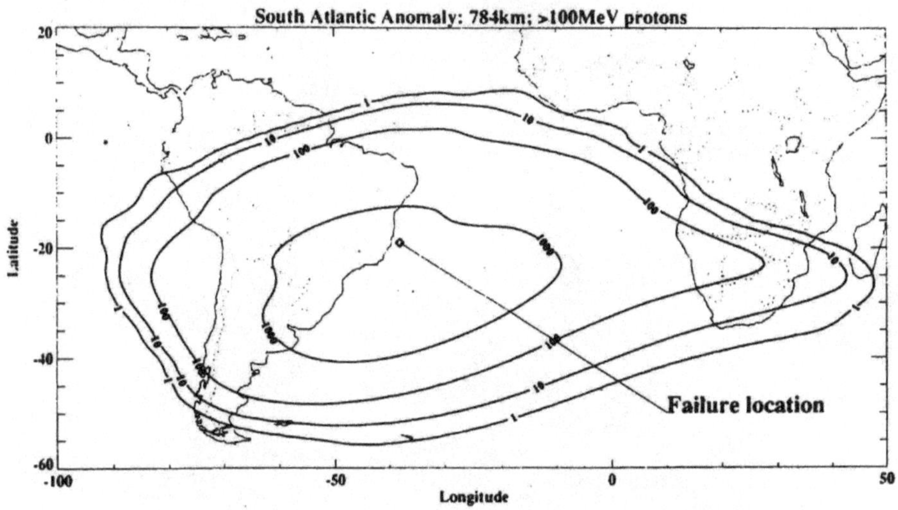

Figure 16. Location of the PRARE on ERS-1 latch-up failure, from [16].

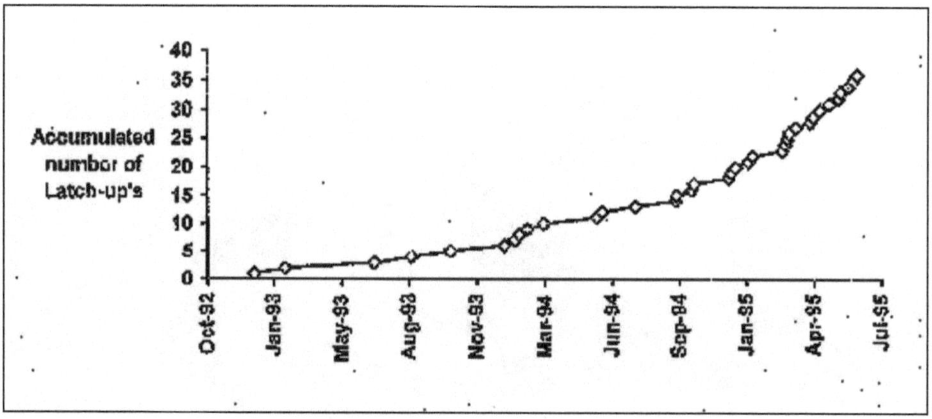

Figure 17. Number of latch-ups versus time on FREJA, from [17].

Another well documented LEO case is the one of the joint Swedish-German project Freja (601 x 1756, 63°) which experienced a number of SEE effects including false commands (0.5 to 1 per day, in the SAA), telemetry format errors, memory SEUs, and finally latch-ups on the CPU boards. The latch-ups were detected in the housekeeping data through records of switches to redundant units. The anomalies were studied by ESA [17], and the latch-ups were reproduced on flight model spares using high energy proton beams. The origin of latch-up events was traced back to CMOS NS32C016 circuits in the CPU board.

5. The particular case of sensors

The various sensors used in a satellite platform or payload are particularly sensitive to radiation effects because of their exposition and the nature of their detectors which are, by design, very sensitive to charge deposition. The sensors possibly impacted are Earth imaging sensors, attitude and orbit control (AOCS) sensors such as star trackers, and finally all types of astronomy payloads (gamma ray sensors, X-ray CCDs, UV, optical and IR imaging sensors).

Current imagers for ultraviolet, optical and near-infrared wavelengths rely in their great majority on CCD linear or matrix detectors. Increasing use is being made of charge injection devices (CIDs) and CMOS active pixel sensors (APSs) as these do not suffer problems associated with charge transfer between pixels. At other wavelengths of the infrared spectrum, the major sensor material types are InSb, InGaAs, GaAs/GaAlAs, HgCdTe, PtSi and extrinsic silicon.

Space radiation effects on these various types of sensors may be :

- transient signals from proton or heavy ion prompt ionisation, these are single events. Depending on the arrival direction of the particle, or whenever secondaries are created, the effects can involve one or many pixels and can appear as tracks in matrix detectors.
- semi-permanent (because of annealing) or permanent ionising dose or dis placement damage degradation of pixel properties.

5.1 Transient signals

Transient signals are due to the collection of the charge generated by a proton or ion track. They can be seen in many space images, and are particularly associated with radiation belt (and especially the South Atlantic Anomaly for SSO satellites), or solar flare protons. When first discovered on their images above the South-Atlantic by the CNES SPOT-1 project team, they were surnamed "UFOs". Some examples of "UFOs" on SPOT images are shown in Figure 18.

The transients can affect one or many pixels and their manifestation also depends on the nature of the read-out electronics. The illustrations in Figure 19 are part of SPOT images showing such effects. The dashed appearance of some "UFOs" is due to the odd/even scheme applied for the readout registers. This effect is zoomed in Figure 20. The proton has impacted a given pixel, and charge collection has spread to nearby pixels, as shown in Figure 20 by the dimming of pixel over-current on both sides of the original impact. The read-out architecture relies on the separate readout of even and odd pixels. The effect results in a dashed line of pixels.

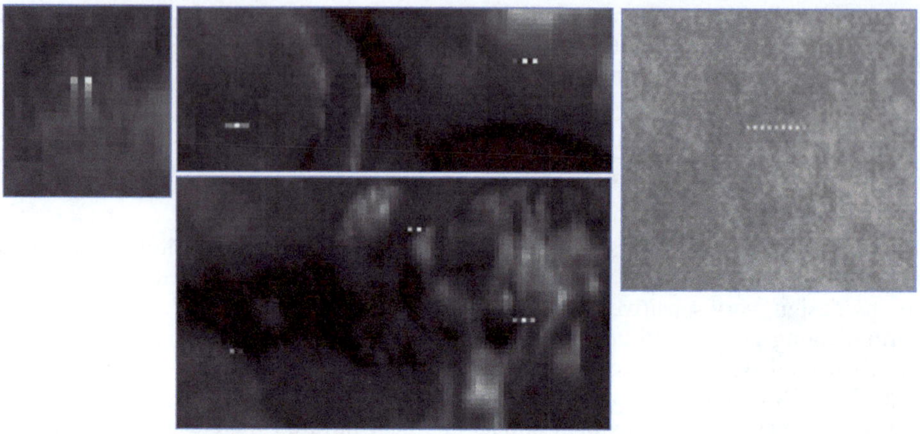

Figure 18. Examples of transient signals on various CNES SPOT images – credits CNES, distribution Spot Image.

For Earth imaging satellites usually orbiting in SSO orbits (600-800 km, 98°), the transients are not such a critical issue because their number remains low and they can be easily filtered in image processing. For higher orbits, and space science or astronomy imagers, more dramatic effects can be observed during solar flares. For SSO orbits, solar flares would have an effect only in the polar zones, where landscape uniformity and reflectance, and also a lesser request for image acquisition, contribute to minor the impact of the phenomenon.

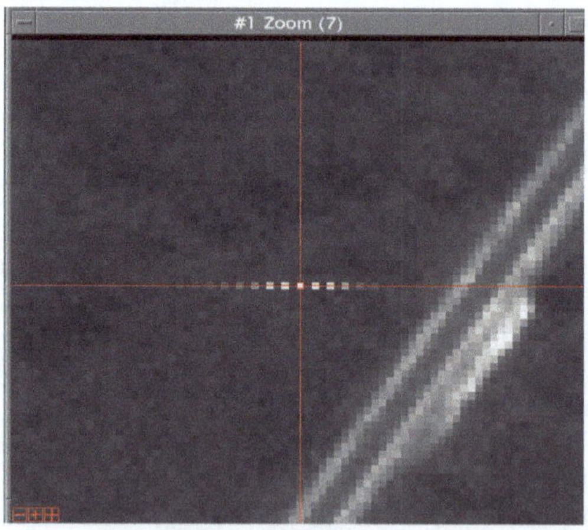

Figure 19. Example of a transient on part of a SPOT5 HRG image – credits CNES, distribution Spot Image.

In some occasions, the transients due to particle tracks can completely blur an image, as illustrated by the Figure 20 showing successive images of the Earth taken during the July 2000 solar flare by the VIS (visible imaging system) on the NASA POLAR satellite.

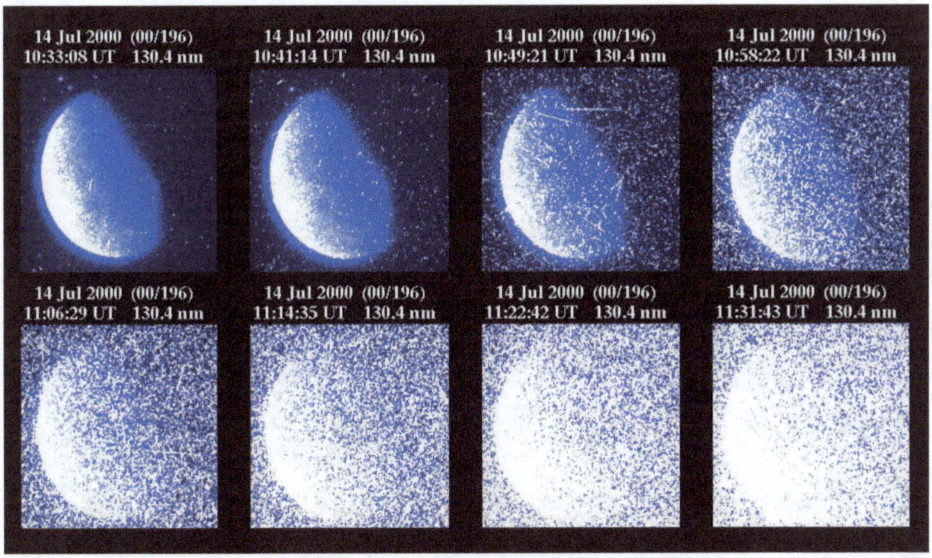

Figure 20. Blurring of the VIS camera on NASA POLAR satellite during July 2000 solar flare – credits NASA.

Figure 21. JPL GALILEO image of Io showing particle transients, image n°PIA00593 [1] – credits JPL.

Effects have also been observed on imagers on board planetary probes, see for example the extraordinary GALILEO image shown in Figure 21 (NASA Photojournal web page, image PIA00593 [18]) and quoted in [6]. The image in Figure 19 shows a Sodium volcanic plume on the Io moon and its scattering of Jupiter's light. White dots are impacts from particles of Jupiter's radiation belts, whose flux is very intense at the level of Io's orbit.

The images in Figure 22 shows the effects of flare protons on SOHO images of the Sun itself. The proton tracks are seen on the detectors a short time after the first images of the flare itself were taken : the imagers are also "filming" the arrival of the protons.

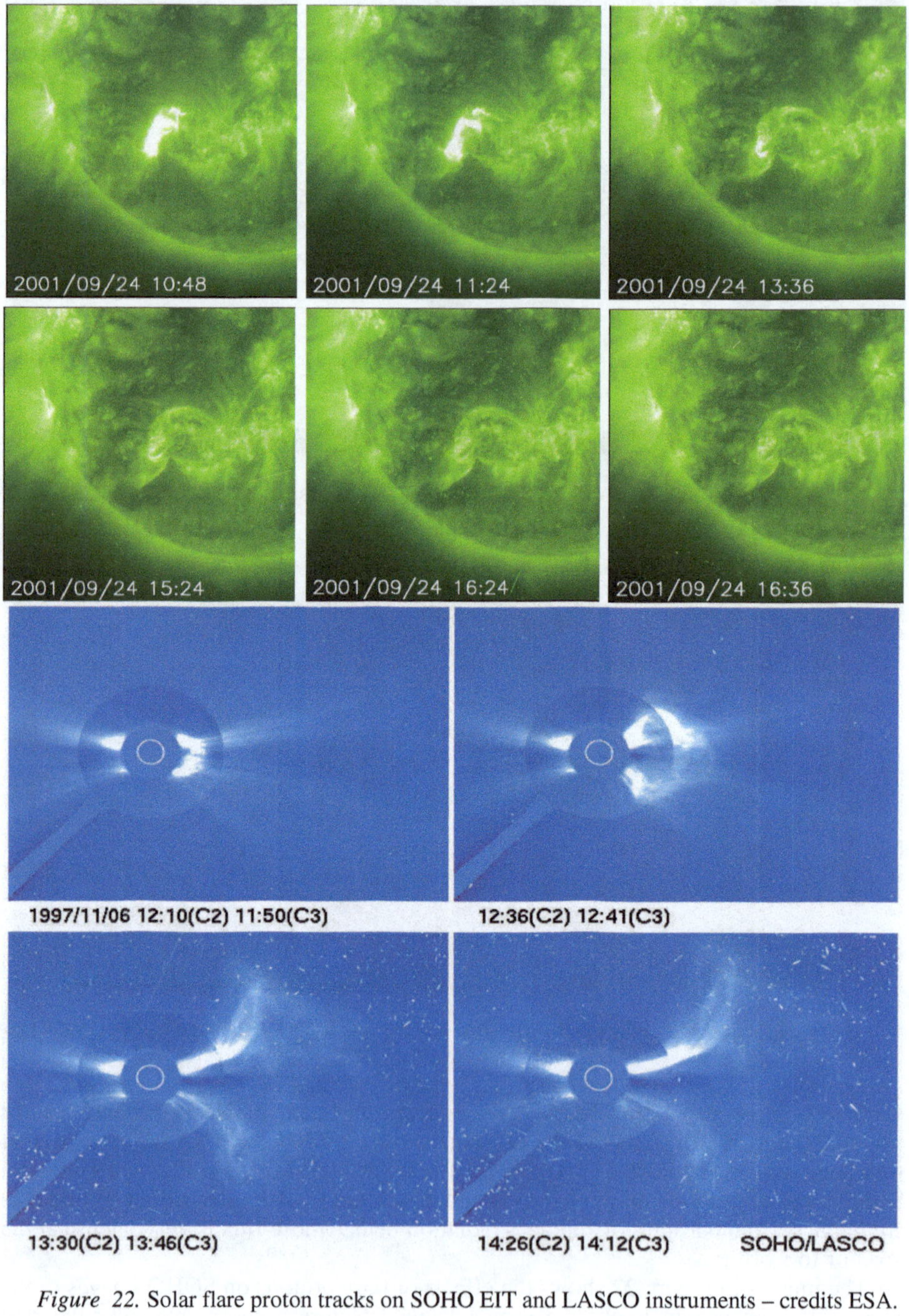

Figure 22. Solar flare proton tracks on SOHO EIT and LASCO instruments – credits ESA.

Sun, Earth and star sensors have been of common use on spacecraft for attitude determination. The growing requirements for fine pointing accuracy tends to spread and privilege the use of star trackers in the last generations of satellites.

Similarly to imaging systems, these sensors use photodetectors such as CCDs or APS. A specific problem with star trackers is due to proton or ion transient signals in the CCD or APS detector. The star pattern recognition algorithm can think that the impacted bright pixels are false stars, and get "lost" if the rate of events is too high. This effect could severely impact on the star tracker's dependability in regions and times of high proton fluxes.

As an example, the star tracker of the CNES/JPL JASON-1 satellite went out of the AOCS loop when the spacecraft passed through the South Atlantic Anomaly. Figure 23 shows examples of read-outs of the CCD matrix (EEV) during acquisitions inside and outside the SAA. Outside the SAA, the sky is "clear", but inside the SAA, one could observe the multiplication of white dots and also the diagonal tracks visualising the travel of grazing protons through the detector. Close-ups of these tracks are shown below the full array images in Figure 24. Figure 25 is a close up of real star images. In the case of JASON (1335 km, 66°), the outage duration was short (a maximum of 20 mn), and predictable, as well as the orbital passes concerned. During these periods, JASON was steered with its gyroscopes without any degradation of mission parameters.

The JPL GENESIS probe uses the same star tracker as JASON. During 21 April 2002 solar event, the star tracker was blinded 4 times by high energy protons.

Switching to redundant units is not an applicable solution in this case, because they will exhibit the same behaviour than the nominal ones. Hardening solutions consist in using radiation improved or hardened CCDs or APS, and trying to design the smarter algorithms, but the signal to noise ratio may be so degraded that this later solution has limits.

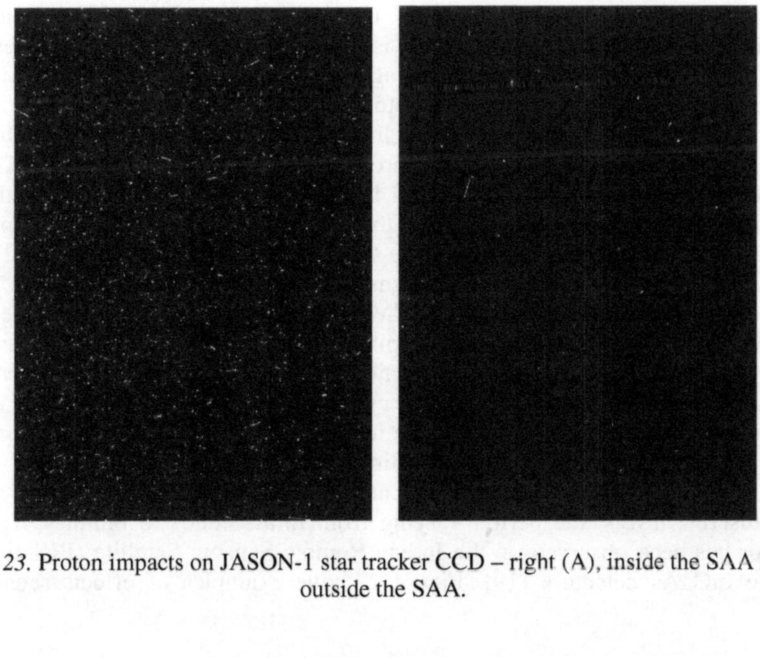

Figure 23. Proton impacts on JASON-1 star tracker CCD – right (A), inside the SAA left (B), outside the SAA.

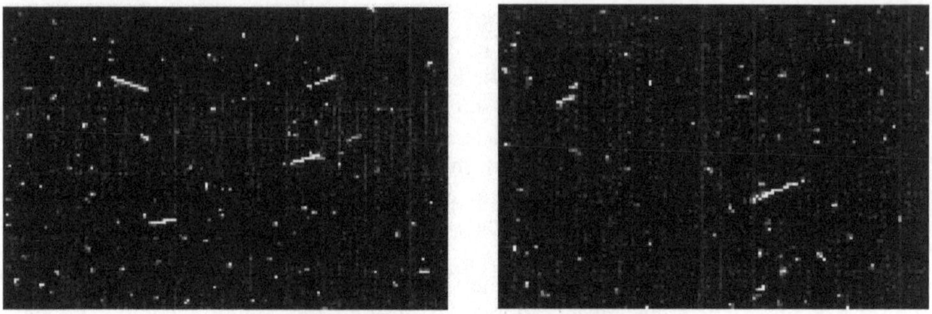

Figure 24. Close-up on image (A) inside the SAA – proton tracks are clearly visible.

Figure 25. Close-up on image (B) – clear star field.

5.2 Permanent or semi-permanent damage

The French Earth observation satellite SPOT 4 (830x830 km, 97°) is using 3 MIR detectors (Middle InfraRed) based on InGaAs photodiodes. This linear array, developed by ATMEL Grenoble, France, is built with ten elementary modules butted together and enclosed in a ceramic package. Each elementary module consists of a two-sided buttable InGaAs die of 300 photodiodes and two 150-element Silicon-based CCD arrays on either side of the photodiode die to multiplex the signal from the photodiodes (Fig. 26). The photo-diodes are reverse-biased (1V) and the device works in vidicon mode i.e. the photo-generated holes (electrons are grounded) reduce the reverse-bias of the diode. A fraction of the known amount the electrons (preload), which are injected into the diode, restore this reverse-bias. The remaining electrons are collected back as an inverted photo signal.

Since they have been launched, the MIR detectors have shown an unexpected behaviour of the dark current. Some pixels have a dark current that increases suddenly. After such a jump, the majority of affected pixels recover stable performances without any degradation except in their dark current. But, when the dark current jump is high (10 to 100 times the initial dark current level) the performances regarding noise as well as temporal stability of the dark current can be affected. The pixel becomes "unstable" with the dark current level periodically switching between several discrete levels, the period varying from milliseconds to minutes. The same behaviour has been observed on the Indian Remote Sensing Satellite (IRS-1C) using the same InGaAs detectors [19]. Figure 27 give examples of effects seen on the pictures.

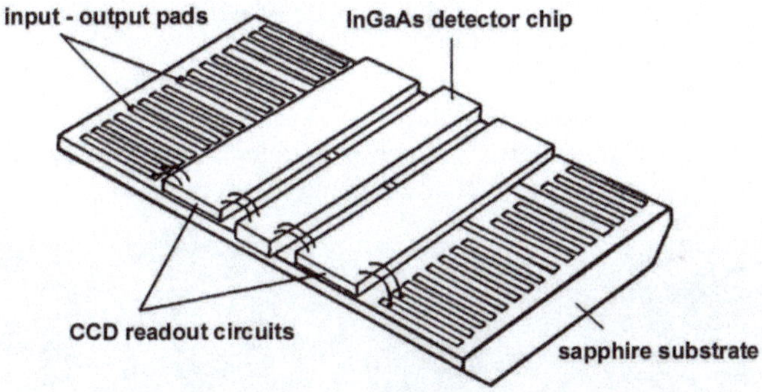

Figure 26. Elementary module of the MIR detector.

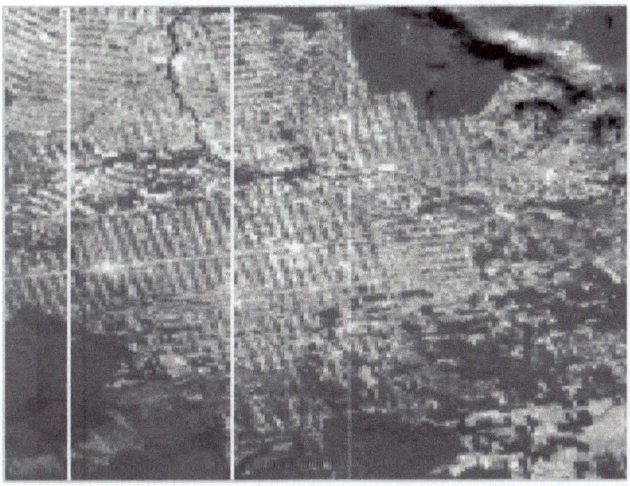

Figure 27. Example of effects : right, "hot pixels" : some pixels are saturated leading to "white" columns, left : example of RTS leading to "white / black" toggle – credits CNES, distribution Spot Image.

These phenomena have been studied by several authors [20]-[24] in the case of silicon detectors. These works have shown that displacement damage induced by protons can lead to the apparition of dark current spikes (or hot pixels) and the apparition of dark current instability. In the last case, the dark current switches between well-defined levels with the appearance of a random telegraph signal (RTS).

Proton tests were done on spare parts. The results showed that the dark current behaviour observed in-flight was compliant with displacement damages induced by protons for extreme dark current values as well as for the unstable pixels. It was also observed that the mean dark current is correlated with the NIEL calculated in InGaAs. Calculations lead to conclude that these high values were due to inelastic interactions. RTS signals were also characterised (Fig. 20). This study resulted in proposing a semi-empirical model prediction of the number of extreme values. This model gives good results in the case of SPOT 4 [25].

Some processing have been set up to correct the proton induced defects. One dark current measurement every 6 days allows to set the list of defective detectors and to compute the on board dark current which is updated every 26 days. Figures 28a and 28b show a MIR picture respectively before and after correction.

a. 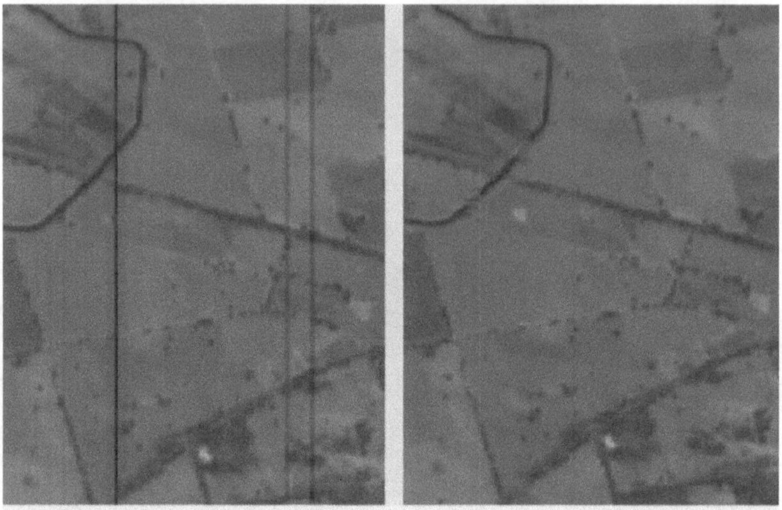 b.

Figure 28. Effects of the processing implemented to correct the MIR pictures – credits CNES, distribution Spot Image.

Besides, an automatic processing of the images dealing with these defective detectors was implemented. It aims a estimating the dark current variation of the detectors from a statistical analysis of the image. According to this variation, the detector will be either recovered by updating its dark current or interpolated.

As far as star trackers are concerned, TID and DDD effects on detectors will increase the mean dark current and alter dark current non uniformity during the mission time in orbit. This would have two effects:

- modifying the uniformity of sensitivity
- and the global Signal / Noise ratio would also decrease.

Thus, it implies that lower magnitude stars would be more hardly detected and the barycentric centre calculation perturbed. The sensor performances in term of bias and noise would be degraded.

6. Dedicated instruments and experiments

6.1 Space environment monitors

The space radiation fluxes can only be measured through in situ observations : there is no other solution than to "go out there" and observe what is going on. All the radiation dimensioning models that we used today are based on actual measurements in space interpolated for completion of the model. In situ measurements are the cornerstone of space environment models. Since the beginning of the space era, both civilian and defence institutions in the USA have spent great efforts in this field. These efforts lead to the internationally used environment specification models and proved invaluable for the world space community. Japanese JAXA has also developed an ambitious program for space environment characterisation and technology assessment.

Apart from large scientific payloads able to characterise the nature and energies of the particles, such as used on US CRRES satellite, medium or small size instruments can be of invaluable interest. Such small size instruments can be proposed on a variety of missions with little dimensioning impact, and could also be used by operational satellites as "black boxes" able to help in post-event analysis if anomalies occur.

We will not undergo here a detailed description of the detection principles used in large scientific payloads. Let us just quote solid state detectors, scintillators, mass spectrometry, electrostatic filters. Small size instruments, that we will call "monitors" usually rely on the simplest principles to implement, i.e. solid state detectors. These detectors are fully depleted Si junctions of various thickness for addressing various energy ranges. They can be used in "telescope" mountings for improving energy resolution.

In Europe, ESA, CNES and other institutions are developing small size instruments and trying to embark them as often as possible on various missions. ESA developed the SREM (standard radiation environment monitor), presently being flown on STRV-1c, PROBA-1 (LEO mission), INTEGRAL, ROSETTA, and the MRM (miniaturised radiation monitor, based on an APS detector). CNES developed, from an original design flown by CESR on XMM, a family of instruments (ICARE) which were flown on space stations MIR and ISS, SAC-C (LEO/SSO mission) and STENTOR (aimed for GEO orbit but the launch failed). In the USA, Amptek Inc. proposes the CEASE (compact environment anomaly sensor) on a commercial basis.

6.2 Technology experiments

The goal of technology experiments is to give feedback from the "real world" to radiation effects on components models. These dedicated experiments help in developing new models or more finely tune older ones. They can also provide advanced information on enabling technology whose use is not common at the current date on operational satellites. As an example, CNES had flown the IBM LUNA-C 16-Mbit DRAMs on the "SPICA" version of ICARE on space station MIR, before these same components were used on SPOT-5 SSR.

Technology experiments allow to quantify in details the observed effects. For example, upsets on operational SSRs may not be dated and localised very precisely, and the SEU or MBU signature may not be recorded. A dedicated experiment could address all these parameters. Technology experiments allowed to evidence all the effects of interest in the past and today (SEU, SEL, SEB, ELDRS,…). The problem at stake is obviously, to quantify the relevance of the ground tests, the flight behaviour estimation techniques, and the exact rating of design margins.

Here again, the USA have lead this field with major technology initiatives such as MPTB (Microelectronics and Photonics Test Bed), APEX (Advanced Photovoltaic and Electronics Experiments), a variety of piggy-back payloads, and today the SET (Space Environment Test Bed) initiative within the NASA LWS program (Living With a Star). These projects are in most of the cases opened to international collaboration : for example, TIMA Laboratory (Grenoble, France), CNES and NASA/GSFC were able to fly a joint artificial neural network processing board on MPTB. In Japan, a strong technology program is also developed by JAXA. In Europe, CNES has been working in this field since more than 15 years, through initiatives such as a collaboration with RKK-ENERGIA on MIR (access time to space 6 months), the SPOT-4 technological passenger (PASTEC) and the ICARE instruments which associate environment monitoring and component test board. An example of such results from ICARE on SAC-C are shown in Figure 29. These results are representative of actual SEU effects on operational SSRs (see for example Figure 15). ESA is currently planning future developments.

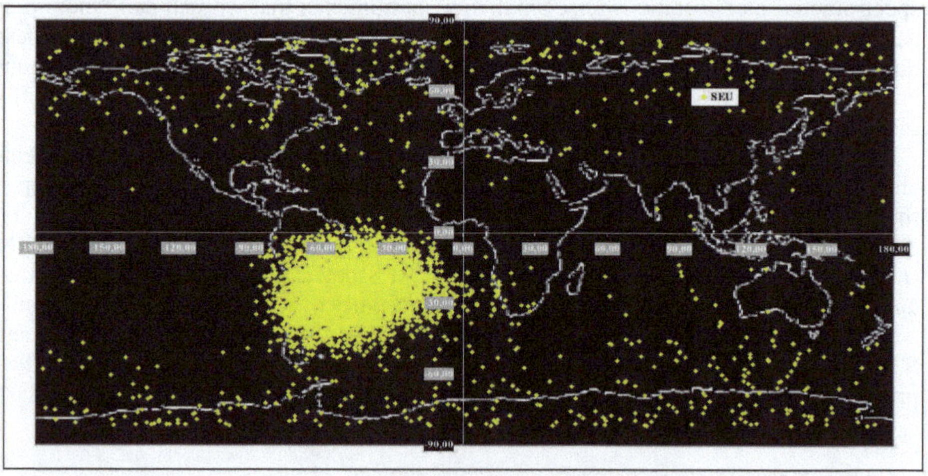

Figure 29. Results from ICARE / SAC-C component test board : cumulated upset map for 7 memory types, 29 memory samples, from 256 k-bit SRAM to 64 M-bit DRAM – CNES / ONERA results.

References

[1] "Spacecraft system failures and anomalies attributed to the natural space environment", NASA reference publication 1390, August 1996.

[2] H.C. Koons, J.E. Mazur, R.S. Selesnick, J.B. Blake, J.L. Roeder, P.C. Anderson, "The impact of the space environment on space systems", 6th Spacecraft Charging Technology Conference, AFRL-VS-TR-20001578, 1 September 2000

[3] http://www.sat-index.com/

[4] E. Vergnault, R. Ecoffet, R. Millot, S. Duzellier, L. Guibert, J.P. Chabaud, F. Cotin, "Management of radiation issues for using commercial non-hardened parts on the Integral spectrometer project", 2000 IEEE Radiation Effects Data Workshop, p68.

[5] E.J. Daly, F. van Leeuwen, H.D.R. Evans, M.A.C. Perryman, "Radiation belt and transient solar magnetospheric effects on Hipparcos radiation background", IEEE. Trans. Nucl. Sci., vol 41, no 6, p2376, Dec. 1994.

[6] P.D. Fieseler, S.M. Ardalan, A.R. Frederickson, "The radiation effects on Galileo spacecraft systems at Jupiter", IEEE. Trans. Nucl. Sci., vol 49, no 6, p2739, Dec. 2002.

[7] A.R. Frederickson, J.M. Ratliff, G.M. Swift, "On-orbit measurements of JFET leakage current and its annealing as functions of dose and bias at Jupiter", IEEE Trans. Nucl. Sci., vol 49, no 6, p2759, Dec. 2002.

[8] G.M. Swift, G.C. Levanas, J.M. Ratliff, A.H. Johnston, "In-flight annealing of displacement damage in GaAs LEDs : a Galileo strory", IEEE Trans. Nucl. Sci., vol 50, no 6, p1991, Dec. 2003.

[9] http://eosweb.larc.nasa.gov/HPDOCS/misr/misr_html/darkmap.html

[10] R. Ecoffet, M. Prieur, M.F. DelCastillo, S. Duzellier, D. Falguère, "Influence of the solar cycle on SPOT-1,-2,-3 upset rates", IEEE Trans. Nucl. Sci., vol 42, no 6, p1983, Dec 1995.

[11] R. Harboe-Sorensen, E. Daly, F. Teston, H. Schweitzer, R. Nartallo, P. Perol, F. Vandenbussche, H. Dzitko, J. Cretolle, "Observation and analysis of single event effects on-board the SOHO satellite", RADECS 2001 Conference Proceedings, p37.

[12] C. Poivey, J. Barth, J. McCabe, K. LaBel, "A space weather event on the Microwave Anisotropy Probe", RADECS 2002 Workshop Proceedings, p43.

[13] D.C. Wilkinson, S.C. Daughtridge, J.L. Stone, H.H. Sauer, P. Darling, "TDRS-1 single event upsets and the effect of the space environment", IEEE Trans. Nucl. Sci., vol 38, no 6, p1708, Dec. 1991.

[14] D.R. Croley, H.B. Garrett, G.B. Murphy, T.L. Garrard, "Solar particle induced upsets in the TDRS-1 attitude control system RAM during the October 1989 solar particle events", IEEE Trans. Nucl. Sci., vol 42, no 5, p 1489, October 1995.

[15] C.M. Seidleck, K.A. LaBel, A.K. Moran, M.M. Gates, J.M. Barth, E.G. Stassinopoulos, T.D. Gruner, "Single event effect flight data analysis of multiple NASA spacecraft and experiments ; implications to spacecraft electrical design", RADECS 1995 Conference Proceedings, p581.

[16] L. Adams, E.J. Daly, R. Harboe-Sorensen, R. Nickson, J. Haines, W. Schafer, M. Conrad, H. Griech, J. Merkel, T. Schwall, R. Henneck, "A verified proton induced latch-up in space", IEEE Trans. Nucl. Sci., vol 39, no 6, p1804, Dec. 1992.

[17] B. Johlander, R. Harboe-Sorensen, G. Olsson, L. Bylander, "Ground verification of in-orbit anomalies in the double probe electric field experiment on Freja", IEEE. Trans. Nucl. Sci., vol 43, no 6, December 1996, p2767.

[18] http://photojournal.jpl.nasa.gov/catalog/PIA00593.

[19] A.S. Kirankumar, P.N. Babu, and R. Bisht, "A study of on-orbit behaviour of InGaAs SWIR channel device of IRS-1C/1D LISS-III camera", private communication.

[20] P.W. Marshall, C.J. Dale, and E.A Burke, "Proton-induced displacement damage distributions in silicon microvolumes", IEEE Trans. Nucl. Sci., Vol.37, no. 6, pp 1776-1783, 1990.

[21] C.J. Dale, P.W. Marshall, and E.A Burke, "Particle-induced spatial dark current fluctuation in focal plane arrays", IEEE Trans. Nucl. Sci., Vol.37, no. 6, pp 1784-1791, 1990.
[22] G.R. Hopkinson, "Space Radiation effects on CCDs", Proceedings ESA Electronics Components Conference, ESA SP-313, p301, 1990.
[23] G.R. Hopkinson, "Cobalt60 and proton radiation effects on large format, 2D, CCD arrays for an earth imaging application", IEEE Trans. Nucl. Sci., Vol.39, no. 6, pp 2018-21025, 1992.
[24] G.R. Hopkinson, "Radiation-induced dark current increases in CCD's", RADECS 93, IEEE Proc., pp 401-408, 1994.
[25] S. Barde, R. Ecoffet, J. Costeraste, A. Meygret, X. Hugon, "Displacement damage effects in InGaAs detectors : experimental results and semi-empirical model prediction", IEEE Trans. Nucl. Sci., Vol 47, no 6, pp 2466-2472, December 2000.

Multi-level Fault Effects Evaluation

L. Anghel, M. Rebaudengo, M. Sonza Reorda, M. Violante

Laboratoire TIMA, Grenoble, France
Politecnico di Torino, Dip. Automatica e Informatica, Torino, Italy
lorena.anghel@imag.fr
{maurizio.rebaudengo, matteo.sonzareorda, massimo.violante}@polito.it

Abstract. The problem of analyzing the effects of transient faults in a digital system is very complex, and it may be addressed successfully only if it is performed at different steps of the design process. In this work we report and overview of fault injection, discussing which techniques are available that can be used starting from the early conception of the system and arriving to the transistor level.

1. Introduction

Soft Errors caused by particle hits are a serious problem for modern complex integrated circuits. As the dimensions and operating voltages are decreasing to satisfy consumer needs for low power consumption and greater functionality at higher integration densities, sensitivity to transient phenomena increases dramatically. Smaller interconnect features and higher operating frequencies lead to an increase of errors generated by violations of the timing specifications. The rate of intermittent faults coming from process variation and manufacturing residuals is growing with every process generation. Smaller transistor size and low power voltages result in a higher sensitivity to neutron and alpha particles. When energetic particles like neutrons from terrestrial atmosphere or alpha particles originating from impurities in the packaging material strike sensitive regions in the semiconductor, the resulting transient phenomena can alter the state of the system resulting in a soft error that in turn may lead to failures in the field.

Soft Errors in RAM memories have traditionally been a much higher concern than soft errors in combinational circuits since memories contain by far the largest number and density of bits susceptible to particle strikes. In the next decade, technology trends as smaller feature size, lower voltage levels, and higher frequencies are foreseen to cause an increase in the soft error failure rates in combinational parts. Analytical models predict that the soft error rate in combinational circuits will be comparable to that of the memory element by 2011 [1] [2].

Mitigating the transient faults threat in a cost effective manner requires identifying the most sensitive parts of a design and select the most effective solutions for each of these parts in terms of sensitivity reduction and hardware/power/time penalty. Efficient soft error analysis methods and simulation tools are required to make feasible this complex task without impacting adversely the design time of new products and especially of complex SoCs.

R. Velazco et al. (eds.), Radiation Effects on Embedded Systems, 69–88.

Fault Injection is a consolidated technique used for the dependability evaluation of computer-based systems. It is defined as the deliberate insertion of faults into an operational system to observe its response [3]. Several approaches are available for performing this action, which we can organize into the following categories by taking into account the *injection method*: simulation-based techniques [4], software-implemented techniques [5][6][18][19], and hybrid techniques, where hardware and software approaches are applied jointly to optimize the performance [20][21]. Other criteria of classification are based on the abstraction level and on the domain of representation that is used to model the Design Under Test (DUT). As far as the abstraction level is concerned, we can identify four classes:

1. *System*: the components of the DUT are processor, memory modules, input/output devices, etc.
2. *Register transfer*: the components of the DUT are registers, arithmetic and logic units, bus structures, etc.
3. *Logic*: the components of the DUT are combinational gates, or simple memory elements.
4. *Device*: the components of the DUT are simple integrated devices.

Conversely, as far as the domain of representation is concerned, we have the following options:

1. *Behavioral*: the relation between DUT's input stimuli and DUT's output responses is coded.
2. *Structural*: the topology of the DUT is described in terms of components and connections among components.
3. *Physical*: the equations describing the principle functions governing the components are described.

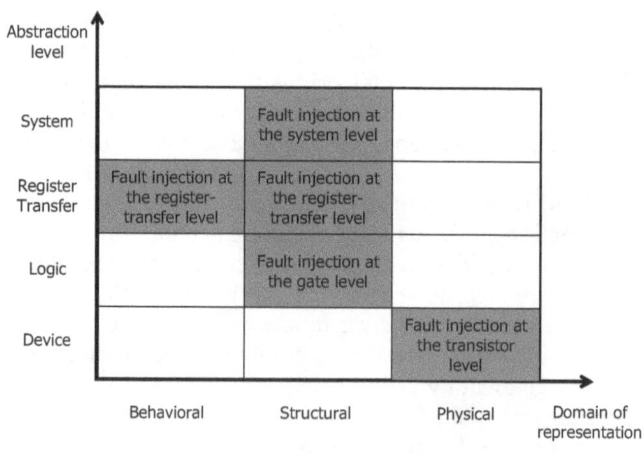

Figure 1. Abstraction level/domain of representation.

By composing the abstraction level with the domain of representation, we obtain the matrix reported in Figure 1, which provides also a key for reading the rest of this chapter. Section "Fault injection at the transistor level" describes how faults can be injected while working at the physical domain with device-level abstraction. Section "Fault injection at the gate and register-transfer level" describes how faults can be injected while working either at the behavioral or structural domain while adopting the register-transfer or the logic-level abstraction. Finally, section "Fault injection at the system level" presents an approach suitable for injecting faults at the structural domain with system-level abstraction.

It is worth to note that fault-injection analysis performed at each level allows a designer to obtain necessary information to forecast the dependability of any complex circuit and to take some counter measures before the circuit is sent to manufacturing.

In the particular case of soft errors, the probability of a complex system to fail within the duration of a certain application runtime is highly depending on several (conditional) probabilities:

1. Probability of the failure, assuming that an error provoked a different system machine state. This can be evaluated by system- and/or Register Transfer-level analyses. Note that there are cases where a erroneous machine state not necessarily lead to a system failure (e.g. error in an invalid tag).
2. Probability that the error appeared in the system machine state was induced:
   a. Either by a direct hit in the latches (bit flip in a memory cell) that had propagated into a machine state. This case can be evaluated by combined device-, logic- and register-level fault injection analyses.
   b. Or by a strike in the combinational logic propagating to the latches and to the machine states or directly to the machine states. Again this can be evaluated by device-, logic- and register-level fault injection analyses to determine the probability that a given type of voltage pulse occurs on the outputs of each gate, and then device- or gate-level analysis to determine the probability of propagation up to a flip-flop input.
3. Probability of the original event (e.g. particle impact). This can only be evaluated with respect to the type of event (radiation, process variation inducing intermittent faults, etc) and to the operational circuit environment. This probability is usually obtained by fault injection at device level.

All these probabilities are dependent on a lot of parameters and the analysis results strongly depend on the accuracy of the circuit description at each level.

Before entering into the details of the possible fault-injection approaches we provide the reader with some background details in section 2. Moreover, we state the assumptions behind this work in section 3.

2. The FARM model

A good approach to characterize a fault injection environment is to consider the FARM classification proposed in [19]. The FARM attributes are the following:

1. *F*: the set of *faults* to be deliberately introduced into the system.
2. *A*: the set of *activation trajectories* that specify the domain used to functionally exercise the system.
3. *R*: the set of *readouts* that corresponds to the behavior of the system.
4. *M*: the set of *measures* that corresponds to the dependability measures obtained trough the fault injection.

The FARM model can be improved by also including the set of workloads *W*, i.e., the set of input stimuli that produce resource demands on the system.

The measures M can be obtained experimentally from a sequence of fault-injection case studies. An injection campaign is composed of elementary injections, called *experiments*. In a fault injection campaign the input domain corresponds to a set of faults F and a set of activations A, while the output domain corresponds to a set of readouts R and a set of measures M.

The single experiment is characterized by a fault *f* selected from F and an activation trajectory *a* selected from A with a workload *w* from W. The behavior of the system is observed and constitutes the readout *r*. The experiment is thus characterized by the triple <*f, a, r*>. The set of measures M is obtained in an injection campaign elaborating the set of readouts R for the workloads in W.

2.1 Fault injection requirements

The FARM model can be considered as an abstract model that describes the attributes involved in a fault-injection campaign, but it does not consider the fault-injection environment, (i.e., the technique adopted to perform the experiments). The same FARM set can be applied to different fault injection techniques. Before presenting the techniques described in this chapter, we focus on the parameters that should be considered when setting up a fault injection environment: intrusiveness, speed, and cost.

In the following the DUT is always composed of at least one processor, connected with some memory modules and possible ad-hoc developed circuits, running a given application. Fault injection is not necessarily performed on the whole DUT. Sometimes only a portion of it undergoes dependability analysis.

2.2 Intrusiveness

The intrusiveness is the difference between the behavior of the DUT and that of the same system when it is the object of a fault injection campaign. Intrusiveness can be caused by:

1. The introduction of instructions or modules for supporting fault injection: as an effect, the sequence of executed modules and instructions is different with respect to that of the DUT when the same activation trajectories are applied to its inputs.
2. Changes in the electrical and logical setups of the DUT, which result in a slowdown of the execution speed of the system, or of some of its components; this means that during the fault injection campaign the DUT shows a different behavior from the temporal point of view; we will call this phenomenon *time intrusiveness*.

3. Differences in the memory image of the DUT, which is often modified by introducing new code and data for supporting the fault injection campaign.

It is obvious that a good faultinjection environment should minimize intrusiveness, thus guaranteeing that the computed results can really be extended to the original DUT.

2.3 Speed

A fault injection campaign usually corresponds to the iteration of a high number of fault injection experiments, each focusing on a single fault and requiring the execution of the DUT in presence of the injected fault. Therefore, the time required by the whole campaign depends on the number of considered faults, and on the time required by every single experiment. In turn, this depends on the time for setting up the experiment, and on the one for executing the application in presence of the fault.

The elapsed time for a fault injection campaign can be reduced and optimized adopting the techniques described in the following subsections.

2.4 Speeding-up the single fault-injection experiment

The speed of a fault-injection experiment is computed considering the ratio between the time required by the normal execution (without fault injection) and the average elapsed time required by a single fault injection experiment. The increase in the elapsed time is due to the operations required to initialize the experiment, to observe the readouts, to inject the fault, and to update the measures.

Since in a given time the number of possible experiments is limited, a crucial issue when devising a fault-injection environment is the computation of the list of faults to be considered. One challenge is to reduce the large fault space associated with highly integrated systems, improving sampling techniques and models that equivalently represent the effects of low-level faults at higher abstraction levels.

2.5 Fault list generation

The fault list should be representative enough of the whole set of possible faults that can affect the system, so that the validity of the obtained results is not limited to the faults in the list itself. Unfortunately, increasing the size of the fault list is seldom a viable solution due to the time constraints limiting the maximum duration of the fault-injection experiment. In general, the goal of the fault list generation process is to select a representative sub-set of faults, whose injection can provide a maximum amount of information about the system behavior, while limiting the duration of the fault-injection experiment to acceptable values.

2.6 Cost

A general requirement valid for all the possible target systems is that the cost of the fault injection environment must be as limited as possible, and negligible with respect to the cost of the system to be validated.

We can consider as a cost the following issues:

1. The hardware equipment and the software involved in the fault injection environment.
2. The time required to set up the fault injection environment and to adapt it to the target system.

The first issue is strictly related to the fault injection technique chosen, whereas the second one implies to define a system as flexible as possible that can be easily modified when the target system is changed, and can be easily used by the engineers involved in the fault injection experiments.

3. Assumptions

In this section we report the assumptions in terms of the FARM model, and choices underlying the organization of the fault-injection environments we will present in the following:

1. *Set F.* It is the set of faults to be injected in a fault-injection campaign. First of all, the fault model has to be selected. This choice is traditionally made taking into account from one side the need for a fault model that is as close as possible to real faults, and from the other side the practical usability and manageability of the selected fault model. Based on these constraints, the fault model we selected is the SEU/SET. Each fault is characterized by the following information:
   - *Fault injection time*: the time instant when the fault is first inoculated in the system.
   - *Fault location:* the system's component the fault affects.
   - *Fault mask:* in case the faulty component is an n-bit-wide register, the fault mask is the bit mask that selects the bit(s) that has (have) to be affected by the SEU.
2. *Set A.* Two important issues are related to this point. On the one side it is important to define how to determine an input trajectory to be applied to the target system during each fault injection experiment. Although several solutions have been proposed, their description is out of the scope of this chapter, where we limit our interest to the techniques for performing the fault injection campaign, once the trajectory is known. On the other hand, there is the problem of how to practically apply the trajectory to the system. This issue is particularly critical when considering embedded systems, since they often own a high number of input signals of different types (digital and analog, high- and low-frequency, etc.).
3. *Set R.* This set of information is obtained by observing the DUT behavior during each fault injection experiment, and by identifying the differences with respect to

the fault-free behavior. Note that all the operations involved by the observation task should also be minimally intrusive.

4. *Set M*. At the end of the fault-injection campaign, a proper tool should build a report concerning the dependability measures and fault coverage computed on the whole fault list. Fault coverage is defined with respect to the possible effects of faults:
   - *Effect-less fault*. The fault does not propagate as an error neither as a failure. In this case the fault appeared in the system and remained passive for a certain amount of time, after which it was removed from the system. As an example, let us consider a fault that affects a variable x used by a program. If the first operation the program performs on x, after the fault, is a write operation, then a correct value is overwritten over the faulty one, and thus the system returns in a fault-less state.
   - *Failure*. The fault is propagated within the system until it reached the system's output.
   - *Detected fault*. The fault produced an error that was identified and signaled to the system's user. In this case the user is informed that the task was corrupted by a fault, and the user can thus take the needed countermeasures to restore the correct system functionalities. In systems able to tolerate the presence of faults, the needed countermeasures may be activated automatically. Error detection is performed by means of mechanisms, *error-detection mechanisms*, embedded in the system whose purpose is to monitor the behavior of the system, and to report anomalous situations. When considering a processor-based system, error-detection mechanisms can be found in the processor, or more in general in the hardware components forming the system, as well as in the software it executes. The former are usually known as *hardware-detection* mechanisms, while the latter are known as *software-detection* mechanisms. As an example of the hardware-detection mechanisms we can consider the *illegal instruction trap* that is normally executed when a processor tries to decode an unknown binary string coming from the code memory. The unknown binary string may be the result of a fault that modified a valid instruction into an invalid one. As an example of the software-detection mechanisms we can consider a code fragment the designers inserted in a program to perform a range check, which is used to validate the data entered by the systems' user, and to report an alert in case the entered data is out of the expected range. To further refine our analysis, it is possible to identify three types of fault detections:
     - *Software-detected fault*. A software component identified the presence of an error/failure and signaled it to the user. As an example, we can consider a subprogram that verifies the validity of a result produced by another subprogram stored in a variable x on the basis of range checks. If the value of x is outside the expected range, the controlling subprogram raises an exception.
     - *Hardware-detected fault*. A hardware component identified the presence of an error/failure and signaled it to the user. As an example, we can consider a parity checker that equips the memory elements of a processor. In case a fault changed the content of the memory elements, the checker will identify a parity violation and it will raise an exception.
     - *Time-out detected fault*. The fault forced the processor-based system to enter in an endless loop, during which the system does not provide any output re-

sult. As an example, the occurrence of this fault type may be detected thanks to a watchdog timer that is started at the beginning of the operations of the processor-based system, and that expires before the system could produce any result.

- *Latent fault.* The fault either remains passive in the system, or it becomes active as an error, but it is not able to reach the system's outputs, and thus it is not able to provoke any failure. As an example, we can consider a fault that modifies a variable x after the last usage of the variable. In this case, x holds a faulty value, but since the program no longer uses x, the fault is unable to become active and to propagate through the system.
- *Corrected fault.* The fault produced an error that the system was able to identify and to correct without the intervention of the user.

4. Fault injection at the transistor level

In the case of SEU/SET faults, this type of fault-injection analysis phase evaluates the environment effects on the integrated silicon devices. More specifically, it considers the interactions between the radiation and device and it relates to the understanding of the cosmic particle environment and physical dimension of the circuit elements. Fault injection at this level leads to a probability distribution for the amount of charge deposited on a sensitive circuit.

4.1 Particles that cause soft errors

Cosmic rays are particle that originate from outer space and enter the earth's atmosphere. High-energy neutrons (having energies higher than 1MeV) could induce soft errors in semiconductor devices via the secondary ions produced by the neutrons interactions with silicon nuclei. This form of radiation is known to be a significant source of soft errors in memory elements. Alpha particles are another source of soft errors. This type of radiation comes almost entirely from impurities in packing material, and can vary widely from one type of circuit to another. A third soft error mechanism is induced by low energy cosmic neutron interactions with the isotope B^{10} in IC materials, specifically in borophosphosilicate glass used in insulation layers during IC manufacturing. It was shown in [7] that this is the dominant source of errors in 0.25 micron technologies using BPSG.

4.2 Modeling single event transients in silicon

It is well known that the reverse-biased junction of transistors is the most charge sensitive part of the circuit, especially when the junction is floating or weakly driven [8]. During a particle strike, a cylindrical track of electron-hole pairs with a submicron radius and very high concentration form along the energetic particle trajectory. If the particle-strike is close or traverse a depletion region the electric field rapidly collects carriers, creating a current glitch at that node. Figure 2 shows the charge generation and collection in a reverse biased junction and the resultant current pulse caused by the energetic particle strike. The physics of charge collection have been extensively studied in details through the use of two dimensional (2D) and three dimensional (3D) numerical simulations [8][9].

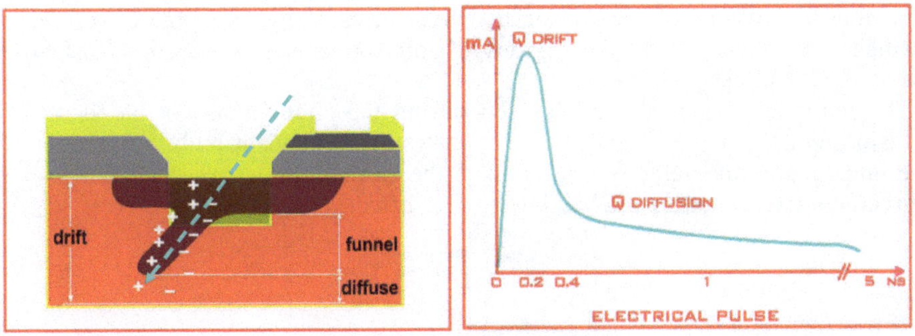

Figure 2. Illustration of a funneling in an n+/p silicon junction following the ion strike and the resulting electrical transient current.

The funnel size and the collection time are highly depending on the substrate doping – the funnel distortion increases as the substrate doping decreases. The collection time usually complete within picoseconds, then an additional phase starts when the diffusion begins to dominate until all the excess carriers have been collected. This second phase is longer than the first one, usually hundreds of nanoseconds.

The transient current pulse duration and shape resulting from a particle strike depends on various factors, such as: the type of the particle, its energy, the angle of the strike, the substrate structure and doping and the cell size. Moreover, for each particle type and energy, the resulting current pulse also depends on the distance of the incident point of the particle to the sensitive node. In general, the farther away from the junction the strike occurs, the smaller the charge that will be collected and the less likely that the event will cause a soft error. The charge collected (*Qcoll*) is not the only parameter that is taken into consideration in analysis. The excess charge plays an important role in the device sensitivity. This sensitivity depends primary on the characteristics of the circuit (effective capacitance of the drain nodes and drain conduction currents) and voltage supply [10]. All theses actors define the amount of charge (*Qcrit*) required to observe a change in the data state, and this parameter is not a constant, being highly dependent on the radiation pulse characteristics and dynamic response of the circuit itself. It is assumed that for a memory element, if a particle strike produces a *Qcoll* less than *Qcrit* the circuit will not experience a soft error.

4.3 Physical level 2D/3D device simulations

The charge injected into the silicon by a particle of a given energy can be found in tables published in the literature. From this charge, it is possible to determine with some margins the charge collected on a sensitive node.

For doing that, designers are modeling a combinational logic cell (or a memory cell) from a given technology library at the device level and numerically simulate the particle impact. Device level simulation of a full cell was made possible only recently

[10], as many improvements have been made on commercial device level simulators. A specific generation module allowing simulation of ions tracks with different locations, directions and ranges has been proposed and are commercially available [11], [12], [13], [14]. The structure of the device is represented in 2D/3D reflecting exactly the layout dimensions of the cell of any given technology (see Figure 3). Mesh refinements are necessary at the regions of interest: channels, junctions, LDD, and around the ion track.

The numerical physical models used during the simulations are based on drift-diffusion models including field and concentration dependent mobility models; they take into account the velocity saturation of the carriers, Shockley Read Hall and Auger recombination models and band narrowing effects.

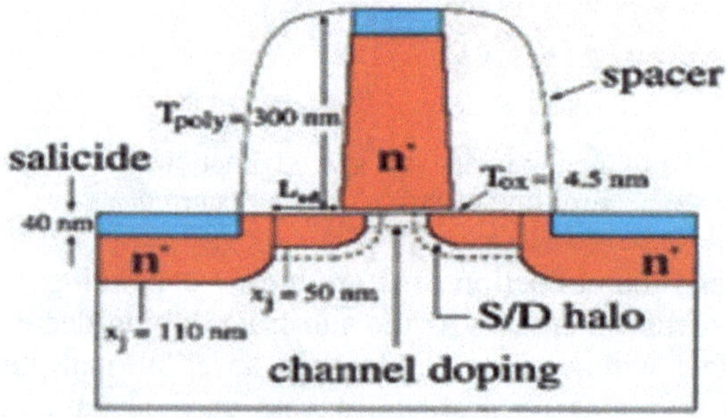

Figure 3. 2D model for a NMOSFET transistor, in 90 nm technology (after 15).

The equations to be solved are the Poisson equations and the current continuity equations, together with the constitutive relationships for current densities (actual drift-diffusions equations). These equations are discretized and solved on a mesh using finite-difference or finite-element techniques. To drift-diffusion models, one should add all effects becoming important in nanometric domain, quasiballistic transport, carrier heating etc. 2D/3D simulation programs are routinely used in the semiconductor industry for device analysis. To these programs we should add heavy ion numerical models, where a heavy ion is modeled by electron hole pairs columns with the track as axis and a Gaussian time distribution [10].

Intensive 2D/3D physical simulations are performed to establish heuristic rules for estimating the percentage of the injected charge that is collected by a sensitive node as well as the timing parameters of the charge collection phenomenon. This, in turn, will allow determining the duration and density of the resulting transient current pulse, while the shape of this pulse is well approximated by a double exponential curve.

Equation 1 gives the transient current obtained after a particle strike [16]:

$$I_{\text{int}}(t) = \frac{Q_{coll}}{\tau_\alpha - \tau_\beta}\left(e^{-\frac{t}{\tau_\alpha}} - e^{-\frac{t}{\tau_\beta}}\right) \tag{1}$$

where Qcoll refers to the amount of charge collected due to the particle strike, the parameter τα, τβ are time constant for the charge collection process and depends on CMOS technology.

2D/3D physical level simulations allow obtaining the corresponding transient current generated by a collision between a specific charged particle and a transistor (or an entire logic cell). It allows designers to identify the most vulnerable nodes of any transistor/logic cells and the minimum critical energy of charged particles. However this evaluation requires important costs in terms of runtime and CPU speed. This type of simulation can be performed manually or semi automatically to a certain extent.

4.4 Transient fault injection simulations at electrical level

This type of fault injection deals with the evaluation of the behavior of a transistor or logic cell by taking into account other device parameters that impact the single event phenomenon immediately following an ion strike. Each circuit element (memory cell or logic gate) must be simulated to determine the magnitude and the shape of the voltage transient that would appear at the cell output.

The transient voltage pulse is a function of the transient current pulses obtained through physical 2D/3D simulations, the particular load of that gate and circuit topology.

To perform electrical level fault simulations, a SPICE electrical masking model is built with necessary technology parameters (as V_{TH}, Tox, V_{DD}). The masking model is used to simulate the degradation of the current pulse as it travels through a gate in the logic circuit.

Electrical masking model refers to two electrical effects that reduce the strength of the current/voltage pulse as it passes through a logic gate.

1. Circuit delays caused by the switching time of the transistors cause the rise and fall time of the pulses to increase, thus reducing their global effective duration.
2. On the other hand, for short duration pulses, the pulse duration may be further reduced as the gate can start to turn off before the output reaches the full amplitude.

For short duration pulses, this second effect is more important, and these effects can cascade along the propagation path through several gates and can eventually degrade the pulse to its extent [17]. The electrical masking model should take into account not only gate delay model and its decomposition in rise and fall times on the input signal and switching voltage to deal with short duration inputs, but also interconnexion delays, as being predominant in the timing model in nowadays and future technologies. Another parameter to take into account is the existence of reconvergent paths in a given design than may have an important effect on the pulse duration.

In the electrical model of DUT is added a current generator parameterized with valued obtained through physical device level simulations. The simulation tool can be any commercial SPICE/analog simulator. The injection points are chosen manually or automatically by means of simulation scripts and commands.

In Figure 4 we show the transient current injection principle. The obtained transient current is injected at the output of the affected gate to obtain the actual voltage pulse. The transient voltage pulse is then propagating through subsequent logic gates and the amplitude and duration is dependent on the total node capacitance (output capacitance of the gate, number of gates connected to the output, parasitic wire load).

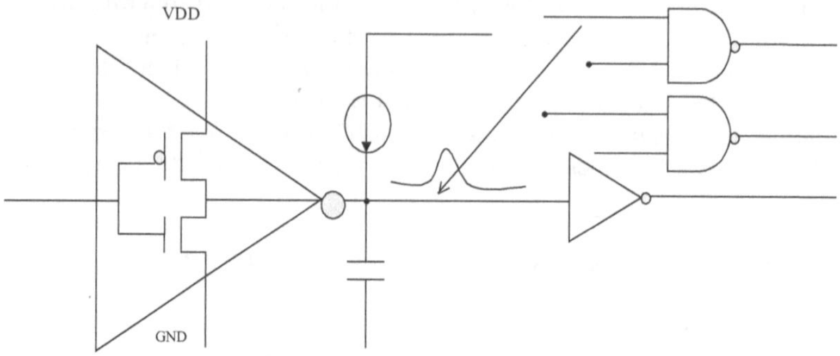

Figure 4. Transient current pulse injections at electrical level.

Although electrical simulation is much faster than 2D/3D simulation, it is still a time-consuming process. Dependability analysis on a complex design through electrical simulations is quite impossible, due to the important number circuit nodes to take into account and the length of each simulation process. However, it is a very powerful tool that helps designers to compute the electrical masking factors while building the complete FIT model calculation.

5. Fault injection at the gate and register-transfer level

This type of fault injection consists in evaluating the DUT response to the presence of faults by means of simulation tools. Fault injection can be implemented in three different ways:

1. The simulation tool is enriched with algorithms that allow not only the evaluation of the faulty-free DUT, as normally happen in VHDL or Verilog simulators, but also their faulty counterparts. This solution is very popular as far as certain fault models are considered: for example commercial tools exist that support the evaluation of permanent faults like the stuck-at or the delay one [24]. Conversely, there is a limited support of fault models like SEU, or SET, and therefore designers have to rely on prototypical tools either built in-house or provided by universities.
2. The model of the DUT is enriched with special data types, or with special components, which are in charge of supporting fault injection. This approach is quite popular since it offers a simple solution to implement fault injection that requires limited implementation efforts, and several tools are available adopting it [25][26][27][28].
3. Both the simulation tool and the DUT are left unchanged, while fault injection is performed by means of simulation commands. Nowadays, it is quite common to find, within the instruction set of simulators, commands for forcing desired values within the model [29]. By exploiting this feature it is possible to support SEUs and SETs, as well as other fault models.

As an example of a simulation-based fault-injection system we describe the approach presented in [29], whose main components are:

1. The model of the DUT (coded in VHDL language) that describes the functions the system under analysis implements.
2. The VHDL simulator, which is used to evaluate the DUT. For this purpose any simulation tool supporting the VHDL language, as well as a set of commands allowing monitoring/changing the values of signals and variables during simulation execution is viable.
3. The Fault Injection Manager that issues commands to the VHDL simulator to run the analysis of the target system as well as the injection of faults.

Simulation-based fault-injection experiments may require huge amount of times (many hours or days) for their execution, depending on the complexity of the model of the DUT, the efficiency of the VHDL simulator adopted, of the workstation used for running the experiments, as well as the number of faults that have to be injected. In order to overcome this limitation, the approach presented in [29] adopts several techniques aiming at minimizing the time spent for running fault injection.

The approach is composed of three steps:

1. *Golden-run execution*: the DUT is simulated without injecting any fault and a trace file is produced, gathering information on the DUT's behavior and on the state of the simulator.
2. *Static fault analysis*: given an initial list of faults that must be injected, by exploiting the information gathered during golden-run execution we identify those faults whose effects on the DUT can be determined a-priori, and we remove them from the fault list. Since the injection of each fault encompasses the simulation of the DUT, by reducing the number of faults that we need to inject we are able to reduce the time needed by the whole experiment.
3. *Dynamic fault analysis*: during the injection of each fault, the state of the DUT is periodically compared with the golden run at the correspondent time instant. The simulation is stopped as early as the effect of the fault on the DUT becomes known, e.g., the fault triggered some detection mechanisms, the fault disappeared from the DUT, or it manifested itself as a failure. Although the operations needed for comparing the state of the DUT with that of the golden run come at a not-negligible cost, the benefits they produce on the time for running the whole experiment are significant. In general, a fault is likely to manifest itself (or to disappear) after few instants since its injection. As a result by monitoring the evolution of the fault for few simulation cycles after its injection, we may be able to stop the simulation execution in advance with respect of the completion of the workload. We can thus save a significant amount of time. Similarly, in case the fault is still latent until few simulation cycles after its injection, it is likely to remain latent, or manifest itself, until the completion of the workload. In this case, the state of the DUT and those of the gulden run are no longer compared, thus saving execution time, until the end of the injection experiment.

In the following section we give more details about the approach introduced in [29].

5.1 Golden run execution

The purpose of this step is to gather information on the fault-free DUT. Given a set of input stimuli (the workload) that will remain constant in the following fault-injection experiments, two sets of information are gathered, one for performing the static fault analysis and one for performing the dynamic fault analysis.

Static fault analysis requires the complete trace of:

1. *Data accesses*: whenever a data is accessed, the time, the type of access (read or write) and the address are recorded.
2. *Register accesses*: whenever a register is accessed, the time, the register name and the type of access are recorded.
3. *Code accesses*: at each instruction fetch, the address of the fetched instruction is stored in a trace file.

The method collects the needed information resorting to ad-hoc modules written in VHDL, called code/data watchers, inserted in the system model. This approach is not intrusive, since code/data watchers work in parallel with the system and do not affect its behavior.

Conversely, for performing dynamic fault analysis the simulation is stopped periodically and a snapshot of the system is recorded. A snapshot is composed of the content of the DUT's memory elements at the current simulation time (i.e., the time instant at which the sample is taken).

This approach is effective because allows gathering information on the system with zero intrusiveness. On the other hand, when addressing very large systems, it could require the availability of large amounts of both memory and disk space. As a consequence, the number of snapshots should be carefully selected.

5.2 Static fault analysis

Faults are removed from an initial fault list according to two sets of rules, which are applied by analyzing the information gathered during golden run execution.

A fault affecting data is removed if it verifies at least one of the following conditions:

1. Given a fault f to be injected at time T at address A, f is removed from the fault list if A is never read again after time T; this rule allows removing the faults that do not affect the system behavior.
2. Given a fault f to be injected at time T at address A, f is removed from the fault list if the very first operation that involves A after time T is a write operation.

Conversely, a fault affecting the code is removed if it verifies the following condition: given a fault f to be injected at time T at address A, f is removed from the fault list if the address A corresponds to an instruction that is never fetched again after time T. This rule identifies faults that do not produce any effect and whose injection is therefore useless.

5.3 Dynamic fault analysis

Dynamic fault analysis is based on the idea of identifying as early as possible the effects of the injected fault during its simulation. As soon as the effects of a fault become evident, the simulation is stopped, potentially saving a significant amount of simulation time.

The fault-injection procedure starts by setting a set of breakpoints in the VHDL code of the system to capture the following situations:

1. *Program completion*: a breakpoint is set so that simulation is stopped after the execution of the last instruction of the program running on the DUT. This mechanism is useful to early stop the simulation of faults that cause a premature end of the simulated application.
2. *Interrupt*: in order to detect asynchronous events, a breakpoint is set to the VHDL statements implementing the interrupt mechanism activation, which is often used to implement hardware and software Error Detection Mechanisms.
3. *Time-out*: the simulation is started with a simulation time much higher than the time required for the golden run program completion. A time-out condition is detected if simulation ends and any breakpoints are reached.

After all the required breakpoints have been properly set, the DUT is simulated up to the injection time, then injection takes place. Injection is done by exploiting the VHDL simulator commands to modify signals/variables in the VHDL source. After injection, the DUT is simulated up to the time instant corresponding to the first snapshot after injection time. Finally, the DUT is compared with the golden run, and the following situations are considered:

1. *No failure*: the state of the DUT is equal to the golden run; two alternatives are possible:
   - When injecting in the data area this implies that the fault effects disappeared from the DUT and that the fault has no effect, the simulation can thus be stopped.
   - When injecting in the code area, if the faulty instruction is never fetched again this means that the fault effects disappeared from the DUT and the simulation can be stopped.
2. The state of the DUT does not match that observed during the golden run; in this case two alternatives are possible:
   - *Failure*: the fault has affected DUT's outputs (thus causing a failure) and simulation can be stopped.
   - *Latent fault*: the fault is still present in the DUT but it did not affect system outputs: further simulations are therefore required.

6. Fault injection at the system level

As an example of system-level fault injection we consider DEPEND [30], which is an integrated design and fault-injection environment.

DEPENDS is suitable for modeling rapidly fault-tolerant architectures and performing extensive fault-injection experiments. The components of the system under analysis, as well as the interaction between components, are described by resorting to a collection of interacting processes, thus obtaining several benefits:

1. It is an effective way to model system behavior, repair schemes, and system software in detail.
2. It simplifies modeling of inter-component dependencies, especially when the system is large and the dependencies are complex.
3. It allows actual programs to be executed within the simulation environment.

According to DEPEND, the designer models the system's behavior by writing a control program in C++ that exploits the objects available in the DEPEND library. The program is then compiled and linked with the DEPEND objects and the run-time environment.

As soon as the compiled and linked model of the system is available, it is executed in a simulated parallel run-time environment where system's behavior is evaluated. During simulations faults are injected, repair schemes are initiated, and reports classifying faults' effects are produced.

The DEPEND library offers to designers complex objects, as well as simpler objects that perform basic tasks as fault injection and results analysis. The complex objected includes:

1. Processors.
2. Self-checking processors.
3. N-modular redundant processors.
4. Communication links.
5. Voters
6. Memories.

6.1 Fault models

When system-level models are considered, and thus few details about the system's implementation are available, abstract fault models should be used. System-level fault models approximate the effects of real faults, like for example permanent or transient faults, by describing how the faults modify the behavior of system's components, while neglecting components' architectures and implementations.

This approach is exploited by DEPEND, which enriches each component in the DEPEND library with a faulty behavior, which is specified alongside the fault-free behavior.

As far as processors are considered, DEPEND assumes that the processor hangs when a permanent or transient fault is discovered. If the fault is transient, it disappears when the CPU is restarted. If the fault is permanent, it is corrected only when the CPU is replaced.

As far as the communication infrastructures are considered, DEPEND simulates the effects of a noisy communication channel by corrupting bits in a message or destroying the message.

Finally, two system-level fault models are available for memory and I/O subsystems. Either a bit of a word is flipped or a flag is raised to represent the error. The fault can be detected with a byte-by-byte comparison or by checksum comparison if the fault is a flipped bit. Otherwise, it can be detected by checking the status of the flag.

6.2 Supporting fault injection

The DEPEND library includes an object, called *injector*, that implements the mechanisms for injecting faults. To use the injector the user specifies the number of components, the distribution of injection time for each component, and the fault routine that specifies the fault model.

Several distributions of injection time are available: constant time, exponential, hyperexponential, and Weibull.

```
initialize()
{
      for all components do
          if (component is OK & On)
              compute and store time to fault
          else
              time to fault is ∞
          end if
      end
}
main()
{
      initialize()
      do forever
          find minimum_time_to_fault among components
          sleep (minimum_time_to_fault-current_time)
          if (sleep not aborted)
              call fault subroutine
              set time to fault of this component to ∞
          end if
      end
}
```

Figure 5. Routine for activating the fault injection.

When initialized, the injector samples from a random number generator to determine the earliest time to fault, sleeps until that time, and calls the fault subroutine. As described in [30], the implementation of DEPEND uses a table where an entry is kept for each system's component. The entry specifies the component's condition (OK, or Failed), the injection status (Injection off, Injection on), the time to fault distribution, and the time to next fault. The algorithm used to determine which is the component for which the injection must start is described in Figure 5.

As soon as a component is repaired or turned on, its time to fault is computed and entered in the table. The injector is then called to take the new component's fault time into account.

6.3 Speeding-up fault injection execution

Although the system level abstracts most of the implementation details of system's components, simulation times may still be a critical issue. When complex systems are addressed and complex workload is considered, the time spent for running fault injection may be prohibitive even at the system level.

To overcome this problem, DEPEND implements three techniques to reduce simulation time:

1. *Hierarchical simulation.* According to this technique, a complex model is partitioned into simpler sub-models. The sub-models are analyzed individually and their results are combined to derive the solution of the entire system. When the interactions among the sub-models are weak, this approach provides valid results.
2. *Time acceleration.* DEPEND objects produces the time of the next event, such as the time when the next fault will arrive, or when the next latent fault will be activated. The sets of these events that affect the systems are kept in a chronologically sorted list. The simulator leaps forward to the time of the event at the head of the list, and resumes processing at the granularity of the system clock until the effect of the event has subsided.
3. *Variance reduction.* DEPEND provides direct control of the simulation engine so that sampling techniques can be implemented.

7. Conclusions

The problem of analyzing the effects of faults in a digital system is very complex, and it may be addressed successfully only if it is performed at different steps of the design process. System-level fault injection can be used to gain an initial feedback about the dependability the adopted architecture provides. Further insights on the system can then be obtained while exploiting register-transfer-level fault injection to investigate the sub-systems that compose the overall system's architecture, as well as gate-level fault injection to understand the behavior of crucial components within sub-systems. Finally, transistor-level fault injection is mandatory to assess the robustness of the adopted manufacturing technology.

As a final remark, we would like to underline that, although invaluable for anticipating dependability issues through the whole design process starting from the earliest stages, fault injection is not the ultimate answer to the problem of validating system's dependability. Further investigations, like those possible through accelerated radiation testing, are still important and should be exploited as well. In conclusion, we state that fault injection is mandatory to anticipate the problems as early as possible, thus minimizing the need for extremely costly accelerated radiation testing experiments that can possibly be exploited late in the design process, only.

References

[1] N. Cohen et al. "Soft Error Considerations for Deep Submicron CMOS Circuit Applications", Int. Electron Devices Meeting, 1999, pp. 315-318.

[2] P. Shivakumar et al, "Modeling the Effect of technology Trends on the Soft Error Rate of Combinational Logic", Int. Conference of Dependable Systems and Networks, 2002, pp. 389-398.

[3] J. Clark, D. Pradhan, Fault Injection: A method for Validating Computer-System Dependability, IEEE Computer, June 1995, pp. 47-56.

[4] T.A. Delong, B.W. Johnson, J.A. Profeta III, A Fault Injection Technique for VHDL Behavioral-Level Models, IEEE Design & Test of Computers, Winter 1996, pp. 24-33.

[5] J. Carreira, H. Madeira, J. Silva, Xception: Software Fault Injection and Monitoring in Processor Functional Units, DCCA-5, Conference on Dependable Computing for Critical Applications, Urbana-Champaign, USA, September 1995, pp. 135-149.

[6] G.A. Kanawati, N.A. Kanawati, J.A. Abraham, FERRARI: A Flexible Software-Based Fault and Error Injection System, IEEE Trans. on Computers, Vol 44, N. 2, February 1995, pp. 248-260.

[7] R. C. Baumann, E. Smith: Neutron Induced B10 Fission as a major Source of Soft Errors in High Density SRAMs, in Microelectronics reliability, vol 41, no2, feb. 2001, pp. 211-218.

[8] C. Hsieh, P. Murley, R. O'Brien: Dynamics of Charge Collection from Alpha Particle Tracks in Integrated Circuits, Proc. 19 th Int'l Reliability Physics Symp IEEE Electron Device, 1981, pp. 2-9.

[9] H. Shone et al, "Time Resolved Ion Beam Induced Charge Collection in Microelectronics", IEEE Trans Nuclear Sci, vol 45, pp. 2544-2549, dec. 1998.

[10] G. Hubert, et al. "SEU Sensitivity of SRAM to Neutron or Proton Induced Secondary Fragments", NSREC 2000, Reno.

[11] Davinci Three Dimensional Device Simulation Program Manual: Synopsys 2003.

[12] Taurus Process/device User's manual, Synopsys 2003.

[13] Athena/Atlas Users's Manual: Silvaco int, 1997.

[14] DESSIS User's Manual, ISE release 6, vol 4, 2000.

[15] Antoniadis et al., "Two-Dimensional Doping Profile Characterization of MOSFET's by Inverse Modeling Using Characteristics in the Subthreshold Region", IEEE Transactions on Electron Devices, Vol. 46, No. 8, August 1999, pp. 1640-1649.

[16] L. Freeman "Critical Charge Calculations for a Bipolar SRAM Array", IBM Journal of research and Development, vol 40, no 1, pp. 119-129, jan. 1996.

[17] M.J. Bellido-Diaz, et al "Logical Modelling of Delay Degradation Effect in Static CMOS Gates", IEEE Proc Circuit devices Syst, 147(2), pp. 107-117, April 2000.

[18] T. Lovric, Processor Fault Simulation with ProFI, European Simulation Symposium ESS95, 1995, pp. 353-357.

[19] J. Arlat, M. Aguera, L. Amat, Y. Crouzet, J.C. Fabre, J.-C. Laprie, E. Martins, D. Powell, Fault Injection for Dependability Validation: A Methodology and some Applications, IEEE Transactions on Software Engineering, Vol. 16, No. 2, February 1990, pp. 166-182

[20] L. T. Young, R. Iyer, K. K. Goswami, A Hybrid Monitor Assisted Fault injection Experiment, Proc. DCCA-3, 1993, pp. 163-174.

[21] P. Civera, L. Macchiarulo, M. Rebaudengo, M. Sonza Reorda, M. Violante, "Exploiting Circuit Emulation for Fast Hardness Evaluation", IEEE Transactions on Nuclear Science, Vol. 48, No. 6, December 2001, pp. 2210-2216.

[22] A. Benso, M. Rebaudengo, L. Impagliazzo, P. Marmo, "Fault List Collapsing for Fault Injection Experiments", Annual Reliability and Maintainability Symposium, January 1998, Anaheim, California, USA, pp. 383-388.

[23] M. Sonza Reorda, M. Violante, "Efficient analysis of single event transients", Journal of Systems Architecture, Elsevier Science, Amsterdam, Netherland, Vol. 50, No. 5, 2004, pp. 239-246.
[24] TetraMAX, www.synopsys.com.
[25] E. Jenn, J. Arlat, M. Rimen, J. Ohlsson, J. Karlsson, "Fault Injection into VHDL Models: the MEFISTO Tool", Proc. FTCS-24, 1994, pp. 66-75.
[26] T.A. Delong, B.W. Johnson, J.A. Profeta III, "A Fault Injection Technique for VHDL Behavioral-Level Models", IEEE Design & Test of Computers, Winter 1996, pp. 24-33.
[27] D. Gil, R. Martinez, J. V. Busquets, J. C. Baraza, P. J. Gil, "Fault Injection into VHDL Models: Experimental Validation of a Fault Tolerant Microcomputer System", Dependable Computing EDCC-3, September 1999, pp. 191-208.
[28] J. Boué, P. Pétillon, Y. Crouzet, "MEFISTO-L: A VHDL-Based Fault Injection Tool for the Experimental Assessment of Fault Tolerance", Proc. FTCS′98, 1998.
[29] B. Parrotta, M. Rebaudengo, M. Sonza Reorda, M. Violante, "New Techniques for Accelerating Fault Injection in VHDL descriptions", IEEE International On-Line Test Workshop, 2000, pp. 61-66.
[30] K. K. Goswani, R. K. Iyer, L. Young, "DEPEND: A Simulation-based Environment for System Level Dependability Analysis", IEEE Transactions on Computer, Vol. 46, No. 1, January 1997, pp. 60-74.

Effects of Radiation on Analog and Mixed-Signal Circuits

Marcelo Lubaszewski[1], Tiago Balen[1], Erik Schuler[1], Luigi Carro[2], Jose Luis Huertas[3]

[1] Departamento de Engenharia Elétrica, UFRGS – Universidade Federal do Rio Grande do Sul, Av. Osvaldo Aranha 103, 90.035-190, Porto Alegre, RS, Brazil
luba@ece.ufrgs.br, tiago.balen@ufrgs.br, eschuler@eletro.ufrgs.br

[2] Instituto de Informática, UFRGS – Universidade Federal do Rio Grande do Sul, Av. Bento Golçalves 9500, 91501-970, Porto Alegre, RS, Brazil
carro@inf.ufrgs.br

[3] Instituto de Microelectrónica de Sevilla, Centro Nacional de Microelectrónica, Av. Reina Mercedes 41012, Sevilla, Spain
huertas@imse.cnm.es

Abstract. Radiation effects translated into Single Event Transients (SETs) and Single Event Upsets (SEUs) are dealt with, in this chapter, in the realm of analog and mixed-signal circuits. First of all, we revisit concepts and methods of the analog testing field looking for techniques that may help mitigating, at the system level, SETs and SEUs in these circuits. Then, two mixed-signal case studies are presented. The first case study investigates the effects of SEUs in a new kind of analog circuit: the Field Programmable Analog Arrays (FPAAs). Some FPAA devices are based on SRAM memory cells to implement the user programmability. For this reason the effect of radiation in such circuits can be as dangerous as it is for FPGAs. BIT-flip experiments are performed in a commercial FPAA, and the obtained results show that a single BIT inversion can result in a very different configuration of that previously programmed into the device. The second case study is focused on ΣΔ A/D Converters. A MatLab-based model of such converter is built and a series of fault injection experiments is performed. The results show that the ΣΔ converter can be used in radiation environment, if its digital part is protected. Such protection can be achieved by adopting some design directives. This chapter ends by proposing the use of on-line analog test methods, in particular self-checking circuits, that can be applied to detect SET and SEU faults during the circuit operation, therefore allowing the design of self-recovering systems.

1. Introduction

State-of-the-art integrated circuits (with growing functionality, signal heterogeneity, lower dimensions, higher performances, less power consumption) are becoming widely used in space environment in order to meet spacecraft requirements such as

R. Velazco et al. (eds.), Radiation Effects on Embedded Systems, 89–119.

size, weight, power and cost. However, new integration technologies increase the vulnerability of the devices due to smaller sizes of transistors and of capacitances.

In space, integrated circuits are subjected to hostile environments composed of a variety of particles. The charged particles can hit the ICs resulting in non-destructive or destructive effects according to the charge intensity and to the hit location. The design of analog and mixed-signal circuits for space application needs to consider the radiation effects and to qualify these circuits for such applications.

Space radiations can produce Single Event Upsets (SEU) and Single event Transients (SET) in the electronics embedded into space applications. SEUs and SETs as well as their effects on digital circuits have been widely studied from more than a decade [19], [27] and several techniques to implement fault tolerant digital circuits have been devised in the last years, e.g. [4], [7], [17]. However, few works concern the problem of SEU in analog circuits [1], [16], [26] since this type of device is usually considered robust.

In this chapter, we revisit concepts and methods of the analog testing field looking for techniques that may help mitigating, at the system level, radiation effects in analog and mixed-signal circuits. The analog testing problem is addressed in section 2.

A particular class of analog circuits, which has recently been introduced to the market, is the Field Programmable Analog Arrays (FPAAs). This kind of circuit is based on configurable analog blocks and is able to implement several analog functions that, depending on the device model, can achieve a relatively high degree of complexity. They boost design flexibility, allow the fast prototyping and offer some interesting features for applications such as adaptive control, instrumentation, etc. These features can be very useful when the environmental variables can assume a wide range of values and the system must respond properly to these variations. This kind of device can be possibly applied to space exploration missions, where it could exist the need to re-calibrate the sensor conditioning circuits of spacecrafts to correct errors or improve system performance, for example. Some FPAAs may have their programmability based on SRAM memory blocks. For this reason the same problem observed in the SRAM-based FPGAs (Field Programmable Gate Arrays), concerning radiation effects, can be extended to FPAAs. In section 3, a set of practical experiments are performed in order do identify how bit-flips in the programming data (which can occur due to radiation exposure) affect the implemented analog circuit.

Another special class of analog circuits, often present in complex integrated systems, is the Analog to Digital (A/D) converter. More specifically, A/D converters based on sigma-delta ($\Sigma\Delta$) modulation are very popular nowadays. These converters are specially insensitive to circuit imperfections and component mismatch, and provide a means of exploiting the enhanced density and speed of scaled digital VLSI circuits. We can note several uses for $\Sigma\Delta$ A/D converters in the space environment. These kind of applications introduce the possibility of the circuitry to suffer from radiation effects. In order to verify the behavior of both analog and digital parts of the converter under radiation exposure we develop, in section 4, a MatLab model of the converter, inserting faults that emulate the radiation effects. In section 4 we also study several implementations of digital filters in order to minimize the radiation effects.

Finally, before concluding the chapter, in section 5 we revisit some of the existing analog on-line test techniques that can be used to recover the system from the transient effect of space radiations. In this section the design of analog self-checking circuits based on partial replication and on balance checking are addressed.

2. Analog testing

Although most electronic circuits are almost entirely digital, many include at least a small part that is essentially analog. This is due to the need to interface with the real physical world, that is analog in nature. Therefore, transducers, signal conditioning and data converter components add to the final circuit architecture, leading to ever more complex mixed-signal chips with an increasing analog-digital interaction.

Nevertheless, as illustrated in Fig. 1, the development of reliable products cannot be achieved without high quality methods and efficient mechanisms for testing Integrated Circuits (ICs). Practical analog testing solutions are still lagging behind their digital counterparts. Analog and mixed-signal testing has traditionally been achieved by functional test techniques, based on the measurement of circuit specification parameters. However, measuring such parameters is a time consuming task, requires costly test equipment and does not ensure that a device passing the test is actually defect-free. Then, to ensure the quality required for product compe-titiveness, one can no more rely on conventional functional tests: a move is needed towards methods that search for manufacturing defects and faults occurring during the lifetime of a circuit.

Moreover, to achieve acceptable fault coverages has become a very hard and costly task for external testing: test mechanisms need to be built into integrated circuits early in the design process. Ideally, these mechanisms should be reused to test for internal defects and environmental conditions that may affect the operation of the systems into which the circuits will be embedded. This would provide for amplified payback. To give an estimate about the price to pay for faults escaping the testing process, as shown in Fig. 2, fault detection costs can increase by a factor of ten, when moving from the circuit to the board level, then from the board to the system level, and lastly, from the final test of the system to its application in the field.

Compared to digital logic, analog circuits are usually made up of much fewer elementary devices that interface with the external world through a much smaller number of inputs and outputs. Thus, the difficulties of testing analog circuits do not reside in sizing, but in the precision and accuracy that the measures require. Additionally, analog circuits are much more sensitive to loading effects than digital circuits. The simple flow of an analog signal to an output pin may have an important impact on the final circuit topology and behavior.

Digital signals have discrete values. Analog signals, however, have an infinite range, and good signal values are referred to certain tolerance margins that depend on process variations and measurement inaccuracies. Absolute tolerances in analog components can be very large (around 20%), but relative matching is usually very good (0.1% in some cases). Although multiple component deviations may occur, in general, analog design methods promote that deviations cancel each other's effect, placing design robustness on the opposite direction of fault detection. Additionally, simulation time very quickly gets prohibitive for multiple component deviations.

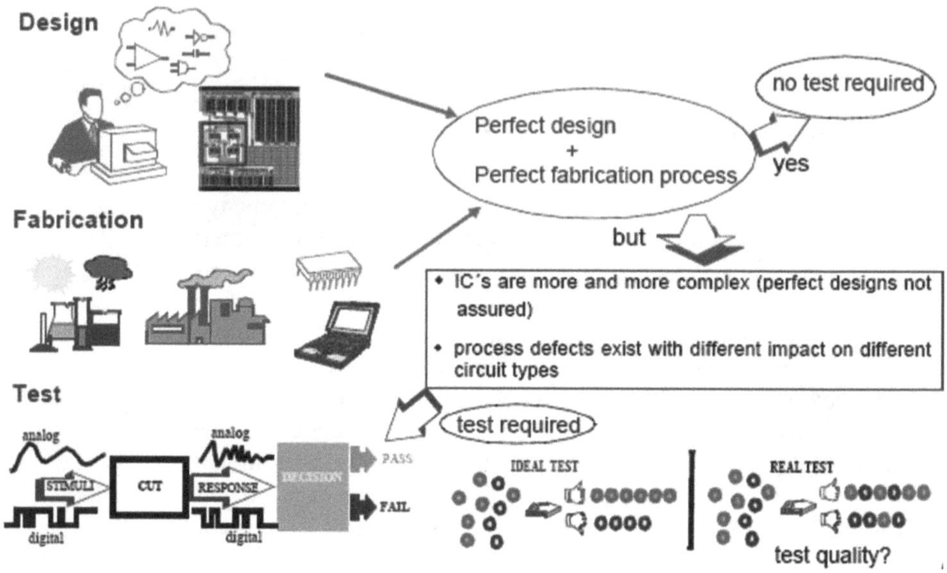

Figure 1. Main steps in IC realization.

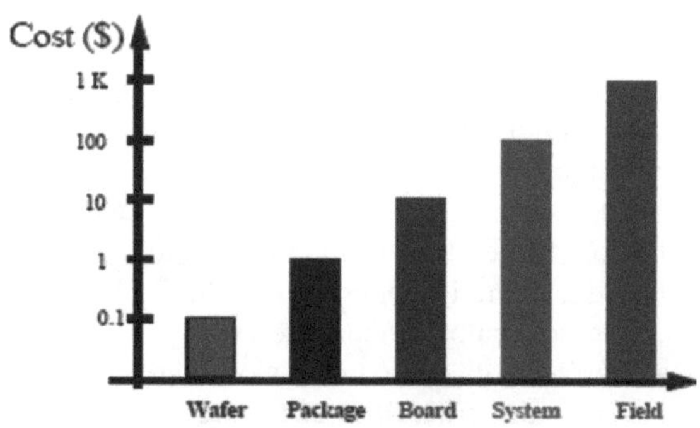

Figure 2. Test economics.

For the reasons above, modeling the circuit behavior is far more difficult in analog than in digital circuits. Furthermore, the function of analog circuits cannot be described by closed-form expressions as in Boolean algebra, that allows for the use of very simple fault models such as the widely accepted digital stuck-at. Instead, the behavior of an analog circuit depends on the exact behavior of a transistor, whose model requires a set of complex equations containing lots of parameters. As a consequence, it turns difficult to map defects to suitable fault models and thus, to accurate simulate the circuit behavior in presence of faults.

Therefore, although defects are absolutely the same for digital and analog circuits as shown in Fig. 3, fault modeling is a much harder task in the analog case. This is also due to a larger number of possible misbehaviors resulting from defects that may affect a circuit dealing with analog signals.

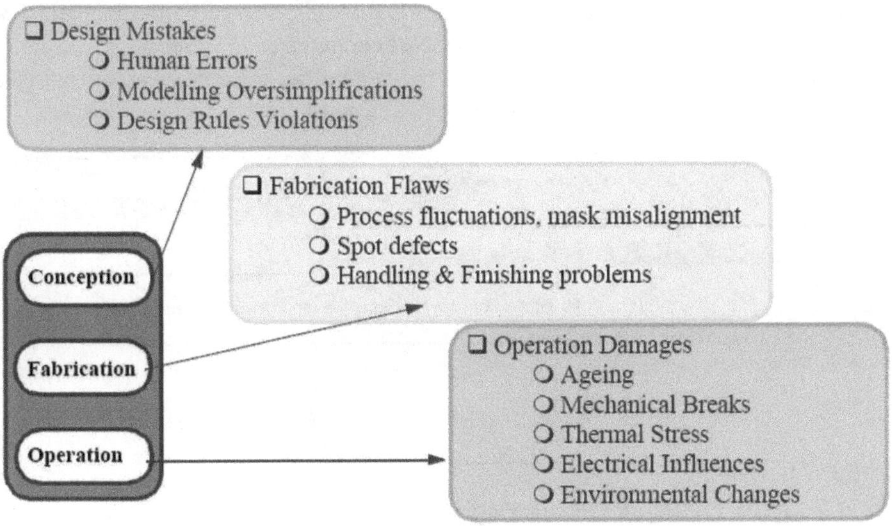

Figure 3. Failure mechanisms and defects.

The most intuitive fault model is the one that simply translates the various facets of the expected behavior of a circuit to a number of parameters that shall conform for the circuit to be considered fault-free (Fig. 4). These parameters are obviously extracted from the circuit design specification and measuring all of them equals to checking the whole circuit functionality, at a cost that may approach the one of full device characterization.

In the context of analog and mixed-signal circuits, several analogies can be drawn between the concepts and methods used in the fields of testing and of mitigation of space radiation effects.

First of all, the class of analog and mixed-signal circuits is essential in space applications as they are the only means to interface sensors and transducers with the dominating digital electronics. Therefore, it must be as hard to mitigate radiation effects in analog and mixed-signal circuits, as it is to test these electronic circuits when compared to digital parts.

Secondly, efficient functional tests capable of detecting realistic defects are the key to ensure high quality manufacturing tests and, since they check the system function, they may also play an important role to detect undesirable radiation effects concurrently to the execution of the application in space. Effects may be seen as soft faults in the case of Single Event Transients, and as hard faults in the case of Single Event

Upsets. Therefore, it will be possible to mitigate the former by filtering functions and the latter by voting or self-correction techniques, for example.

Finally, if for analog testing to embed mechanisms for fault prevention, detection, diagnosis and/or correction into circuits may provide an amplified payback due to reuse along the circuit lifetime, in the case of space applications those mechanisms are the only way to perform remote (from earth) or self-repair in very expensive equipment that would not recover the desired function otherwise.

Within this context, next sections investigate the effects of radiation in Field Programmable Analog Arrays (FPAAs) and ΣΔ A/D converters to propose, at the end of the chapter, the reuse of existing analog testing techniques for concurrent error detection to mitigate SETs and SEUs in mixed-signal circuits.

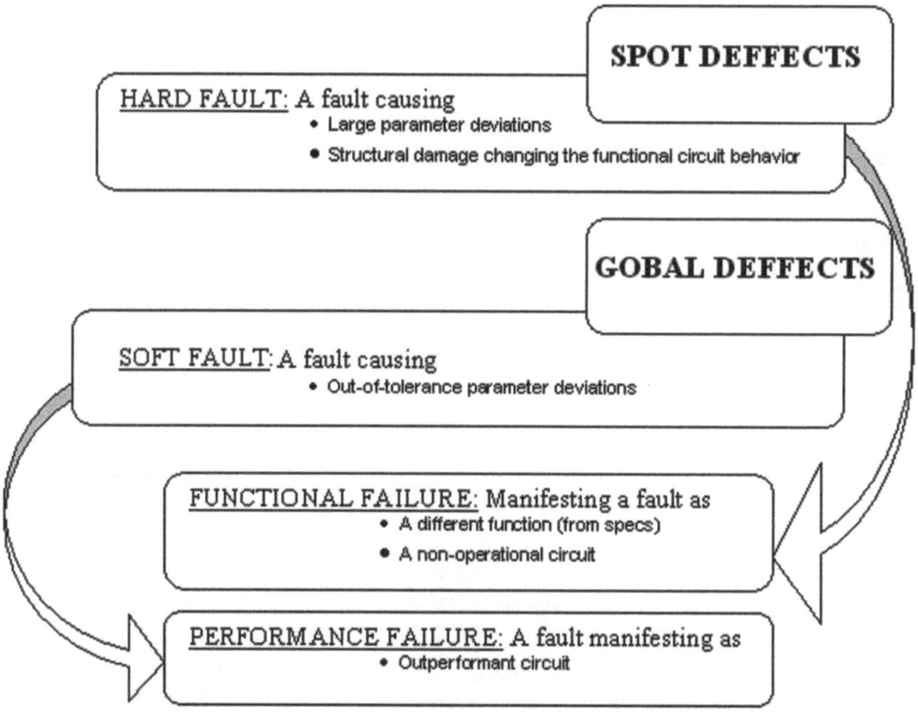

Figure 4. Faults and failures.

3. First case study: SRAM-based FPAAs

Field Programmable Analog Arrays (FPAAs) are devices based on configurable analog blocks. They provide to the analog world the same advantages as their digital counterparts, Field Programmable Gate Arrays (FPGAs), provide to digital circuits. They boost design flexibility, allow the fast prototyping and offer some interesting features for applications such as adaptive control and instrumentation and evolvable analog hardware [12], [28]. These features can be very useful when the environmental variables can assume a wide range of values and the system must respond properly to these variations. As an example one can consider the avionics applications where the external temperature and pressure can vary significantly in few minutes of flight. Another possible application is in the space exploration missions, where could exist the need to re-calibrate the sensor conditioning circuits of spacecrafts to correct errors or improve system performance for example.

For all these reasons FPAAs have becoming an important platform to analog circuit prototyping, thus it is important to ensure the correctness condition for the analog functions implemented into these devices.

Some FPAA architectures are based on SRAM memory to implement the user programmability. In this case, the same as for SRAM-based FPGAs, a SEU can affect the programming memory and change the device configuration, which may lead to a system failure. A SET in an FPAA can change momentarily the value of analog signals in the circuit depending on the biasing current, the stored charge and the voltage of the signal present in the affected nodes. If we consider aeronautical and space applications, in which the incidence of cosmic radiation are more intense, SEU and SET affecting analog circuits (specially the SRAM-based FPAAs) could become a non-rare problem.

Several works propose solutions to mitigate the SEU and SET effects in SRAM-based FPGAs. The most popular techniques are based on hardware and time redundancy. In the TMR (Triple Modular Redundancy) approach [7] the application circuit is tripled and a circuit called "voter" identifies (as correct) the inputs that have the same value, and transfers the value to the output. In time redundancy approaches the register cells are doubled or tripled and the computation of a given signal is made in two or more different instants of time and a majority voter selects the correct output of the registers [17]. This way, a SET with duration smaller than the time delay between the loads of the redundant registers shall not affect the system operation.

The implementation of an analog majority voter is not trivial as it is for the digital circuits. Such voter must be able to select the correct inputs and transfer the correct signal to its output without degradation, which implies in the use of a not simple signal comparison scheme and an analog MUX with the same bandwidth, dynamic range and SNR (Signal to Noise Ratio) of the application circuit. Furthermore, differently from digital circuits, the time redundancy technique is not easily applicable to the analog parts since the analog signal is continuous and in several applications the signal is not sampled nor processed in the digital domain.

3.1 The effects of SEU in SRAM-based FPAAs

A typical FPAA structure comprises Configurable Analog Blocks (CABs), I/O blocks, an interconnection network and memory registers for device programming. In some FPAAs the programming memory is based on SRAM blocks [2], thus it is possible to conclude that the programmable analog hardware can be as susceptible to SEU as its digital counterpart.

Usually the programmability of FPAAs is allowed through switches that set the routing and the values of components within the array. The state of such switches is defined through the value of a BIT-stream stored in a shift register that is serially loaded during the device configuration. This shift register can be based on SRAM-type memory cells. Therefore, the impact of a charged particle in one or more transistors of the memory cells can result in a bit-flip of the previous stored BIT-stream, which may change the state of a switch used in the circuit and modify parameters like values of components or the routing in or between the CABs of the FPAAs. In some cases a SEU in the programming memory can result in a very different configuration from that previously programmed, which can be harmful to the system operation. Fig. 5 illustrates such event considering a typical SRAM-based FPAA architecture.

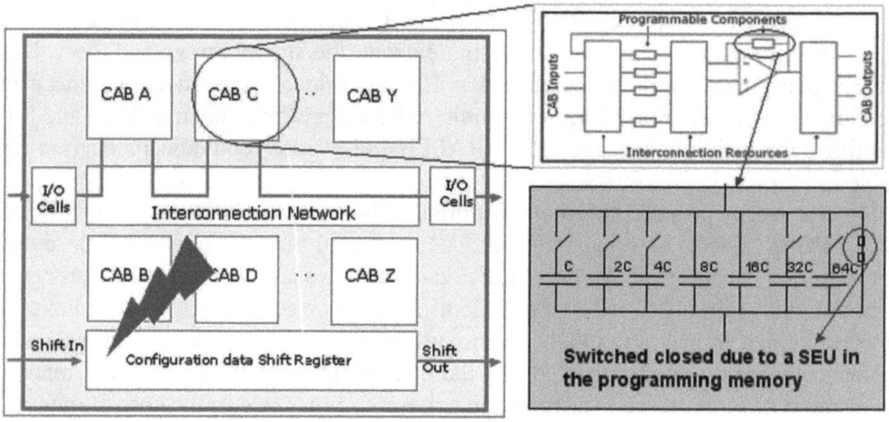

Figure 5. Variation in the programmed circuit due to a SEU in the programming memory of a SRAM-Based FPAA.

In Fig. 5 typical FPAA and CAB architectures are shown. Each CAB is composed of an analog programmable component array, local and global interconnection switching blocks, wiring, and an Output Amplifier (OA) with global and local programmable feedback loops. The components within the array can be implemented as simple wires, passive or active components or other more complex parts. In general, the programmable parameters of CABs are gain of amplifiers, values of resistors and capacitors, as well as setting global and local feedback loops. In some FPAA models the resistors are implemented through the switched-capacitor technique and the value of its resistance may be programmed by the value of the capacitor and its switching frequency. The value of the capacitors (either switched or static ones) is programmed through a bank in which a set of programmable switches is responsible to connect or disconnect the capacitors in order to configure the desired value.

In the example of Fig. 5 a previous programmed circuit is set with a capacitor whose relative value is 32C. If a SEU occurs in the programming memory cells its possible that a component used in the circuit implementation suffers a change in its value. In this case a bit-flip was considered in one of the switches which implements one programmable capacitor bank, provoking a short in the programmed capacitor.

One can see that a SEU can be catastrophic when using SRAM-based FPAAs since, in some cases, the correct functioning of a single switch is crucial to the system operation. Besides modifications in the programmed values, a bit-flip in the memory cells can result in the disconnection of components, connections of undesirable components in the circuit (parasites) and even an interruption in the signal path. Such interruption can invalidate the affected analog blocks or even the entire system in which the FPAA is inserted.

If one ore more BITs of the original configuration are modified during the operation of the circuit, the only way to restore the original configuration is reloading the configuration BIT-stream into the FPAA memory cells. In some FPAA models this reload can be done in fractions of milliseconds.

3.2 Fault injection experiments

In order to perform some experiments to study the effect of BIT-flips in the programming memory of FPAAs a commercial device is used. The device studied in this work is the AN10E40 from Anadigm Company [2], a switched capacitor FPAA. It has 20 CABs distributed in a 4x5 array. Each CAB can be connected to any other CAB or to one of the 13 I/O cells through an interconnection network. The block diagram of the AN10E40 is shown in Fig. 6, in which one can see the global wiring surrounding the 20 CABs. This network comprises horizontal and vertical buses organized in 5 rows and 6 columns, each one composed of two wires. Besides the global buses, the connections between the CABs can be made by means of local interconnections.

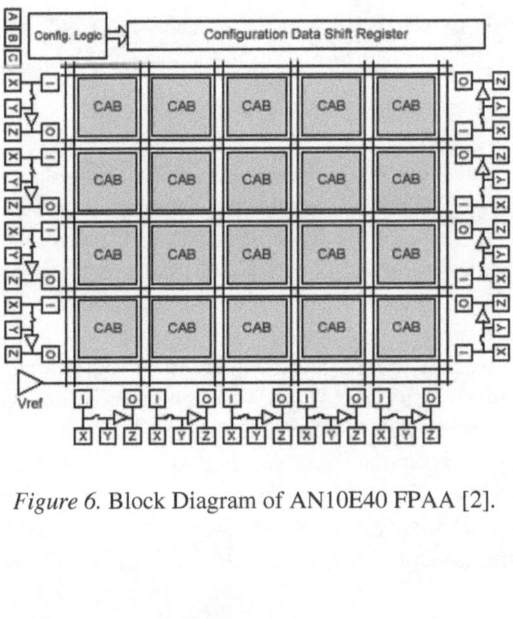

Figure 6. Block Diagram of AN10E40 FPAA [2].

A schematic view of the AN10E40 CAB is depicted in Fig. 7 [2]. Each CAB has 5 capacitor banks that can implement a programmable capacitor or a programmable resistor (switched-capacitor). The values of the programmable components of the CAB cannot be changed directly by the user, since the programming and constructive details are unknown. In order to configure the desired circuit into the FPAA one needs to use the Anadigm Designer Software, which provides a set of pre-built IP modules. The user can link these modules and set parameters like block gain, central frequency of filters, integration constants and thresholds of comparators, for example.

However in one of the data directories of the programming software it is possible to find the files that contain the default BIT-stream for each module available in the programming library. Figure 8 shows the default BIT-stream for 2 of the IP-modules of the programming library (a simple gain stage and a rectifier).

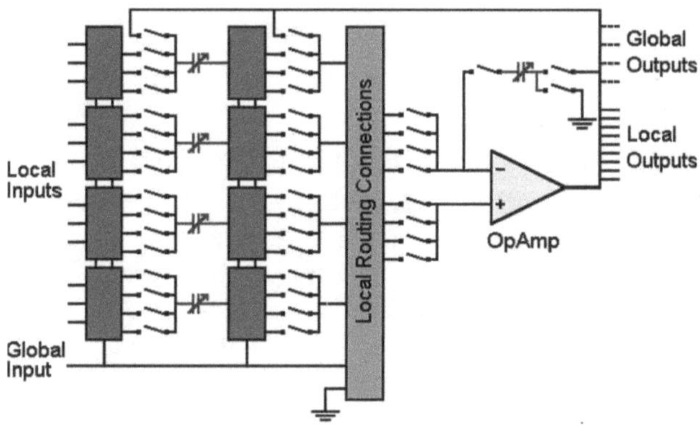

Figure 7. Schematic view of the AN10E40 CAB [2].

IPmodule	Bytes
Simple gain stage	003f c040 0022 ff24 1000 0ff3 fc00 0018 2270 01c0 0805 2090 8000
Rectifier	003f c040 0082 ff20 1000 0ff0 0000 0080 2a70 01c0 0000 2090 9500

Figure 8. Examples of default BIT-stream of IP-modules.

In the example of the Fig. 8 both considered analog IP-modules are built using only one CAB of the FPAA. Each CAB comprises 208 programmable switches [2], which means that each one of these blocks is programmed with a stream of 208 BITs. There are other analog modules built with 2 or 3 CABs, therefore one single analog function can be programmed with up to 624 BITs. As there are 20 CABs in this device the total amount of memory dedicated to the CABs programmability is 4160 BITs. The other programmable resources of the device (routing, IO, programmable voltage reference and clock divisions) are set by means 2704 switches, therefore the whole BIT-stream of the AN10E40 is composed of 6864 BITs [2].

The fault injection experiments are carried out by modifying the default BIT-stream of the IP-modules. For this purpose a copy of the files that contain the BIT-stream was made in such a way that two module libraries are now available, one with the fault-free BIT-stream an the other with single BIT-flips in the configuration BIT-stream. This approach does not allow to inject faults in the BITs that program the global interconnections or the IO cells of the device because the modified files only comprise the BIT-stream of the CABs switches.

The analog module used in this experiment is a sinewave oscillator whose default BIT-stream of both CABs used in the module implementation and the expect output signal are shown in the first block of Fig. 9 (where f_{osc} is the oscillation frequency and A is the amplitude of the signal). Each digit of the BIT-stream above is a hexadecimal representation therefore comprising 4 BITS. Faults were injected in only one of the CABs that comprise the oscillator by changing (inverting) the values of one BIT at a time of the CAB. A total of 208 faults were injected. Examples of the implication of single flips in the considered BIT-stream are shown in the second and third blocks of Fig. 9.

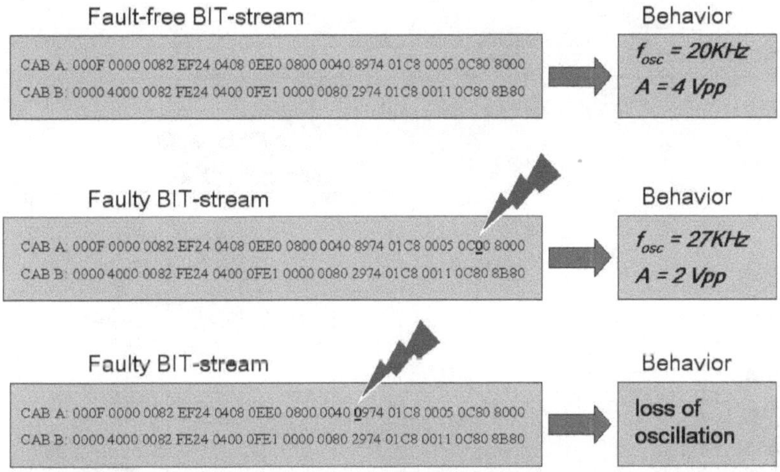

Figure 9. Modified BIT-stream of oscillator and resultant behaviors.

3.3 Experimental results

In order to facilitate the detection of errors caused by SEU in the programming memory an error detection circuit was built using the internal resources of the FPAA. Such error detection circuit is based on a very selective Band-Pass (BP) filter that attenuates the signal if the frequency of the oscillator (tuned with the center frequency of the filter) is different from that previously programmed. The output of the BP filter is rectified and filtered in order to generate a DC level, which is compared to a reference window. Variations in the amplitude or frequency of the oscillation signal will generate a DC signal whose amplitude is out of the considered window, thus being detected by the comparators. Figure 10 shows the block diagram of the error detection circuit while Fig. 11 shows the block representation for this scheme in the Anadigm Designer software environment.

The oscillator frequency and amplitude are programmed as 20KHz and 4Vpp respectively. The filter central frequency and gain are programmed as 20 KHz and 0 dB. The reference window is set as ±10mV around the DC level generated in the output of the rectifier/low-pass block. With this scheme variations of ±10mV in the amplitude and ±100Hz in the oscillator frequency are detected by the evaluation circuit.

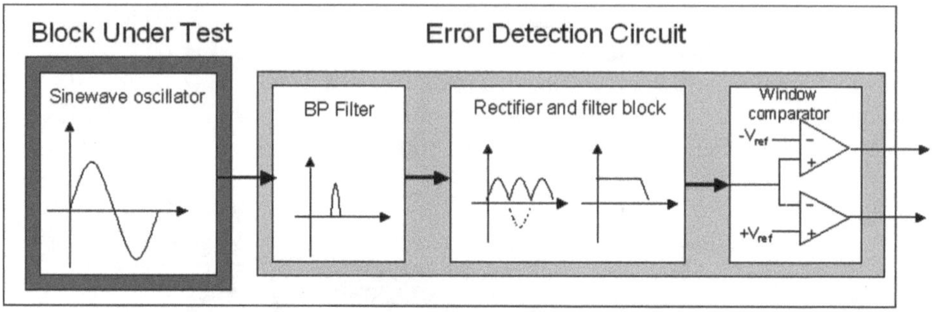

Figure 10. Oscillator and error detection circuit.

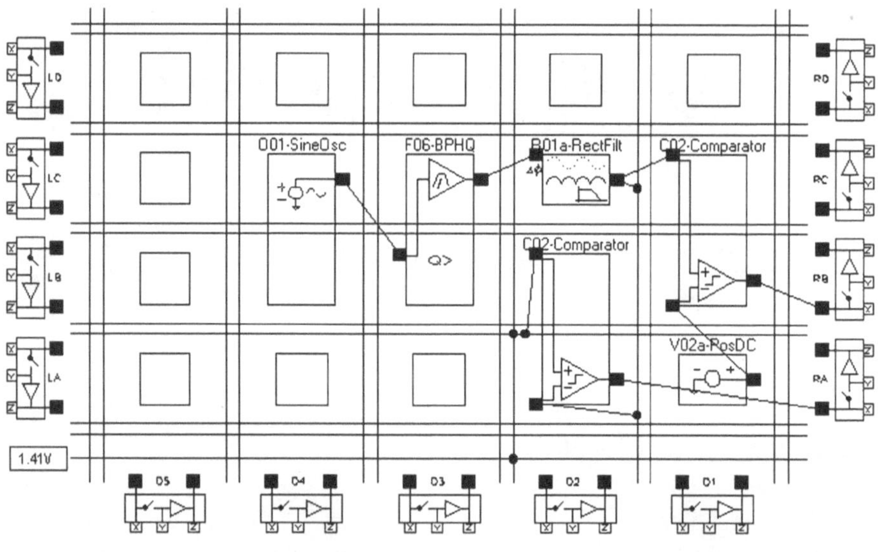

Figure 11. Anadigm Designer software with the programmed circuit representation.

For a total of 208 injected BIT-flips only 57 faults affect the functional behavior of the circuit in a way that the amplitude or frequency of the oscillator deviates from the nominal programmed value. The low rate of errors caused by the injected faults is due to the fact that only part of the CAB resources is used to implement the oscillator. Figure 12 shows the schematic of the oscillator according to the Anadigm IP-Modules Manual [3] and Fig 13 shows the possible implementations of this scheme by using the CAB depicted in Fig. 7 (the comparator in Fig. 12 is a control block and is not considered in this fault injection experiment). It is possible to see that these implementations do not use neither all the programmable components nor the entire local routing of the CABs. For this reason a SEU in a BIT that control some component or branch not used in the circuit can maintain unaltered the functionality of the programmed blocks.

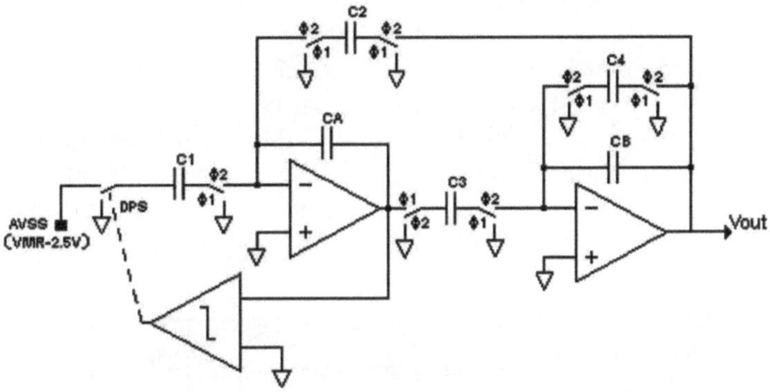

Figure 12. Oscillator schematic (two CABs) [3].

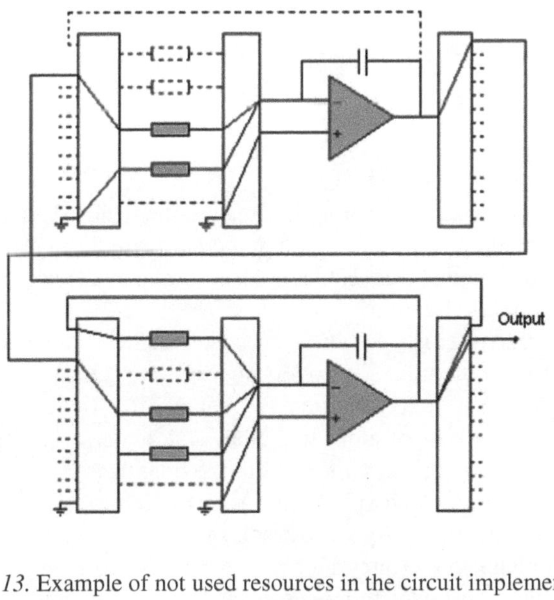

Figure 13. Example of not used resources in the circuit implementation.

4. Second case study: ΣΔ A/D converters

In recent time, oversampling A/D and D/A converters have become popular due to their high resolution for medium-to-low speed applications, such as high-quality digital audio, voice-band communications, wireless communications and others.

A/D converters based on sigma-delta (ΣΔ) modulation combine sampling at rates well above the Nyquist rate with negative feedback and digital filtering in order to exchange resolution in time for that in amplitude. Figure 14 shows the block diagram of the ΣΔ A/D Converter. First the analog signal is modulated into a simple code, a single-bit word, at a frequency much higher than the Nyquist rate (quantization). The decimation stage transforms the modulated signal to longer words at lower word rates. The digital filter stage is designed to remove noise and interference. We will describe and explain every circuit individually in following sub-sections.

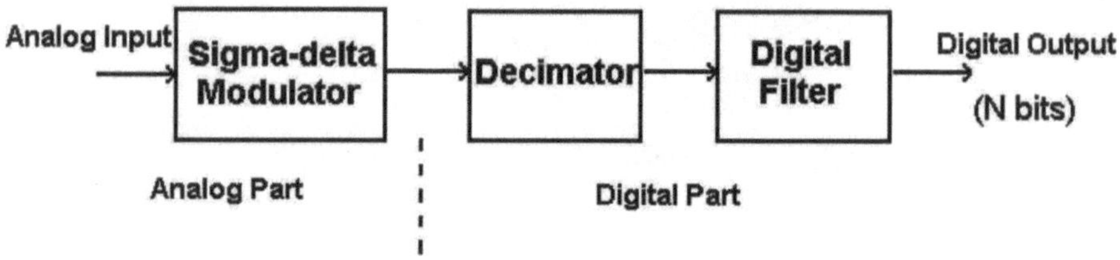

Figure 14. Block diagram of the Sigma-Delta A/D Converter.

4.1 The ΣΔ modulator

Quantization of amplitude and sampling in time are the heart of all digital modulators. The analog signal is modulated into a simple code, a single-bit word, at a frequency much higher than the Nyquist rate. Hence, this system provides a good approximation of a PCM (Pulse Code Modulation).

The quantization error (e) is the difference between the input and output values, that can be treated as white noise. The oversampling rate (OSR) is defined as the ratio of the sampling frequency f_s to the Nyquist frequency $2f_0$.

The ΣΔ modulator consists of an analog filter and a coarse quantizer enclosed in a feedback loop. The input of the circuit is fed to the quantizer via an integrator, and the quantized output is fed back and subtracted from the input. This feedback forces the average value of the quantized signal to track the average input. Any difference between them accumulates in the integrator and eventually corrects itself. The feedback loop acts to attenuate the quantization noise at low frequencies, while emphasizing the high-frequency noise. Since the signal is sampled at a frequency which is much higher than the Nyquist rate, high-frequency quantization noise can be removed without affecting the signal band.

ΣΔ modulators are specially insensitive to circuit imperfections and component mismatch since they employ only a simple two-level quantizer, and this quantizer is embedded within a feedback loop.

Higher order ΣΔ modulators, containing more than one integrator, offer the potential of increase resolution up to 16-20 bits. However, modulators containing more than two integrators suffer from potential instability owing to the accumulation of large signals in the integrators. An architecture where several first-order modulators are cascaded in order to achieve performance that is comparable to that of higher order modulators is one way to overcome the stability problem.

Figure 15 shows the basic 1-bit ΣΔ modulator first-order loop, where the filter consists of a single integrator, the quantizer is a comparator and a 1-bit D/A converter.

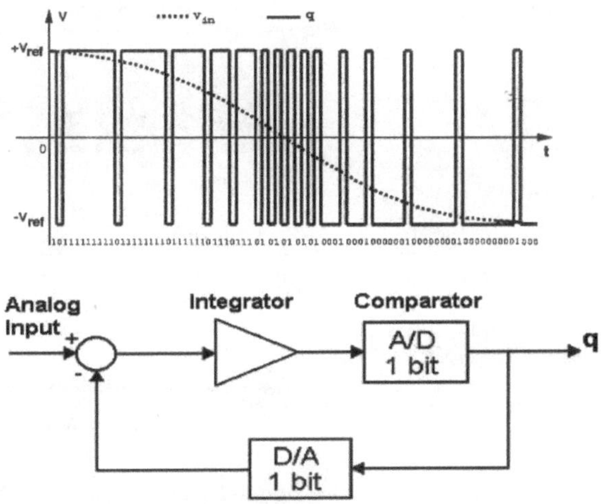

Figure 15. The Sigma-Delta Modulator.

4.2 A MatLab model for the ΣΔ A/D converter

For this case study we developed a MatLab model of the ΣΔ A/D converter based on [21] and [8]. To this end it is convenient to use the discrete-time equivalent circuit of Fig. 16, where the integrator has been replaced by an accumulator. Herein, the input X is a discrete-time sequence of analog samples developing an output sequence of binary-value samples Y. At each discrete instant of time, the circuit forms the difference (Δ) between the input and the delayed output, which is accumulated by a summer (Σ) whose output is quantized by a comparator (one-bit A/D converter). Figure 17 shows the input and output waveforms of the ΣΔ quantizer.

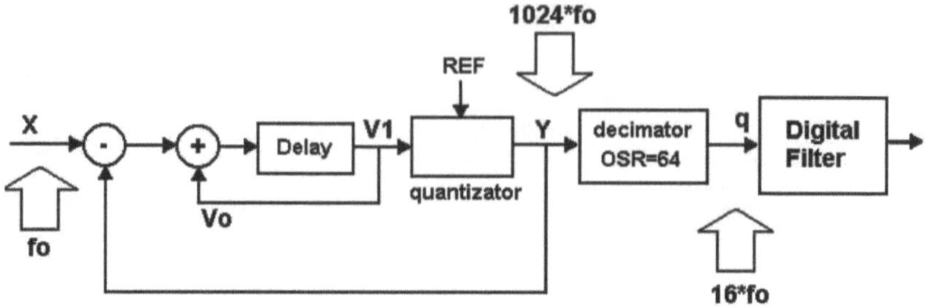

Figure 16. The ΣΔ quantization MatLab model.

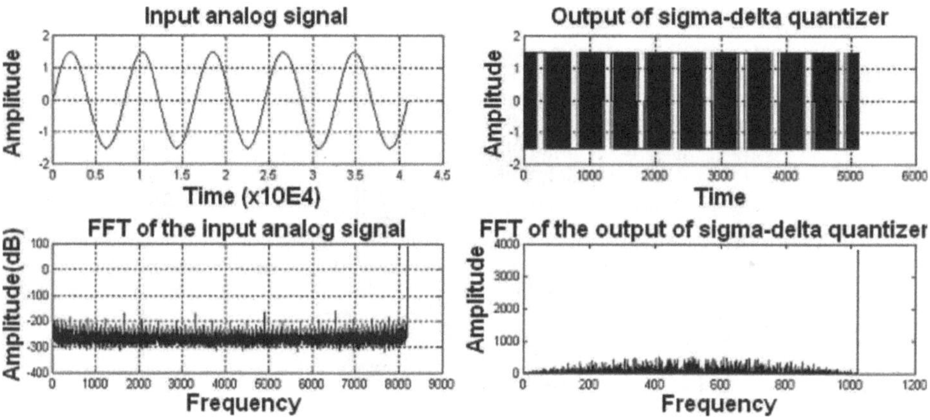

***Figure** 17.* Input and output waveforms of the sigma-delta quantizer MatLab model.

4.2.1 The digital blocks: decimator/filter

The output of the ΣΔ modulator represents the input signal together with its out-of-band components, modulation noise, circuit noise and interference. So, it is necessary a stage (fully digital) that serves to remove all the out-of-band components and re-sample the signal at the Nyquist rate. The decimation stage transforms the modulated signal to longer words at a lower word rate and a digital filter stage is designed to remove noise and interference. We describe and explain every individual circuit in the following.

4.2.1.1 The decimation stage

Decimation is an important component of oversampling analog-to-digital conversion. It transforms the digitally modulated signal from short words occurring at high sampling rate to longer words at the Nyquist rate.

A convenient filter for decimating ΣΔ modulation has a frequency response based on *sinc* functions. The simplest of these decimators is the accumulate-and-dump circuit. If its input samples are x_i occurring at rate f_s and output samples are yk occurring at f_D, then the decimation ratio N is the integer ratio of the input frequency to the output frequency. Figure 18 shows an implementation of the sinc function.

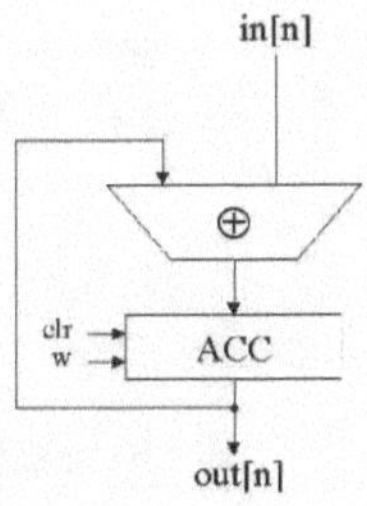

Figure 18. Implementation of the sinc function.

For this case study we developed a MatLab model for the *sinc* filter. The decimation rate is 64, downsampling the quantized signal to 8 times the Nyquist rate (i.e., $16*f_0$). Figure 19 shows the frequency response of the implemented filter and the filtered quantized signal.

4.2.1.2 The lowpass digital filter stage

Since the decimation stage does not provide sufficient out-of-band attenuation, a lowpass digital filter stage is necessary to eliminate the out-of-band quantization and the high frequency components of the signal. Its circuit can usually be very simple. For this case study we consider three different implementations of a lowpass digital filter, to find out which of them suits better our application. We describe and explain each of these implementations in the following.

4.2.2 FIR filter

First, we consider a FIR (Finite Impulse Response) filter implementation. There are several advantages in using FIR filters, since they can be designed with exact linear phase and the filter structure is always stable with quantized filter coefficients. However, in most cases, the order N_{FIR} of a FIR filter is considerably high. In general, the implementation of the FIR filter requires approximately N_{FIR} multiplications per output sample.

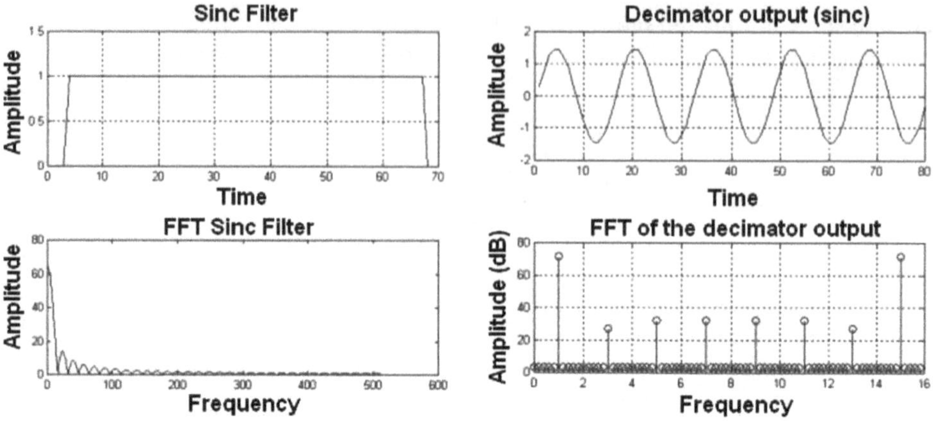

Figure 19. Frequency response of the implemented sinc filter and output signal of the modelled modulator.

Using MatLab we designed a lowpass FIR filter based on [21] to eliminate the out-of-band components of the quantized signal. First the filter order is estimated from the given specifications (f_p=1, f_s=2, Fs=16) and then the coefficients of the filter's transfer function are determined using the estimated order and the filter specifications. For this case the lowpass filter order estimated is 22. Figure 20 shows the direct form structure of the implemented filter and Figure 21 shows its frequency response.

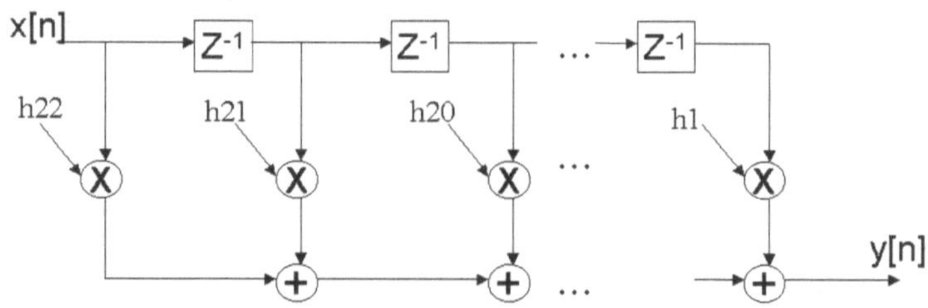

Figure 20. Direct form structure of the FIR filter implemented.

4.2.3 IIR filter

For an IIR (Infinite Impulse Response) digital filter, the computation of the nth output sample requires the knowledge of several past samples of the output sequence, or in other words, requires some form of feedback. An Nth-order IIR digital filter transfer function is characterized by 2N + 1 coefficients and, in general, requires 2N + 1 multipliers and 2N two-input adders for implementation. In most cases, the order N_{FIR} of a FIR filter is considerably higher than the order N_{IIR} of an equivalent IIR filter meeting the same specifications. IIR filters are also usually computationally more efficient.

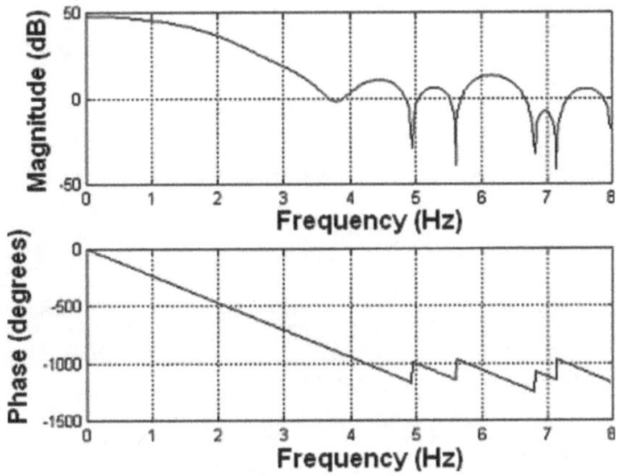

Figure 21. Frequency response of the lowpass FIR filter implemented.

In the case of IIR design, the most common practice is to convert de digital filter specifications into analog lowpass prototype filter specifications, to determine the analog lowpass filter transfer function meeting these specifications, and then transforming it into the desired digital filter transfer function.

For this case the lowpass filter order estimated is 6. Figure 22 shows the filter's frequency response.

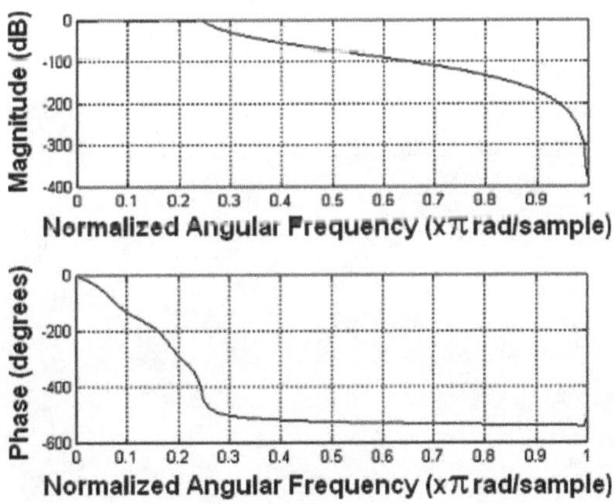

Figure 22. Frequency response of the lowpass IIR filter implemented.

Therefore, using MatLab we designed a cascade IIR filter, transforming the 6th order lowpass IIR filter in a three-stage cascade filter using a second-order IIR DFIIt structure in each stage.

4.2.4 IIR filter with the delta operator

Reduction of finite-word-length effects in digital filter has been studied in a large number of papers during the past few decades. Recently, delta operation realizations have gained interest due to their good finite-word-length performance under fasting sampling [11], [14].

The delta operator is defined as:

$$\delta = \frac{Z-1}{\Delta} \tag{1}$$

where z is the forward shift operator and Δ is proposed to be the sampling interval. The parameter Δ can be considered as a free parameter allowing values from 0 to 1 and it can be used for optimizing the roundoff noise of the filter.

The implementation of δ^{-1} operation is shown on Fig. 23. In filter realizations the delays are replaced with δ^{-1} blocks.

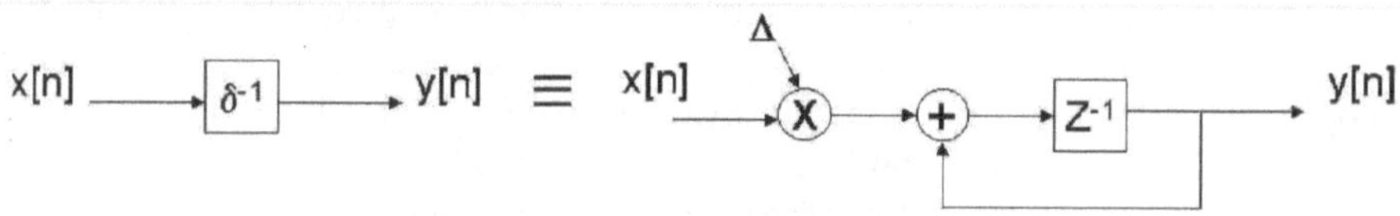

Figure 23. Implementation of the delta operation.

Different delta operator realized filter structures have been compared [11], [14] and the direct form II transposed (DFIIt) structure has been found to perform better. Therefore, DFIIt structure was chosen for comparison of the implementation complexity of delta and delay realizations. A second-order implementation of the delta DFIIt structure is shown in Fig. 24.

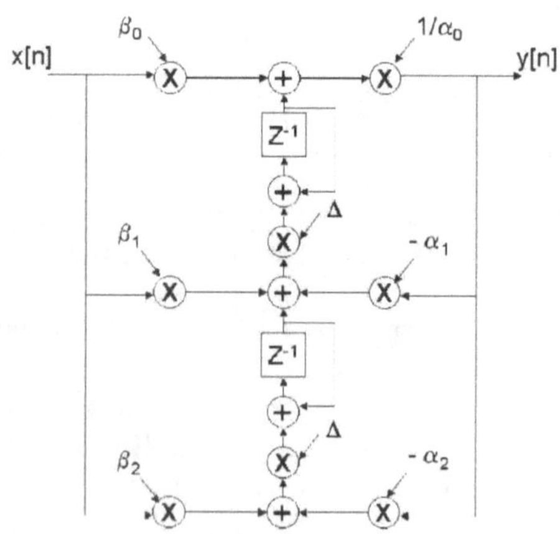

Figure 24. A second-order IIR Delta DFIIt structure.

Therefore, using MatLab we designed a cascade IIR filter with the delta operator, a three-stage cascade filter using a second-order IIR delta DFIIt structure in each stage. The coefficients of the delta operator were recalculated choosing the optimal parameter Δ equal to one [14].

4.3 Radiation effects in sigma-delta converters

4.3.1 Fault injection

In order to verify the behavior of the converter under radiation exposure, various faults were injected into the MatLab model. The effect of a SEU can be a bit flip in a memory cell or a transient in the combinational logic or in the analog part. When a radiation particle hits a register in the converter digital part, a bit-flip can occur. This is the most vulnerable part of the circuit. This bit-flip can be modeled in MatLab as a random change in the register value. The converter process is stopped, a random bit is flipped, and the process continues at this point. In the analog part, a particle-hit effect can be modeled as a change in the capacitor storage. Again, the converter process is stopped, the capacitor storage is changed to a random value, and the process continues. The point and the instant that the particle hits the converter are fundamental for the resulting consequence. A critical point, for example, is the most significant bit of the last register in the cascade. If this memory cell is hit at determined instant, the result can be disastrous. Next section will show the radiation effect over the modulator and the digital filters.

4.3.1.1 Modulator

The analog part of the ΣΔ converter is the modulator. The effect of a particle hit in an analog circuit is a current pulse. This current pulse can charge a capacitor to a random value. To simulate the effect of radiation over the modulator, we change the storage value of capacitor of the integrator stage of the modulator at a random interval of time. The result is shown in Fig. 25. It can be seen that this fault is entirely suppressed by the digital filters. The current pulse in the modulator, in this way, can be modeled as a high-frequency component that is filtered by the low-pass filters inherent to the ΣΔ modulator. This part, concluding, does not need to be protected against radiation.

4.3.1.2 Decimator

The sinc filter has only one memory element, the accumulator. In the converter designed, the decimator performs an addition of 64 values serially. A radiation particle that hits the most significant bit of the register at the last cycle of the sum can change the result significantly. Figure 26 shows the effect of a particle hit in the decimator and the resulting signal. Again, here the radiation effect can be modeled as a high-frequency component that is filtered by the low-pass filter next.

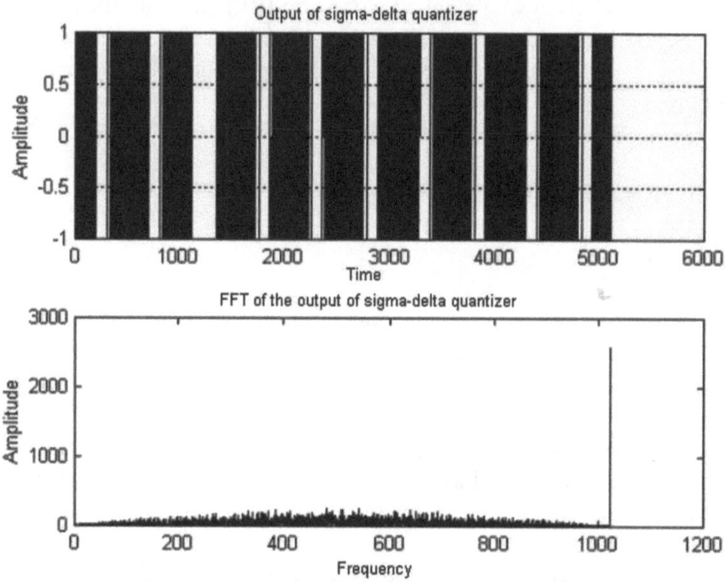

Figure 25. Output waveform of the ΣΔ quantizer with injected faults on the analog part of the converter.

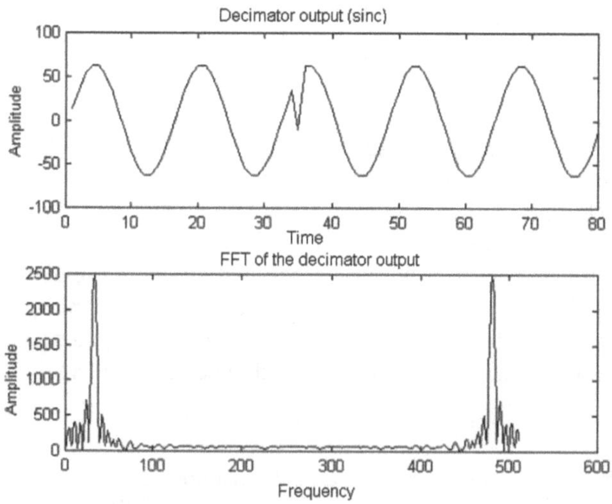

Figure 26. Output signal of the modeled modulator filtered by the decimator stage with injected faults.

4.3.1.3 FIR filter

The low-pass filter at the end of the converter built as a FIR filter has 22 registers. Figure 27 shows the effect of a radiation particle hitting the first register of the cascade. As it can be seen, there is no significant error caused by this fault. A bit-flip is not relevant to the signal process, because there are 22 taps in the filter.

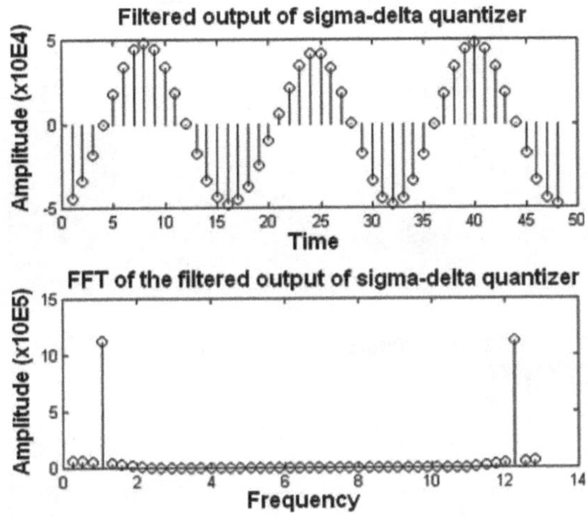

Figure 27. Output of the sigma-delta quantizer filtered by the lowpass FIR filter implemented with injected faults.

4.3.1.4 IIR filter, shift operator

Substituting the FIR filter by an IIR filter whit shift operator and injecting faults in it results in the signal of Fig. 28. The impact of a radiation particle here is disastrous. The fundamental frequency cannot be distinguished anymore.

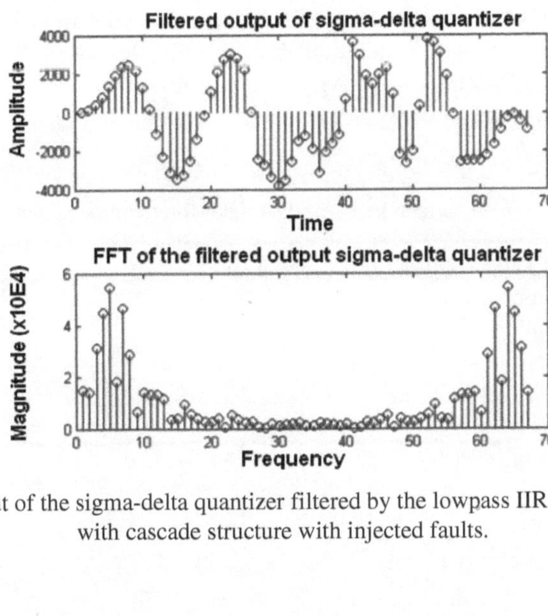

Figure 28. Output of the sigma-delta quantizer filtered by the lowpass IIR filter implemented with cascade structure with injected faults.

4.3.1.5 IIR filter, delta operator

The delta operator now substitutes the shift operator. Figure 29 shows the result signal with an injected fault. The result is better than the filter with shift operator. The signal is reconstructed in only one cycle.

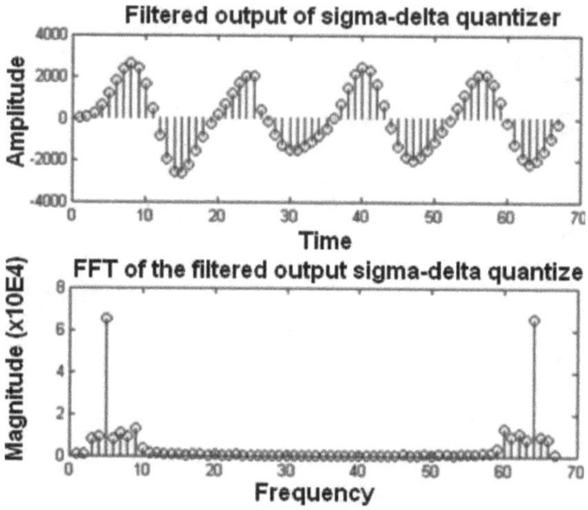

Figure 29. Output of the sigma- delta quantizer filtered by the lowpass IIR filter delta operator implemented with cascade structure with injected faults.

4.3.2 Comparisons between different implementations of the low-pass filter

The FIR filter has less sensitivity to the radiation particle hit, as shown above. However, it needs more silicon area to be implemented. The IIR filter with shift operator (IIR Z) has high sensitivity to radiation, but occupies less area. The IIR filter with delta operator (IIR Δ) has little sensitivity and requires an equivalent area to the IIR Z. Table 1 shows a comparison between these three implementations.

Table 1. Comparison between the digital filters implementations.

Filter Implementations	FIR	IIR Z	IIR Δ
Registers	21	6	6
Multipliers	22	15	15
Adders	21	9	15
Logic Cells (Altera)	1353	1077	1089
Bits	8	16	13

5. Analog self-checking design applied to SEUs and SETs mitigation

From the very first design, any circuit undergoes *prototype debugging*, *production* and periodic *maintenance tests* to simply identify and isolate, or even replace faulty parts. Those are called *off-line tests,* since they are independent of the circuit application and need, in the field, that the application is stopped before the related testing procedures can be run. Off-line tests target the detection of *permanent faults*, like interconnect opens and shorts, floating gates, etc, that can be produced by defects resulting from manufacturing and circuit usage.

In high safety systems, such as space, avionics, automotive, high speed train and nuclear plants, poor functioning cannot be tolerated and detecting faults concurrently to the application becomes also essential. The *on-line detection* capability, used for checking the validity of undertaken operations, can be ensured by special mechanisms, such as self-checking hardware. On-line tests target the detection of *transient faults*, that will appear due to intermittent phenomena, such as electromagnetic interference or space radiations.

In previous sections, we studied the effects of SETs and SEUs in FPAAs and in ΣΔ A/D converters. In this section we revisit some of the existing analog on-line test techniques, the self-checking circuits, that can be used to recover the system from the transient effect of space radiations.

In digital self-checking circuits, the concurrent error detection capability is achieved by means of functional circuits, which deliver encoded outputs, and checkers, which verify whether these outputs belong to error detecting codes. The most usual codes are the parity, the Berger and the double-rail code. The general structure of a self-checking circuit is shown in Fig. 30.

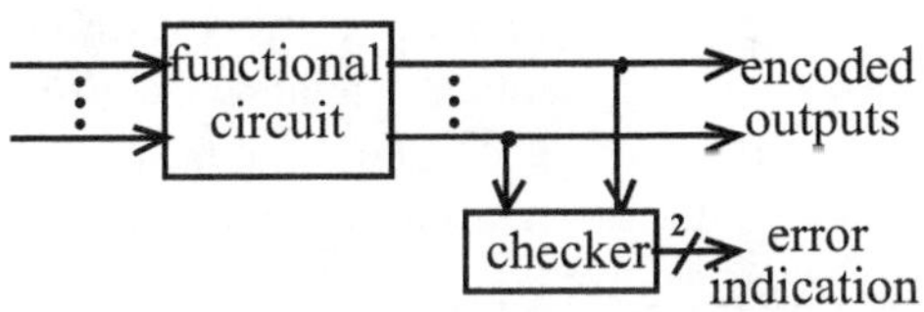

Figure 30. Self-checking circuit.

Most often, self-checking circuits are aimed at reaching the *totally self-checking goal*: the first erroneous output of the functional circuit results in an error indication in the checker outputs.

Similarly to digital self-checking circuits, the aim of designing *analog self-checking circuits* is to meet the totally self-checking goal. This is possible since analog codes can also be defined, for example the differential and duplication codes [15]. A tolerance is required for checking the validity of an analog functional circuit and this is taken into account within the analog code.

The nodes to be monitored by an analog checker are not necessarily those associated with the circuit outputs, due to commonly used feedback circuitry. In addition, the most important difference is that the input and output code spaces of an analog circuit have an infinite number of elements. Therefore, the hypothesis considered for digital circuits becomes unrealistic, since an infinite number of input signals might be applied within a finite lapse of time. In order to cope with this problem, the self-checking properties are redefined for the analog world in [22].

In the last years, the self-checking principle has been applied to on-line testing analog and mixed-signal circuits, including filters and A/D converters [9], [18]. Of particular interest to this work are the following techniques employed for concurrent error detection: *partial replication* of modular architectures, e.g. filters based on a cascade of biquads [13] and pipelined A/D converters [23], and *balance checking* of fully differential circuits [20].

The partial replication principle is shown in Fig. 31 and illustrated in Fig. 32 for the case of a multistage pipelined A/D converter. Since the converter is built from a *cascade of identical functional modules*, the on-line testing capability can be ensured by an additional *checking module* identical to the converter stages and a multiplexing system. The *multiplexing system* must be such that the outputs of every stage can be compared against the outputs of the checking module, when the latter receives the same input as the former. The *control* gives the sequence of testing that evolves sequentially from the first (1) to the last (L) stage, and then restarts.

Figure 33 illustrates the principle of balance checking applied to fully differential circuits and Fig. 34 applies the principle to an integrated filters. In a correctly balanced fully differential circuit, the operational amplifier inputs are at virtual ground. But, in general, transient faults, deviations in passive components and hard faults in operational amplifier transistors corrupt this balance. In [20], an analog checker is proposed which is capable of signalling balance deviations, i.e. the occurrence of a common-mode signal at the inputs of fully differential operational amplifiers. This same technique was used for on-line testing A/D converters in [18] and in [10]. To improve accuracy of concurrent error detection in fully differential circuits, [24] presented a novel analog checker that dynamically adjusts the error threshold to the magnitude of the input signals. This analog checker was used in [25] to validate a new analog on-line testing approach based on circuit state estimation.

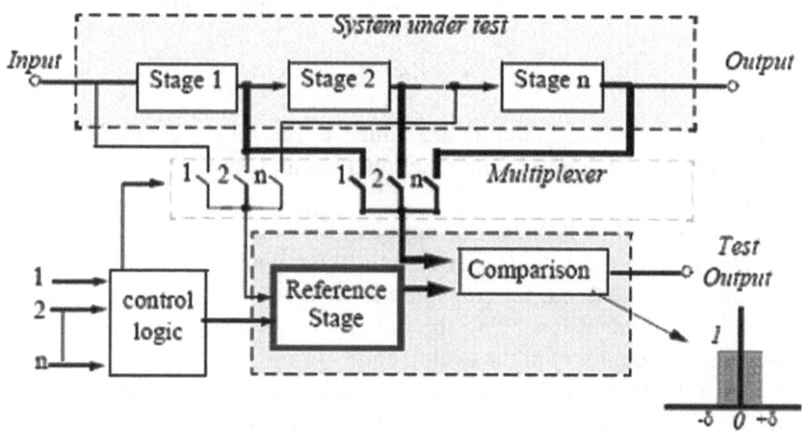

Figure 31. Principle of partial replication.

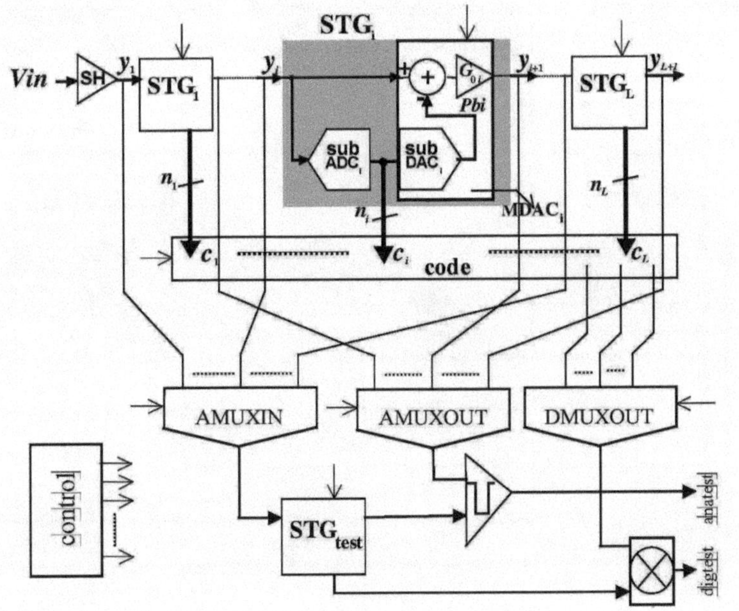

Figure 32. Pipelined A/D converter with on-line test capability.

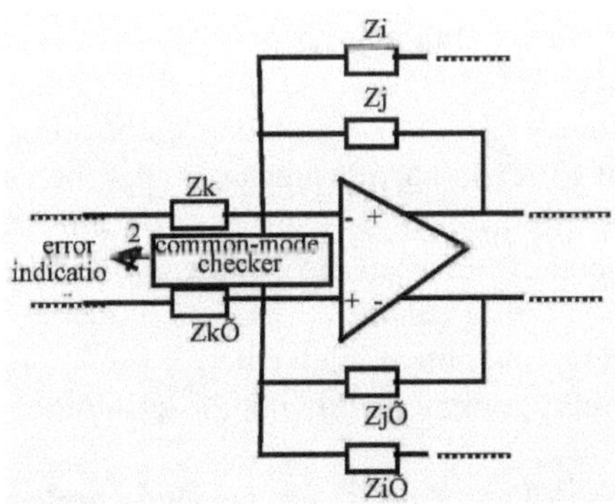

Figure 33. Generic stage of a self-checking fully differential circuit.

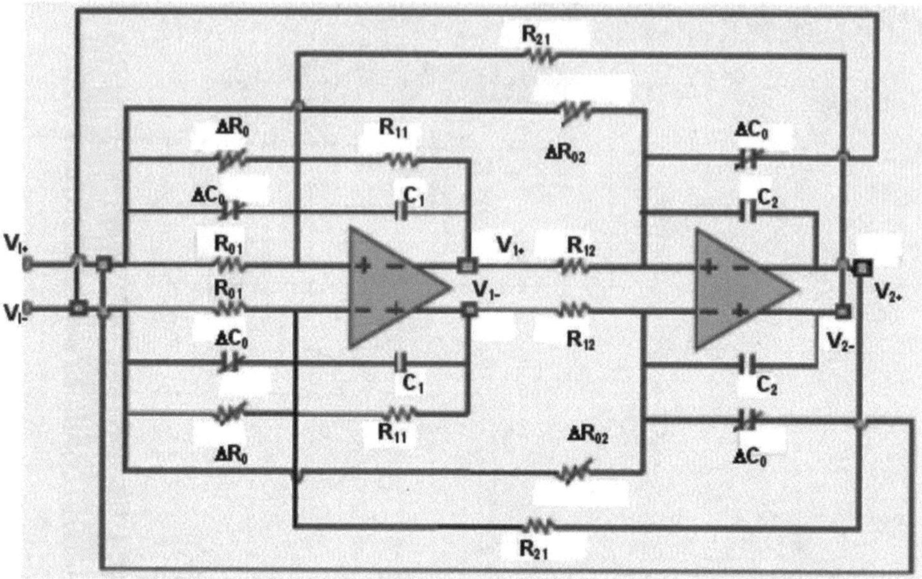

Figure 34. A self-checking fully differential filter (checker not shown).

Note that the partial replication approach perfectly matches the case of FPAAs, bringing additional flexibility to the reference stage (Fig. 31) that can now mimic the behaviour of different stages and is no more limited to circuits that are made up of a cascade of identical modules (Fig. 32). The balance checking approach is more difficult to use in the case of FPAAs, since it precludes that the device has been designed using a fully differential approach and already embeds common-mode checkers at the operational amplifier inputs. In case an error indication occurs, this may be due to a SET or a SEU. In case the error persists, a SEU has probably changed the FPAA programming and the analog function implemented is no more the original one. The FPAA can therefore be reprogrammed allowing the system to recover and be back to a state of good operation.

In the case of the ΣΔ A/D converter, the balance checking approach would certainly be the most suitable technique for the modulator, if the digital stages of the converter did not implicitly filter the occurrence of SETs. For the digital parts of the ΣΔ converter, as already mentioned in this chapter, time and hardware redundancy solutions exist that can properly mitigate space radiation effects.

6. Concluding remarks

The analog and mixed-signal parts were not receiving much attention concerning SET and SEU so far. In fact, the size of transistors in analog circuits does not shrink as faster as in digital ones. Furthermore, in digital CMOS circuits there is current flowing only during state transitions while in the analog parts there are biasing currents that in most cases are greater than the current generated for a SET (considering the double exponential model [27]), therefore the effects of such transient could be not significant in analog circuits.

However, recent studies have shown that those conclusions may not apply for some classes of analog and mixed-signal devices. For example, analog programmable components, the FPAAs, can have their programmability based in SRAM memories. This fact can make the SEU problem in FPAAs as critical as it is for FPGAs.

In order to investigate the effects of BIT-flips in the programming memory of FPAAs a commercial device was used and a series of BIT-flip faults were injected by modifying the value of the default BIT-stream of a programmable analog module. The experiments show that a single BIT inversion can result in a very different configuration of that previously programmed, and in some cases the consequences can be very armful to the system. The experiments also show that a SEU affecting a memory cell that controls a not used resource can result in a correct functional behavior and the fault can remain undetectable.

This way it is possible to conclude that SEU faults are now a matter of concern also in the analog domain given to a feature that is not more exclusive to the digital parts: the programmability.

We have also investigated the radiation effects on a ΣΔ A/D converter. A MatLab model of the converter was developed to allow fault injection and evaluation of the circuit behavior under the effect of SETs and SEUs.

From the experimental results reported, the main conclusion that can be drawn is that the converter can be used in radiation environment if its digital part is protected. An efficient protection form is to use the delta operator substituting the shift operator in the low-pass filter. The delta operator applied to IIR filters is less sensitive to radiation effect that those constructed with shift operators. Another type, the FIR filter, has a minimum sensitivity, but requires more area to be implemented. This chapter showed that the IIR Δ filter has an equivalent area that the IIR Z filter. The converter analog part and decimator do not need to be protected, because the faults caused by radiation can be seen as high frequencies components. These high frequencies are filtered by the next stage of the converter.

Finally, knowing the effects of space radiations in FPAAs and in ΣΔ A/D converters, we have revisited some of the existing on-line test techniques, the analog self-checking circuits, and proposed ways to mitigate SETs and SEUs in such mixed-signal circuits.

References

[1] Adell, P., Schrimpf, R.D., Barnaby, H.J., Marec, R., Chatry, C., Calvel, P., Barillot, C. and Mion, O. "Analysis of Single-Event Transients in Analog Circuits". IEEE Transactions on nuclear Science, Vol. 47, No. 6, December 2000.

[2] Anadigm Company, "Anadigm AN10E40 User Manual", 2002, www.anadigm.com.

[3] Anadigm Company, "Anadigm Designer IP Module Manual", 2002, www.anadigm.com.

[4] Anghel, A., Alexandrescu, D., Nicolaidis, M. "Evaluation of a Soft Error Tolerance technique based on Time and or Hardware Redundancy". Proc. of IEEE Integrated Circuitsand Systems Design (SBCCI), Sept. 2000, pp. 237-242.

[5] Aziz, P. M., Sorensen, H. V., Spiegel, J.; *An Overview of Sigma-Delta Converters*; IEEE Signal Processing Magazine; 1996.

[6] Boser, B. E., Wooley, B. A.; *The Design of Sigma-Delta Modulation Analog-to-Digital Converters*; IEEE Journal of Solid-State Circuits, Vol. 23, No 6; Dec 1988.

[7] Carmichael, C. "Triple Module Redundancy Design Techniques for Virtex Series FPGA". Xilinx Application Notes 197, v1.0, Mar. 2001.

[8] Carro, L., De Nale, L., Jahn, G.; Conversor Analógico/Digital Sigma-Delta; Relatório de Pesquisa, Departamento de Engenharia Elétrica da Universidade Federal do Rio Grande do Sul. Setember, 1999.

[9] Chatterjee, A., 1991, Concurrent error detection in linear analog and switched-capacitor state variable systems using continuous checksums, in: *International Test Conference*, Proceedings, pp. 582-591.

[10] Francesconi, F., Liberali, V., Lubaszewski, M. and Mir, S., 1996, Design of high-performance band-pass sigma-delta modulator with concurrent error detection, in: *International Conference on Electronics, Circuits and Systems*, Proceedings, pp. 1202-1205.

[11] Goodal, R.M., Donoghue, B. J.; *Very High Sample Rate Digital Filters Using the Delta Operator*; IEEE Proceedings; Vol 40, No 3; June, 1993.

[12] Hereford, J., Pruitt, C. "Robust Sensor Systems using Evolvable Hardware". NASA/DoD Conference on Evolvable Hardware (EH'04), p. 161, 2004.

[13] Huertas, J.L., Vázquez, D. and Rueda, A., 1992, On-line testing of switched-capacitor filters, in: *IEEE VLSI Test Symposium*, Proceedings, pp. 102-106.

[14] Kauraniemi, J., Laakso, T. I., Hartimo, I., Ovaska, S.; *Delta Operator Realization of Direct-Form IIR Filters*; IEEE Transactions on Circuits and Systems II: Analog and Digital Signal Processing, Vol 45, No 1; January, 1998.

[15] Kolarík, V., Mir, S., Lubaszewski, M. and Courtois, B., 1995, Analogue checkers with absolute and relative tolerances, *IEEE Transactions on Computer-Aided Design of Integrated Circuits and Systems* **14**(5): 607-612.

[16] Leveugle, R., Ammari, A. "Early SEU Fault Injection in Digital, Analog and Mixed Signal Circuits: a Global Flow". Proceedings of the Design, Automation and Test in Europe Conference and Exhibition (DATE'04), pp.1530-1591, 2004.

[17] Lima, F., Carro, L., Reis, R. "Designing Fault Tolerant Systems into SRAM-based FPGAs". Proc. of Design Automation Conferece (DAC'03), pp. 250-255, 2003.

[18] Lubaszewski, M., Mir, S., Rueda, A. and Huertas, J.L., 1995, Concurrent error detection in analog and mixed-signal integrated circuits, in: *Midwest Symposium on Circuits and Systems*, Proceedings, pp. 1151-1156.

[19] Messenger, G.C. "A summary Review of Displacement Damage from High Energy Radiation in Silicon Semiconductors and Semiconductors Devices". IEEE Transactions on nuclear Science, Vol. 39, No. 3, June 1992.

[20] Mir, S., Lubaszewski, M., Kolarík, V. and Courtois, B., 1996, Fault-based testing and diagnosis of balanced filters, *KAP Journal on Analog Integrated Circuits and Signal Processing* **11**:5-19.

[21] Mitra, S. K.; *Digital Signal Processing - A Computer-Based Approach;* Ed. McGraw-Hill Irwin; 2ª edição; 2001.

[22] Nicolaidis, M., 1993, Finitely self-checking circuits and their application on current sensors, in: *IEEE VLSI Test Symposium*, Proceedings, pp. 66-69.

[23] Peralías, E., Rueda, A. and Huertas, J.L., 1995, An on-line testing approach for pipelined A/D converters, in: *IEEE International Mixed-Signal Testing Workshop*, Proceedings, pp.44-49.

[24] Stratigopoulos, H.-G.D. and Makris, Y., 2003, An analog checker with dynamically adjustable error threshold for fully differential circuits, in: *IEEE VLSI Test Symposium,* Proceedings, pp. 209-214.

[25] Stratigopoulos, H.-G.D. and Makris, Y., 2003b, Concurrent error detection in linear analog circuits using state estimation, in: *International Test Conference,* Proceedings, pp. 1164-1173.

[26] Turflinger, T.L. "Single-Event Effects in Analog and Mixed-Signal Integrated Circuits". IEEE Transactions on nuclear Science, Vol. 43, No. 2, pp. 594-602, April 1996

[27] Yang, F.L., Saleh, R.A. "Simulation and Analysis of Transient Faults in Digital Circuits". IEEE Journal of Solid-State Circuits, Vol. 27, No. 3, March 1992.

[28] Znamirowski, L., Paulusinski, O.A., Vrudhula, S.B.K. "Programmable Analog/Digital Arrays in Control and Simulation". Analog Integrated Circuits and Signal Processing, 39, 55–73, Kluwer Academic Publishers 2004.

Fundamentals of the Pulsed Laser Technique for Single-Event Upset Testing

Pascal Fouillat[1], Vincent Pouget[1], Dale McMorrow[2], Frédéric Darracq[1], Stephen Buchner[2], Dean LEWIS[1]

[1]IXL – Université Bordeaux 1 – UMR CNRS 5818
351 Cours de la Libération – 33405 – Talence - France
fouillat@ixl.fr, pouget@ixl.fr, darracq@ixl.fr, lewis@ixl.fr
[2]Naval Research Laboratory, Code 6812
Washington, DC 20375
mcmorrow@ccs.nrl.navy.mil
[3]QSS Group, Inc., Seabrook, MD 20706; 301-286-5019
sbuchner@pop500.gsfc.nasa.gov

Abstract. This paper describes the use of a pulsed laser for studying radiation-induced single-event effects in integrated circuits. The basic failure mechanisms and the fundamentals of the laser testing method are presented. Sample results are presented to illustrate the benefits of using a pulsed laser for studying single-event upsets in memories.

1. Introduction

Electronic systems operating in space are exposed to radiation in the form of energetic charged particles, such as protons and heavy ions. As those charged particles pass through the semiconductors and insulators that make up the integrated circuits (ICs), they liberate electrons from the constituent atoms. Under certain conditions, the liberated electrons can disturb the normal operation of the IC, causing a variety of different and potentially harmful effects that have been classified as Single-Event Effects (SEEs). Included among those effects are the change in state of a latch or memory cell, referred to as a Single-Event Upset (SEUs), and temporary voltage spikes at internal circuit nodes termed Single-Event Transients (SETs) [1,2,3]. SEEs are of great concern because they may lead to a loss of information, to physical failures or, in the worst case, to a total loss of control of the spacecraft. In space, heavy ions responsible for SEUs and SETs originate from solar flares and cosmic rays, or from proton-induced reactions in low earth orbit. More recently, ions resulting from the interaction of atmospheric neutrons with the silicon and boron atoms in electronic circuits themselves have become a threat for modern advanced technologies, even at ground level.

SEEs disturb the proper functioning of many different kinds of digital and analogue circuits. Early observations of SETs were confined to digital architectures where the effect of the deposited energy in a combinatorial logic gate caused a static bit error once the transient signal reached a sequential logic element, such as a latch, a memory or a register. Clearly, SETs that propagate to and are captured by data storage elements cannot be distinguished from SEUs occurring when particles directly strike

R. Velazco et al. (eds.), Radiation Effects on Embedded Systems, 121–141.

those same memory elements. Therefore, the overall radiation-induced error rate in digital circuits consists of contributions from both SETs and SEUs.

SEE testing of ICs is essential in order to evaluate, understand, and mitigate their sensitivity. The pulsed picosecond laser has become an important tool for the evaluation of single-event effects (SEEs) in microelectronic circuitry [4,5,6,7,8]. In its common implementation, the pulsed laser technique is based upon the excitation of carriers in a semiconductor material using tightly focused, above-bandgap optical excitation. Carrier generation is governed primarily by Beer's law absorption, such that each absorbed photon generates a single electron-hole pair, and the injected carrier density decreases exponentially with distance from the surface of the material. In recent years the pulsed laser has been used successfully in a range of investigations of SEE phenomena, including interrogation of the spatial and temporal aspects of single-event upset and single-event latch up (SEL) in a variety of digital circuits [4,5,7], investigation of the basic charge-collection mechanisms of individual transistors [4,10], and most recently as an essential tool for unraveling the complex SET response of bipolar linear circuits [8,11].

This paper describes the laser testing technique, which has developed into a powerful diagnostic tool for investigating and characterizing the details of SEEs in ICs. It can be considered as a complementary tool to the classical approach, which involves testing with particle accelerators. The first section describes the fundamentals of the laser testing technique. The second section introduces the actual experimental implementations encountered both in the academic and industrial worlds. The last section illustrates the power of the laser method for SEU and SET testing as well as its actual limitations when compared directly to the classical testing with particle accelerators.

2. Fundamentals of the laser testing technique

2.1 Classification of the laser testing techniques

Laser scanning techniques are commonly used for imaging purposes in a variety of scientific domains [12]. In the microelectronics domain, a number of interesting methods for integrated circuit (IC) testing based on laser scanning have been developed in the last decade. Since the operator can interact with the device under test (DUT) in two different ways, either electrically or optically, different combinations of the two can be brought into play, either for stimulating the DUT or analysing its electrical or optical response (see Table 1). Among all these methods, the pump methods consist in mapping the variation of an electrical parameter induced by the localized interaction of an optical beam with the semiconductor material (or the metal layers) of the DUT. Some of these pump methods have gradually migrated from laboratories to industrial applications. This is especially the case for techniques using continuous-wave (cw) laser beams, like the Optical Beam Induced Current (OBIC) technique, and the Optical Beam Induced Resistance Change (OBIRCH) technique, which are available today through commercial systems for defect localization [13].

Recent progress in the development of commercial pulsed laser sources allows adding the extreme temporal resolution of ultra-short laser pulses to the laser scanning techniques for IC testing. Because the clock frequencies of very large scale integration (VLSI) circuits are increasing so rapidly, within a few years ultra-short laser pulses might become the only available probe for time-resolved internal electrical parameter measurement. Moreover, a pulsed source makes the implementation of discrete scanning techniques that can lead to sub-wavelength spatial resolution much easier. This is essential for testing deep sub-micron technologies.

Table 1. Classification of laser testing techniques.

DUT STIMULATION	 ANALYSIS	TECHNIQUE
Electrical	Electrical	Electrical testing (IDDQ, ...)
Electrical	Optical	Probe (Reflectometry, ...)
Optical	***Electrical***	***Pump (Fault injection, SET, SEU...)***
Optical	Optical	Pump-Probe (fs acoustic, ...)

It is also important to note that the constant increase in the number of metal layers in modern devices is not a definitive limitation for the techniques that require interaction of the laser pulse with the semiconductor material. Indeed, the opacity of the front-side metal layers can be easily bypassed today via backside testing, i.e. via focusing the beam through the device substrate. Although this approach requires state-of-the-art sample preparation techniques, it offers complete accessibility of the beam to the active areas of the DUT while leaving it totally functional (Lewis, 2001; McMorrow, et al., 2004).

2.2 Modeling the laser generation rate

Considering a pulsed laser beam focused on the surface of a semiconductor device and neglecting second-order effects such as non-linear absorption and wavefront curvature correction to Beer-Lambert's law, the injection of excess carriers induced in the semiconductor by the laser pulse can be well modeled by the following expression for the generation rate of electron-hole pairs [14]:

$$g_{las}(r,z,t) = \frac{2\alpha T E_L}{\pi^{\frac{3}{2}} \omega_0^2 E_\gamma \tau_{las}} \frac{\omega_0^2}{\omega(z)^2} e^{-\frac{2r^2}{\omega(z)^2}} e^{-\alpha z} e^{-\frac{t^2}{\tau_{las}^2}}$$

$$\omega(z) = \omega_0 \sqrt{1 + \left(\frac{z}{z_{sc}}\right)^2} \tag{1}$$

The lateral profile is described by a Gaussian function in accordance with the fundamental transverse propagation mode of the laser beam. The ω_0 parameter is defined as the beam-waist located on the surface of the semiconductor. The spreading of the beam within the semiconductor is governed by the confocal length z_{sc}. In accordance with Beer-Lambert's law, the intensity of the laser light decays exponentially with distance due to the absorption of the light. The parameter α is the optical absorption coefficient of the semiconductor. E_L is defined as the energy of the laser pulse, $E\gamma$ represents the energy of a photon and T is the energy transmission coefficient of the surface of the semiconductor. The parameter T may include interference effects in the oxide layer. Because the circuit response is much slower than the propagation time of the laser pulse through the structure, the coupling between spatial and temporal variations is neglected and the temporal profile of the generation rate simply reproduces the temporal profile of the pulse. It is assumed to be Gaussian with a width of 1 ps in typical experimental conditions [6].

Figure 1 (left) presents the spatial distribution of hole-electron pairs (time-integrated generation rate) induced by a laser pulse for a wavelength of 0.8μm and a transmitted energy of 8 pJ. The beam is assumed to be focused down to a spot size of 1 μm, as we can expect when using a *100x* microscope objective lens, near the theoretical limit imposed by the wavelength. Note that the logarithmic color scale exaggerates both beam width and spreading. Actually, for the wavelength considered, most of the energy is absorbed before the beam spreading becomes significant.

It is worth noting that some second order effects can be included in the generation rate models, at least under particular conditions. For instance, the non-linear-optical two-photon absorption (TPA) mechanism, which may occur at high laser intensities associated with ultra-short laser pulses or when using near band-gap wavelengths [6,15] can be analytically described for slowly diverging beams (i.e. when the effect of the energy deposited beyond the confocal length is negligible). In such a case, the generation rate is described as follows [16]:

$$
\begin{aligned}
G_{las}(r,z,t) &= G_0 U(r,z,t) + \frac{r_{TPA}}{2} G_0 U(r,z,t)^2 \\
G_0 &= \frac{2\alpha T E_L}{\pi^{\frac{3}{2}} \omega_0^2 E_\gamma \tau_{las}}, \quad r_{TPA} = \frac{2\beta T E_L}{\alpha \pi^{\frac{3}{2}} \omega_0^2 \tau_{las}} \\
U(r,z,t) &= \frac{e^{-\frac{2r^2}{\omega_0^2}} e^{-\alpha z} e^{-\frac{t^2}{\tau_{las}^2}}}{1 + r_{TPA}\, e^{-\frac{2r^2}{\omega_0^2}} e^{-\frac{t^2}{\tau_{las}^2}} \left(1 - e^{-\alpha z}\right)}
\end{aligned}
\qquad (2)
$$

In this expression, β is the non-linear absorption coefficient and the contribution of the non-linear mechanism to the generation rate is governed by the r_{TPA} coefficient. Note that setting this coefficient to *0* in this model leads to the linear expression of equation (1) in which the beam spreading is neglected. This model should be used only for small contributions of the TPA mechanism. Keeping this limitation in mind, it is important to notice from equation (2) that for small values of z, i.e. near the surface, the TPA mechanism increases the generation rate with respect to the linear model. Obviously, when integrating over a virtual infinite depth of semiconductor, the total number of generated carriers is less than in the linear case since it takes globally more than one photon to create an electron-hole pair.

Other effects, such as the spatial variation of the absorption coefficient with doping concentration, could also be analytically included on a case-by-case basis. The free-carrier absorption is usually modeled by a global quantum efficiency factor that may also depend on the structure. However, a rigorous treatment of more complex effects such as self-absorption (i.e. the tail of the pulse is absorbed by the carriers generated by the front of the pulse) would necessitate a numerical approach. These effects are usually negligible in silicon for wavelengths shorter than 0.85µm but they may have an important contribution to the shaping of the generation rate for near bandgap wavelengths [17].

2.3 Laser generation rate versus heavy-ion

We have seen that a pulse of laser light focused on a semiconductor can induce a localized transient generation of electron-hole pairs according to the photoelectric effect when the photon energy is greater than the semiconductor bandgap energy. The particle-induced generation of electron-hole pairs involves a coulomb interaction, which is intrinsically different from the photoelectric effect responsible for the generation of electron-hole pairs by laser light.

The heavy-ion induced generation rate is completely defined by the energy and the initial LET of the ion. For SEE testing, the variation of the ion energy is usually assumed to be negligible and, as a consequence, the initial LET constitutes the chief variable parameter. One commonly used model for describing the heavy-ion induced generation rate in numerical device simulations is the Gaussian cylindrical model, an approximation to the actual charge distribution that cannot be expressed in closed form:

$$g_{ion}(r,z,t) = \frac{1}{\pi^{\frac{3}{2}} r_0^2 \tau_{rad}} \frac{L_i}{E_p} e^{-\frac{r^2}{r_0^2}} e^{-\frac{t^2}{\tau_{rad}^2}} \tag{3}$$

The L_i parameter represents the initial LET at the surface of the irradiated semiconductor. The r_0 parameter defines the radius of the cylindrical generation rate and is usually arbitrarily fixed to a typical value of 0.1µm. The E_p parameter represents the average energy (3.6eV in Si) to create a hole-electron pair. Again, the temporal aspect of the propagation can usually be neglected and the temporal variation of the generation rate is a global Gaussian profile. Its duration, τ_{rad}, includes the time of flight across the structure for the ion and the secondary electrons as well as the relaxation time of the generated carriers. It is on the order of 1ps.

Figure 1 right presents the charge track induced by a 275MeV Fe ion. Note that since the model does not include the ion energy variation, the generation rate is uniform along the propagation direction. This figure reveals the differences in structure between ion-induced (right) and laser-induced (left) charge tracks when the same quantity of photo carriers is generated in the material. The attenuation of the laser beam with depth into the semiconductor cannot be neglected, in contrast to the ion for which a longitudinal quasi-uniform ion track is assumed.

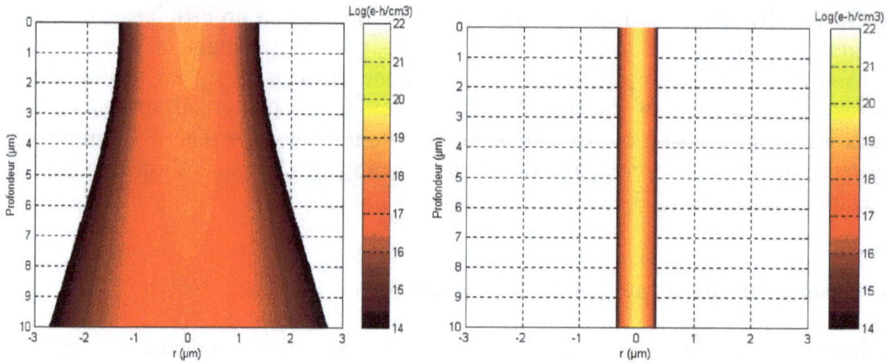

Figure 1. Carriers density generated in silicon versus depth and distance from the centre of the impact : (left) by a laser pulse of 8pJ with a wavelength of 800nm modeled with equation (1) and by a 275 MeV Fe ion with an (right) LET of 24 MeV·cm2/mg and modeled with equation (2).

Concerning the radial distributions, Fig. 2 shows that the laser-induced carrier distribution has a wider lateral extension and a much lower peak value than the ion track. This implies that the amount of charge collected at an electrode of a device by the funneling effect (junction electrical field deformation due to high concentrations of free carriers) will be lower than in the case of an ion impact. A more accurate model has been developed in [16].

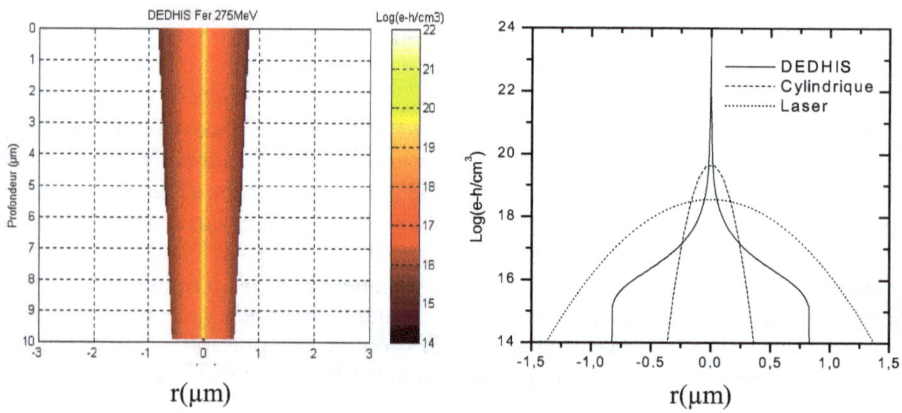

Figure 2. left) Modeling the Fe Ion (275MeV) with DEDHIS (Pouget, 2000) and right) lateral profiles of the induced carrier densities for the models from [1], [2] and DEDHIS.

Although these figures represent particular cases, they are, nevertheless, representative of the differences in radial distribution and peak density. However, despite the differences in structure between ion-induced and laser-induced tracks, numerous experimental investigations as well as numerical simulations have shown that these two kinds of excitation produce similar transient effects in integrated circuits [4,8].

3. Pulsed laser systems for ICs testing

3.1 The basic principle of laser testing

The basic principle of laser testing consists of generating a transient current or voltage able to disturb the normal functioning of the DUT, such as illustrated in Fig. 3.

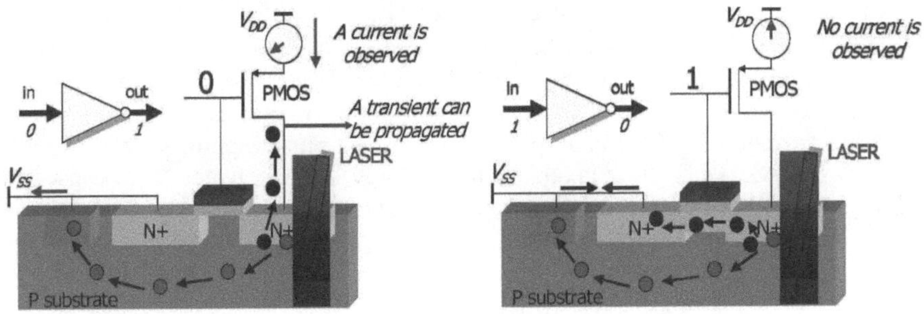

Figure 3. left) a photocurrent is generated through the power lines when striking the OFF transistor, right) no photocurrent is measured in the power line when striking the ON transistor.

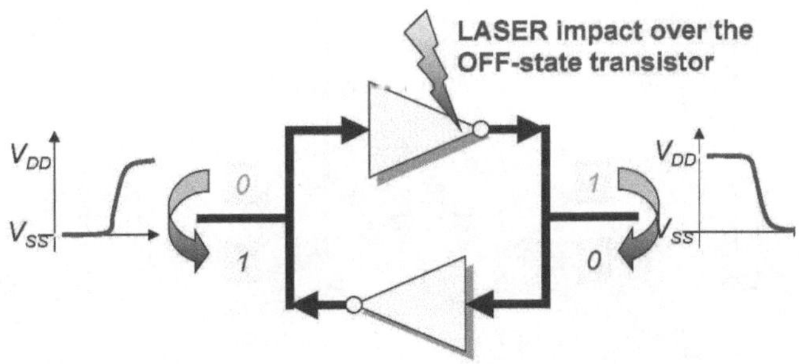

Figure 4. Upsetting a memory cell (the access transistor are not presented) with a laser impacting a blocked transistor in a NOT operator.

For simplicity, the approach is illustrated using the basic structure of an inverter. By irradiating the NMOS drain of the off transistor (the NMOS on the left side of Fig. 3), a photocurrent is initiated between its drain and substrate, with photo-generated holes flowing to the ground line via the NMOS body and photo-generated electrons flowing to the power line via the PMOS transistor, which is initially in its on-state. A latch is formed by connecting two such inverters together, as illustrated in

Fig. 4. By gradually increasing the laser pulse energy, one eventually reaches a point at which the critical value of photo-current produces a transient voltage of sufficient magnitude to cause the latch to switch its state. This mechanism is a classical Single Event Upset (SEU). If the laser light is focused on the drain of the on-state transistor (NMOS in the right side of Fig. 3), no photocurrent is observed in the power line, and the transient voltage will not upset the gates connected to its output.

3.2 Experimental set-up description

Figure 5 presents the primary elements of the laser system facility at IXL laboratory [18]. The laser source is a Ti:Sapphire (Ti:SA) oscillator (model Tsunami from Spectra-Physics) pumped by a 10W cw laser (model Millenia Xs from Spectra-Physics). The oscillator delivers 100fs or 1ps pulses at a frequency of 80MHz. The pulse wavelength is tunable in the red-NIR region from 730nm to 1000nm. This tunability permits adjustment of the penetration depth of the laser light pulse in the semiconductor material [10]. Because, in most cases, a pulse repetition rate of 80MHz is too high for the DUT to relax back to a steady state between two consecutive pulses, a pulse picker is used to reduce this frequency. The pulse repetition rate at the output of the pulse picker is adjustable from single-shot to 4MHz, and the pulse triggering can be synchronized with the DUT clock. Other systems utilize cavity dumping to achieve analogous results.

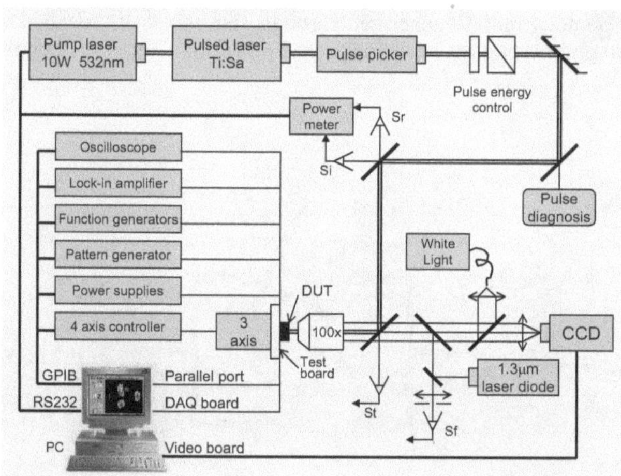

Figure 5. Experimental set-up for ICs testing with a pulsed laser.

For front-side testing of silicon devices, a wavelength of 800nm is usually chosen, giving a penetration depth of approximately 12µm. This is sufficient to ensure efficient photo-generation in the active volume of most modern devices. For backside testing, a longer penetration depth is required to reach the active volume through the silicon substrate. Indeed, the absorption coefficient of silicon drops rapidly for wavelengths above 1000 nm. To induce a sufficient level of photo-generated carriers in the active volume of the device, the best compromise usually lies between 950 and 1000nm, depending on the substrate thickness and doping level.

3.3 Automation

System automation is implemented using a software package called SEEM (for Single-Event Effects Mapper) developed using National Instruments LabWindows/ CVI and Microsoft Visual C++ environments. The user defines a scan window into which the laser strikes the DUT at regular intervals. Different parameters may be measured and stored in such a way that 2D plotting of the scanned area, using a color scale, leads to an SEE sensitivity map of the DUT. Twelve different scanning modes have been implemented but, for most of them, the operations performed for each node of the scanned window consist of three steps: (*i*) positioning of the impact point; (*ii*) laser pulse triggering; and (*iii*) electrical measurements.

3.4 Other systems

Different strategies for laser systems can be found in the literature. Backside testing of complex devices with the test bench described in 3.2 necessitates special preparation of the DUT since the maximum wavelength used cannot reach the sensitive volumes without thinning the device. Using longer wavelengths permits this limitation to be overcome. In this case, however, nonlinear effects are exacerbated, and more care must be taken in quantifying the results [4].

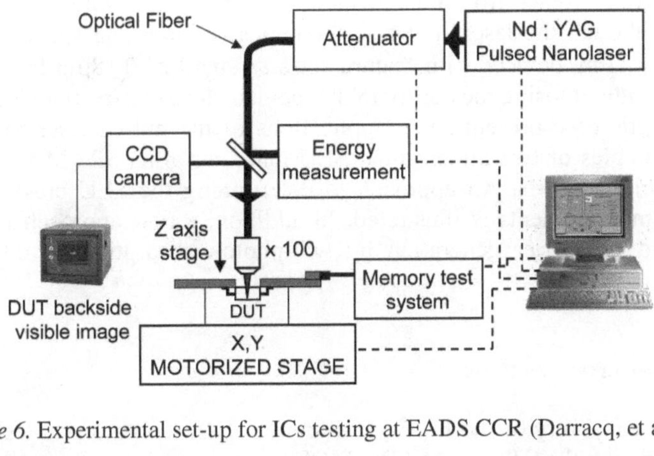

Figure 6. Experimental set-up for ICs testing at EADS CCR (Darracq, et al., 2002).

Figure 6 presents a schematic diagram of the EADS laser test bench designed for SEU testing on COTS components. The laser source is a pulsed Q-Switch Nd:YAG nanolaser (physical dimensions: 6×4×2 cm^3). It has a wavelength of 1064nm and delivers 700ps pulses with a maximum energy of about 5nJ at the surface of the DUT. The repetition rate of the pulses can be adjusted from single shot to 2kHz. For laser safety and experimental convenience, its output is coupled to a multimode optical fiber reducing the beam path in air. A fiber-dedicated motorized attenuator is used to change the laser pulse energy incident on the device. The pulse energy is continuously adjustable, and measured with a calibrated photodiode.

4. Applications of laser systems

The pulsed picosecond laser has become a versatile tool for the investigation and understanding of single-event effects in microelectronic circuitry. During the past 15 years this approach has been applied successfully to the evaluation of single-event effects in a wide variety of circuits and circuit types, including SRAMs, DRAMs, logic circuits, analog-to-digital converters, operational amplifiers, and comparators, to name a few. In addition, the pulsed laser approach has proven invaluable for the investigation of the basic mechanisms of carrier dynamics and charge collection in individual transistors, including GaAs MESFETs, GaAs, InGaAs, and InAs high-electron mobility transistors (HEMTs), GaAs and SiGe heterojunction bipolar transistors (HBTs), and bulk and silicon-on-insulator (SOI) CMOS devices.

One attractive feature of the pulsed laser approach is that spatial and high-bandwidth temporal information is obtained without any radiation damage (total ionizing dose or displacement damage) to the device being tested. An additional, particularly significant feature is that testing typically is performed in air, making the experimental setup straightforward and suitable for high-bandwidth experiments. Further, the "equivalent LET" can be adjusted continuously, simply by increasing or decreasing the incident laser pulse energy. The primary limitations of the pulsed laser approach are interference from metallization and the finite resolution of the focused laser spot. Although the laser spot size theoretically limits the spatial resolution to approximately 1μm, devices with feature sizes as small as 0.18μm have been tested successfully without losing the ability to distinguish close electrical nodes.

In this section we present some applications of the pulsed laser SEE approach, including examples of two-dimensional SEU mapping in an SRAM test chip and in two commercial SRAMs. An approach to determining the SEU cross-section from pulsed laser measurements is illustrated. In addition, a new approach for laser SEE studies, based on carrier generation by two-photon absorption, is introduced and discussed.

4.1 SEU laser cross-sections

To verify the capability of the laser test system at IXL laboratory, a test vehicle has been designed with the 0.8μm AMS BiCMOS technology. It is a basic SRAM cell for which all the dimensions have been stretched in order to understand the cell behavior versus the laser impact location. Figure 7 presents the test vehicle classically organized with 6 transistors. This test vehicle is well suited for a fundamental study of the laser cross-section curves because of the low density of metallic tracks which allows a direct study by the front-side of the DUT.

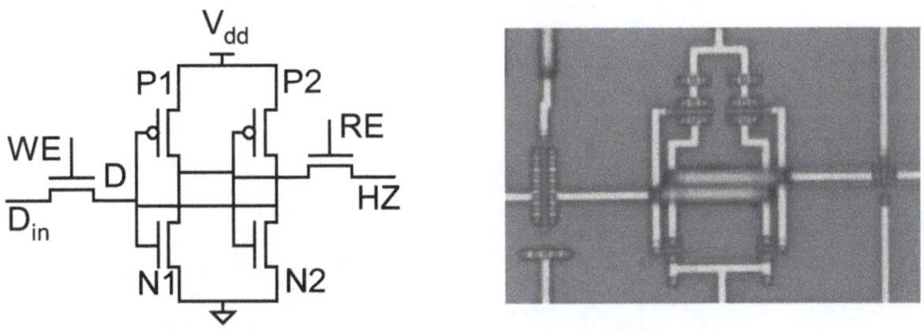

Figure 7. Electrical schematic and picture of the tested basic SRAM cell.

Cross-section extraction procedure. The test procedure is as follows: (i) the cell is initialized with a known state, (ii) a laser energy is selected, (iii) the laser spot is positioned on a given point, (iv) an unique laser pulse is fired, (v) the state of the cell is read and reinitialized if corrupted, (vi) for each strike location that induces an upset, the corresponding point is superimposed on a map of the device with the color of the point representing the laser pulse energy. This procedure is repeated for the same area for different incident laser pulse energies (Fig. 8).

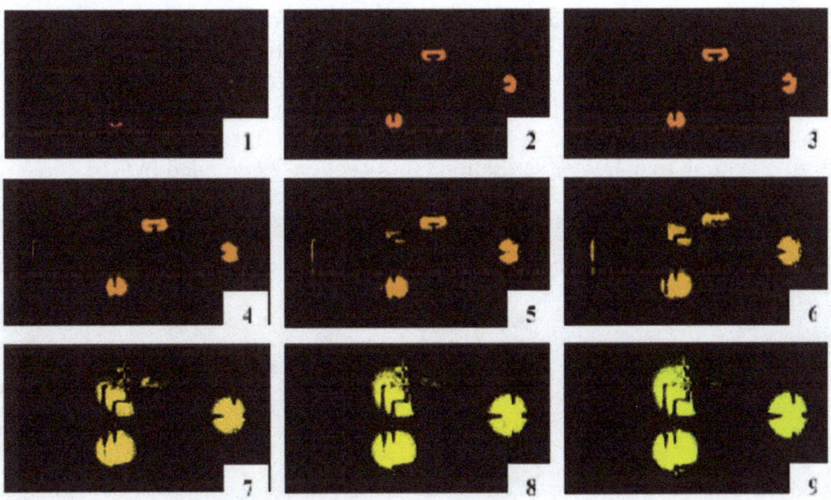

Figure 8. SEU sensitivity mappings for the SRAM cell.

In the second step, by integrating the sensitive (colored) area of each map, we obtain the curve depicted in Fig. 9. The nine points of this curve correspond to the nine SEU maps of Fig. 8. The result is a curve of cross-section versus laser pulse energy, which has a shape similar to that obtained from heavy-ion experiments where the cross-section is plotted versus LET. Experimental laser cross-section curve is then constructed.

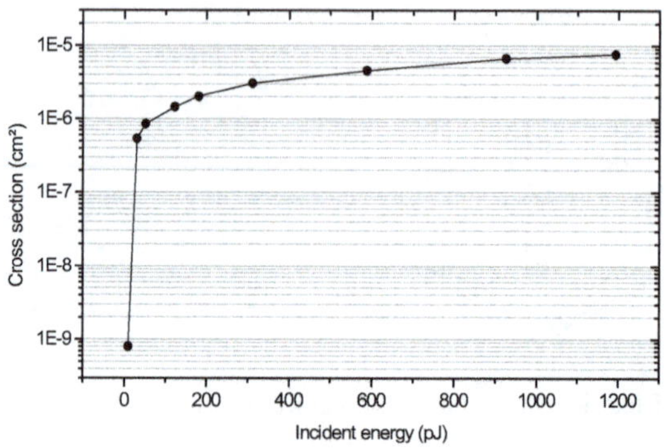

Figure 9. Experimental laser cross-section for the maps of Figure 8.

Influence of the electrical environment. Figure 10 summarizes results for two different electrical conditions. The same color table is used for the 2 pictures. In each picture, the maps for the different energies are superimposed. This allows a global vision of the SEU sensitivity. Some interesting conclusions may be drawn from Fig. 10. For instance, since the red color indicates a greater sensitivity, the well known result that irradiating the drain of the off transistor in a memory cell may lead to an upset is clearly exhibited here. Moreover, not only are the "off" transistors sensitive, but the access transistor on the left of each figure that permits the SRAM cell to be read, also is sensitive to SEU. This transistor, which is in its off-state here (WE=0), also exhibits a small but very active area when its input is at a high level. For this same configuration, a fraction of the n-well contributes to the global sensitivity, leading to a greater saturation cross-section, as detailed in [16].

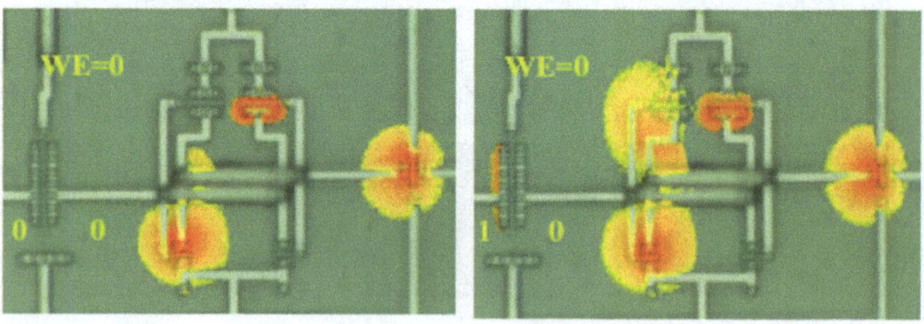

a) Din=0 and D=0 b)Din=1 and D=0

Figure 10. Sensitivity maps for the same internal logic state and a different value at the input of the write transistor. Red points correspond to lower threshold. The step size is 0.2μm. WE is off.

Finally, examination of the laser maps for different configurations constitutes a powerful means of evaluation of the radiation hardness of a given electronic circuits. When performed for different architectures implemented in the same technology, this approach permits the relative sensitivity of each to be determined.

The fundamental limitation of the testing procedure described in this section lies in the fact that the laser pulse must propagate through the metal layers; this can affect the accurate measurement of a laser cross-section. The following section presents an alternate experimental approach, based on backside, through-wafer irradiation, for dealing with very high density SRAMs.

4.2 Laser testing of commercial SRAMs

Laser and heavy ion cross-sections for two commercial SRAMS. This section presents an analysis of the SEU sensitivity of two components performed with the laser test bench at EADS CCR [19]: a 4 Mbit SRAM from Hitachi (HM628512A) and a 1 Mbit SRAM from NEC (μPD431000A). In this case we utilize the same procedure as described in Sec. 4.1 except that the laser wavelength used permits the direct interrogation through the backside of the DUT, without thinning the device.

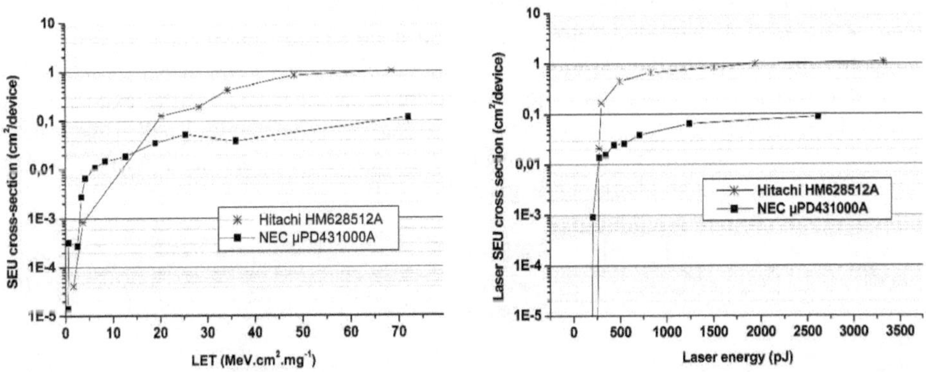

Figure 11. Heavy ion (left) and laser (right) SEU cross-sections for the Hitachi and NEC SRAMs. The laser curve is corrected from the spot size effects.

The cross-section curves obtained from heavy-ion-beam test and laser-beam test are presented in Fig. 11. The laser data is obtained with the laser test bench at EADS Corporate Research Center, developed in collaboration with IXL laboratory. The methodology for backside SRAM SEU sensitivity testing is detailed in [20]. This study focuses on the link between the two results presented in Fig. 11, by relating first the ion cross-section to the laser one, and second, the LET to the laser pulse energy.

Correction for the spot size effect. For a given technology, a threshold energy (*Eth*) is necessary to induce an upset in a memory cell. That is, a laser pulse energy of at least E_{th} is necessary to generate the minimal charge (the critical charge, Qc) required to produce an upset. Figure 12 presents two laser spots on a simplified memory plan composed of regularly spaced memory cells. The spot on the right has an energy *Eth*, so that it induces an SEU in only one cell. On both sides of Fig. 12a, radial projections of the laser intensity are represented for the two laser spots. In this example, the spot on the left has about three times the total energy of the spot on the right, although their radial widths (at half maximum) are identical.

The energy deposited on each memory cell is the laser intensity integrated over the cell area and over the pulse duration. Since the radial distribution of this intensity is a Gaussian function, the spot size remains constant but, the greater the total beam energy, the greater the energy deposited in each cell. Since SEU is a threshold effect, as the laser pulse energy is increased, the number of cells receiving a deposited charge above threshold increases, resulting in a greater number of cells that upset. As a consequence, for the spot on the left in Fig. 12a, several memory cells receive an energy value greater than E_{th}, resulting in multiple upsets (five in this example). The analogous effect is not observed during particle accelerator tests because the radial distribution of radiation-induced carriers is much more confined along the ion path (small spot on Fig. 12a).

Because of this effect, direct application of the laser-generated results leads to an over-estimation of the SEU cross-section and a correction factor is necessary. The correction factor is based on the average number of memory cells inside of which the deposited charge is greater than the critical charge Q_c defined by the experimental threshold energy E_{th}. Let Q_{ij} be the deposited charge in the memory cell ij for a laser beam with a spot size of $2w_0$ centered on (x_0,y_0) onto the memory plan in Fig 12a. The ratio of Q_{ij} to the critical charge Q_c is given by [20]:

$$Q_{ij}/Q_C = \left(E_L/E_{Th}\right)\times\frac{\int_{i\Delta x}^{(i+1)\Delta x}\int_{j\Delta y}^{(j+1)\Delta y}\exp\left(-2\left[\left(x-x_0\right)^2+\left(y-y_0\right)^2\right]/w_0^2\right)dxdy}{\int_{-\frac{\Delta x}{2}}^{\frac{\Delta x}{2}}\int_{-\frac{\Delta y}{2}}^{\frac{\Delta y}{2}}\exp\left(-2\left[x^2+y^2\right]/w_0^2\right)dxdy} \quad (4)$$

For a given laser "impact" on (x_0,y_0) and an energy E_L, this ratio is estimated for each memory cell of the simulated memory plan. Whenever this ratio is greater or equal to one, an upset is added to the total number of upsets induced by the laser pulse. The correction factor is then defined by the average value of the total number of upsets estimated for different "impact" locations. This factor is an increasing function of the laser energy, as shown in Fig. 12b. For this study, $w_0 \approx 3.5\mu m$ and the memory cell dimensions are $\Delta y \approx 6.5\mu m$ and $\Delta x \approx 4.3\mu m$ for the NEC SRAM, and $\Delta y \approx 4.2\mu m$ and $\Delta x \approx 2.3\mu m$ for the Hitachi SRAM.

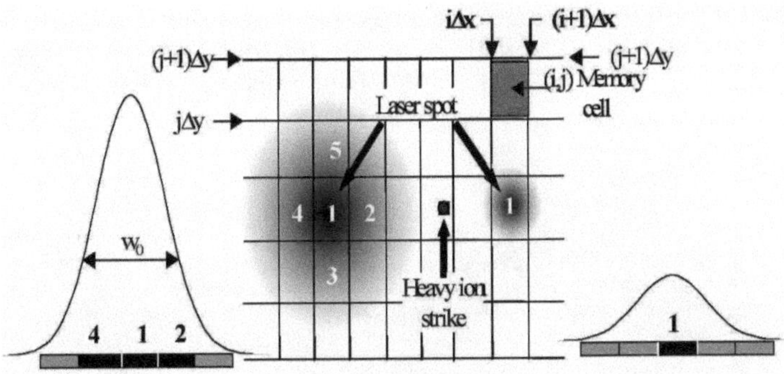

Figure 12a. Influence of the laser spot size.

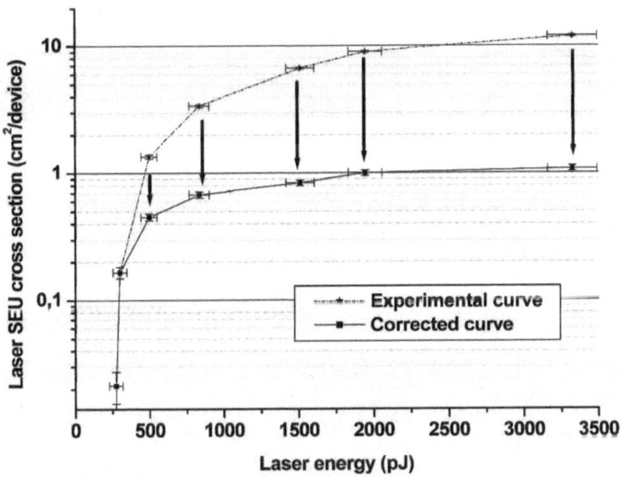

Figure 12b. Experimental and corrected laser cross section versus laser energy for the Hitachi SRAM.

Figure 11 (right) presents the laser test results obtained for Hitachi and NEC SRAMs corrected from the spot size effect. These curves illustrate that the saturation value of the laser-generated cross-sections are comparable to heavy ion saturation values (see Fig. 11 left). Despite these corrections, the cross-section curves cannot be directly compared since the link between heavy-ion LET and laser pulse energy has not been established.

Theoretical investigation of the laser energy versus the ion LET. The complex problem of a possible quantitative relationship between LET and laser energy is a new subject in the literature. As the threshold LET is an important reliability indicator to evaluate the failure rate, the ability to convert laser energy threshold into LET is important.

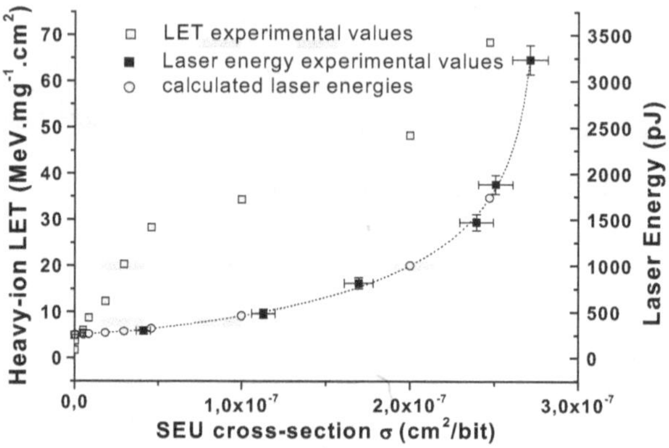

(a)

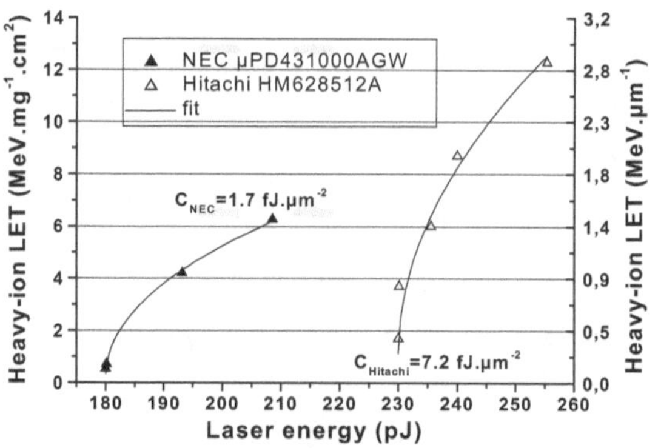

(b)

Figure 13. LET and laser energy versus cross-section for the Hitachi HM628512A (a) and LET versus energy curves (b) for the Hitachi and NEC SRAMs.

Figure 13 shows a plot of experimental LET as a function of laser energy for the two tested SRAMs. These plots come from a comparison of the experimental cross-section obtained during heavy-ion and laser tests (see Fig. 13a). First, LET and energy values are plotted as a function of σ. Then, identifying values of LET and energy that correspond to the same values of σ permits the LET as a function of energy to be plotted, if it is assumed that those two parameters can be identified by their effect on the device. The ability to reproduce the same effect with a laser pulse and with an ion is a basic hypothesis of laser testing. It obviously requires the correction of the cross-section values due to the spot size effect, as discussed in the preceding section. As Fig. 13b shows, the relationship between LET and laser energy appears to be non-linear for SRAMs.

For an energy E and a LET L this relation can be fit to the following expression:

$$L = L_{th} + \sqrt{C(E - E_{th})} \tag{5}$$

where E_{th} is the experimental threshold energy, L_{th} the equivalent threshold LET and C a parameter that depends on the device technology. The extracted values of C for the two SRAMs are given in Fig. 13b. The physical origin of C is undetermined at the present time. We can remark that this parameter has the dimensions of an energy per unit area (L is then expressed in MeV.μm^{-1}), which could be linked to a characteristic energy of the problem. Additional work is required to understand its physical meaning, i.e. the origin of this non-linearity.

To estimate the reliability of a component, the final piece of data needed is the threshold LET, L_{th}. An empirical correlation between laser pulse energy and the threshold LET for SEU and SEL in a range of devices has been observed for nominally 600nm optical pulses [5,7]. From the theoretical perspective, one approach for correlating LET and laser pulse energy is comparison of the total number of carriers produced by each. Assuming that N is the number of carriers produced at the threshold energy E_{th}, the corresponding LET L_{th} is then given by:

$$L_{th} = \frac{E_I}{\rho_{Si}\, l} N \tag{6}$$

where l is the depth along which the carriers causing the upset are produced, ρ_{Si} is the atomic density of silicon, and E_I is the average energy required to produce one electron-hole pair. Table 2 summarizes the values of the reliability indicators obtained during particle accelerator and laser test for the examples presented here. As is evident, the reliability indicators obtained with the laser are in a good agreement with those obtained using a particle accelerator.

Table 2. Reliability indicators L_{th} ,σ_{sat} for the two tested SRAMs.

	SEU Heavy ion test results		***SEU Laser test results***	
	L_{th} (MeV.mg^{-1}.cm^2)	σ_{sat} (cm^2)	Calculated L_{th} (MeV.mg^{-1}.cm^2)	σ_{sat} (cm^2)
NEC	0.3	0.11	0.31 ± 0.10	(0.09±0.01)
Hitachi	0.5	1.1	0.51 ± 0.14	(1.08±0.04)

4.3 Laser SEE based on carrier generation using two-photon absorption

Laser SEE experiments typically are performed under conditions such that higher-order nonlinear-optical effects, such as two-photon absorption and photogenerated free-carrier absorption, are not significant. Recently, a new method of laser-induced carrier generation for SEE applications based on two-photon absorption (TPA) using high peak power femtosecond pulses at sub-bandgap optical wavelengths has been introduced and demonstrated [15,21]. In two-photon absorption, the laser wavelength is chosen to be less than the bandgap of the semiconductor material, such that, at low light intensities, no carriers are generated (no optical absorption). At sufficiently high intensities, however, the material can absorb two or more photons simultaneously to generate a single electron-hole pair [22]. Because carrier generation in the two-photon process is proportional to the square of the laser pulse intensity, significant carrier generation occurs only in the high-intensity focal region of the focused laser beam, as illustrated in Fig. 14, which compares two-photon absorption and single-photon absorption using above band-gap illumination at 800nm. This characteristic allows charge injection at any depth in the structure, and permits three-dimensional mapping of single-event effects, and backside, through-wafer irradiation of devices.

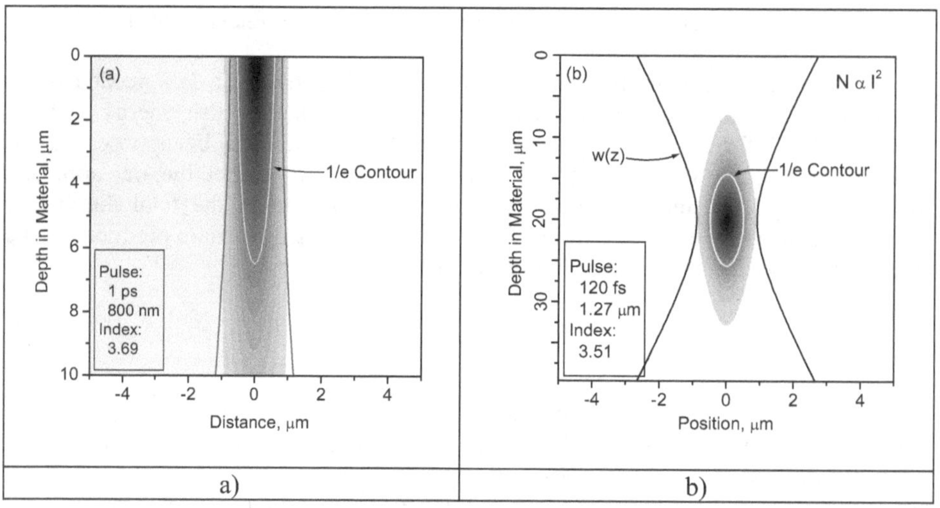

Figure 14. Electron-hole density plots for (a) single-photon and (b) 2-photon excitation processs in silicon at 800 nm and 1.27µm, respectively, as a function of depth (z). Note the difference in scale between (a) and (b).

Motivations for development of the TPA SEE experiment are several-fold. Recent generation technologies are becoming increasingly complex, with multiple metallization layers posing a major impediment to conventional top-side laser SEE testing. Additionally, the advent of flip-chip mounted devices renders both top-side laser testing and conventional heavy-ion testing impractical (or impossible), as is illustrated in Fig. 15. The two-photon absorption method represents a novel approach to SEE interrogation with unique capabilities not exhibited by other techniques. The experimental approach is analogous to that depicted in Fig. 5, with the exception of the laser source and its characteristics. In particular, the efficient implementation of this approach requires the use of intense femtosecond optical pulses ($\Delta\tau \approx 100$fs) with an optical wavelength in the sub-bandgap region of the semiconductor material under investigation (for silicon this involves wavelengths greater than 1.15μm).

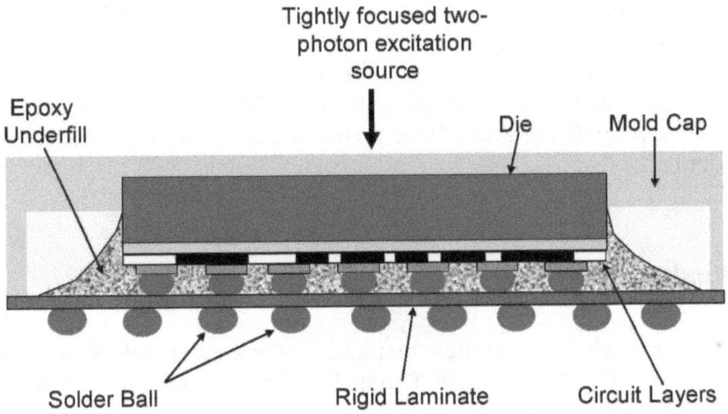

Figure 15. Schematic diagram illustrating one type of flip-chip mounted device. Carriers are introduced through the wafer using the sub-bandgap, two-photon absorption approach.

The diagnostic characteristic of above-bandgap, single-photon excitation is the exponential decrease in carrier density as a function of depth in the material [4]. While this characteristic leads to well defined reproducible conditions for charge injection in the material, it precludes the possibility of carrier injection at a controlled depth. Conversely, when two-photon absorption is the primary means of carrier generation, the optical loss and penetration depth can be deterministically manipulated: because carrier generation is proportional to the square of the laser pulse intensity, the generated carriers are highly concentrated in the high-irradiance region near the focus of the beam (*cf.*, Fig. 14). For a material that is transparent to the incident radiation, the high irradiance region can be directed to any depth in the material by translation of the DUT with respect to the focusing element. This permits both 3-D response mapping within a device, and carrier injection via backside illumination when front surface access to circuit elements is blocked.

To date, the TPA SEE approach has been used for interrogating the single-event transient response of linear bipolar technologies to charge deposited at various depths and locations in the device [21], and the backside, through-wafer approach has been demonstrated for a linear device with a lightly-doped substrate [23] and for a high-density SRAM with a highly-doped substrate [24].

5. Conclusions

We have presented experimental systems for pulsed laser testing of integrated circuits, with a detailed discussion of the fundamentals of photon-induced, and a comparison with heavy ion testing. The complete automation and synchronization of the system allows time-resolved scanning for timing-related investigations in complex VLSI circuits. The system was successfully used for analyzing SEUs in commercial SRAMs, SETs in analogue devices, and time-resolved transient fault injection in an ADC for information propagation imaging and critical phase extraction. The fundamental differences between the laser light and the effects of heavy ions have been highlighted. Next generation of laser tools could mitigate many of the actual limitations, as demonstrated by new techniques recently presented in the literature, such as the use the two-photon absorption (TPA) phenomena to generate photo-carriers in a very small volume whose position can be controlled anywhere inside the device under test.

Acknowledgments

The authors acknowledge, for their respective works, the following supports: Region Aquitaine, CNES, EADS, Defense Threat Reduction Agency (DTRA) and the Office of Naval Research.

References

[1] Adell P., Schrimpf R.D., Barnaby H.J., Marec R., Chatry C., Calvel P., Barillot C., Mion O., "Analysis of single event transient in analog circuits", IEEE Trans. Nucl. Sci., 47, p. 2616, 2000.
[2] Buchner S.P., Baze M.P., Single-Event Transients in Fast Electronic Circuits, IEEE Nuclear and Space Radiation Effects Conference (NSREC) Short Course, Section V, 2001.
[3] Turflinger T. "Single-Event Effects in Analog and Mixed-Signal Integrated Circuits," IEEE Trans. Nucl, Sci., 43, p. 594, 1996.
[4] Melinger J.S., Buchner S., McMorrow D., Stapor W.J., Weatherford T.R., Campbell A.B., "Critical evaluation of the pulsed laser method for single-event effects testing and fundamental studies", IEEE Trans. Nucl. Sci., Vol. 41, p. 2574, 1994.
[5] Moss S.C., LaLumondiere S.D., Scarpulla J.R., MacWilliams K.P., Crain W.R., and Koga R., "Correlation of picosecond laser-iInduced latchup and energetic particle-induced latchup in CMOS test structures," IEEE Trans. Nucl. Sci., Vol. 42, pp. 1948-1956, 1995.
[6] Pouget V. , D. Lewis, H. Lapuyade, R. Briand, P. Fouillat, L. Sarger, M.C. Calvet, "Validation of radiation hardened designs by pulsed laser testing and SPICE analysis", Microelectronics Reliability, vol. 39, pp. 931-935, October, 1999.
[7] McMorrow D., Melinger J., Buchner S., Scott T., Brown R.D., and Haddad N., "Application of a pulsed laser for evaluation and optimization of SEU-hard designs", IEEE Trans. Nucl. Sci., Vol. 47, pp. 559-565, 2000.
[8] Lewis D., Pouget V., Beaudoin F., Perdu P., Lapuyade H., Fouillat P., Touboul A., "Backside laser testing for SET sensitivity evaluation", IEEE Trans. Nucl. Sci., 48, p 2193, 2001.
[9] McMorrow D., Melinger J.S., Knudson A.R, Buchner S., Tran L.H., Campbell A.B., and Curtice W.R., "Charge-enhancement mechanisms of GaAs field-effect transistors: experiment and simulation", IEEE Trans. Nuc. Sci., 45, 1494 (1998).
[10] Melinger J.S., McMorrow D., Campbell A.B., Buchner S., Tran L.H., Knudson A.R., and Curtice W.R., "Pulsed laser induced single event upset and charge collection measurements as a function of optical penetration depth", J. App. Phys., Vol. 84, p. 690, 1998.
[11] Pease R., Sternberg A., Boulghassoul Y., Massengill L., Buchner S., McMorrow D., Walsh D., Hash G., LaLumondiere S., and Moss S., "Comparison of SETs in bipolar linear

circuits generated with an ion microbeam, laser and circuit simulation", IEEE Trans. Nuc. Sci., Vol. 49, pp. 3163-3170, 2002.
[12] Clegg W., Jenkins D.F.L., Helian N., Windmill J.F.C., Fry N., Atkinson R., Hendren W.R., Wright C.D., "A scanning laser microscope system to observe static and dynamic magnetic domain behavior", IEEE Trans. Instr. & Meas., vol. 51, 2002.
[13] Beaudoin F., Haller G., Perdu P., Desplats R., Beauchêne T., Lewis D., "Reliability defect monitoring with thermal laser stimulation: biased versus unbiased", Microelectronics Reliability, vol. 42, 2002.
[14] Pouget V. , PhD. Thesis, Université Bordeaux 1, 2000.
[15] McMorrow D., Lotshaw W.T., Melinger J.S., Buchner S., Pease R., "Sub-bandgap laser induced single event effects: carrier generation via two photon absorption", IEEE Trans. Nucl. Sci., Vol. 49, p. 3002, 2002.
[16] Pouget V., Fouillat P., Lewis D., Lapuyade H., Darracq F., Touboul A, "Laser cross section measurement for the evaluation of single-event effects in iIntegrated circuits", Microelectronics Reliability, vol. 40, October, 2000.
[17] Johnston A.H., "Charge Generation and Collection in p-n Junctions Excited with Pulsed Infrared Lasers", IEEE Trans. Nucl. Sci., 40, p. 1694, 1993
[18] Pouget V., Lewis D., Fouillat P., "Time-resolved scanning of iIntegrated circuits with a pulsed laser: application to transient fault injection in an ADC", IEEE, Trans on Instrumentation and Measurement, vol. 53 (4), pp. 1227-1231, 2004.
[19] Darracq F., PhD. Thesis, Université Bordeaux 1, 2003.
[20] Darracq F., Lapuyade H., Buard N., Mounsi F., Foucher B., Fouillat P., Calvet M-C., Dufayel R., "Backside SEU Laser Testing for Commercial Off-The-Shelf SRAMs", IEEE Trans. Nucl. Sci., vol. 49, p.2977, 2002.
[21] McMorrow D., Lotshaw W.T., Melinger J.S., Buchner S., Boulghassoul Y., Massengill L.W., and Pease R., "Three dimensional mapping of single-event effects using two-photon absorption", IEEE Trans. Nuc. Sci., vol. 50, pp. 2199-2207, 2003.
[22] Van Stryland E.W., Vanherzeele H., Woodall M.A., Soileau M.J., Smirl A.L., Guha S., Boggess T.F., "Two photon absorption, nonlinear refraction and optical limiting", Opt. Eng., Vol. 24, pp. 613-623, 1985.
[23] McMorrow D., Buchner S., Lotshaw W.T., Melinger J.S., Maher M., and Savage M.W., "Demonstration of through-wafer, two-photon-induced single-event effects", IEEE Trans. Nuc. Sci., Vol. 51, pp. 3553-3557, 2004.
[24] McMorrow D., Lotshaw W.T., Melinger J.S., Buchner S., Davis J.D., Lawrence R.K., Bowman J.H., Brown R.D., Carlton D., Pena J., Vasquez J., Haddad N., Warren K., and Massengill L., "Single-event upset in flip-chip SRAM induced by through-wafer, two-photon absorption", IEEE Trans. Nuc. Sci., Vol. 52, in press.

Design Hardening Methodologies for ASICs

Federico Faccio

CERN, Route de Meyrin 385, 1211 Geneva 23, Switzerland
Federico.Faccio@cern.ch

Abstract. Application-Specific Integrated Circuits (ASICs) can be effectively hardened against radiation effects by design, an approach that is commonly known as "Hardening By Design" (HBD). This contribution describes several HBD methodologies that are commonly used in CMOS technologies to protect the circuit from both Total Ionizing Dose (TID) and Single Event Effects (SEE).

1. Introduction

The functionality of CMOS-based Application-Specific Integrated Circuits (ASICs) in a radiation environment is at risk because of both Total Ionizing Dose (TID) and Single Event Effects (SEE). The traditional approach for the development of radiation-hard ASICs relies on a combination of technology-based and design-based techniques. In this approach, the designer uses only qualified radiation-hard CMOS technologies that have been custom developed at high cost to guarantee TID tolerance and Single Event Latchup (SEL) immunity up to a specified level, and typically only needs to take care of Single Event Upset hardening with some design technique. This solution is very expensive in a silicon manufacturing world where foundries for state-of-the-art technologies require an investment in the multi-billion-dollar range.

An alternative approach, based on the use of commercial-grade CMOS technologies, is recently gaining in popularity. This approach relies on the results of a research work dating back to the 80s and demonstrating that radiation effects in laboratory SiO_2 layers decrease quadratically with the oxide thickness, and even more sharply when the thickness is reduced below about 10 nm [1],[2]. When scaling down CMOS processes following what is known as Moore's law [3], the gate oxide thickness of the transistors is also decreased. A study of radiation effects in several technology generations [4],[5] demonstrated that these commercial-grade gate oxides followed the expected behavior, as shown in Fig. 1 [6].

Unfortunately, the isolation oxide thickness does not need to scale down, and despite the introduction of the Shallow Trench Isolation (STI) process around the 0.25 μm technology node this represents the weak point, in terms of TID tolerance, of modern submicron CMOS [7]. Leakage currents induced by charge trapping in the thick oxide at first increase the current consumption of the still functional circuit, until

R. Velazco et al. (eds.), Radiation Effects on Embedded Systems, 143–160.

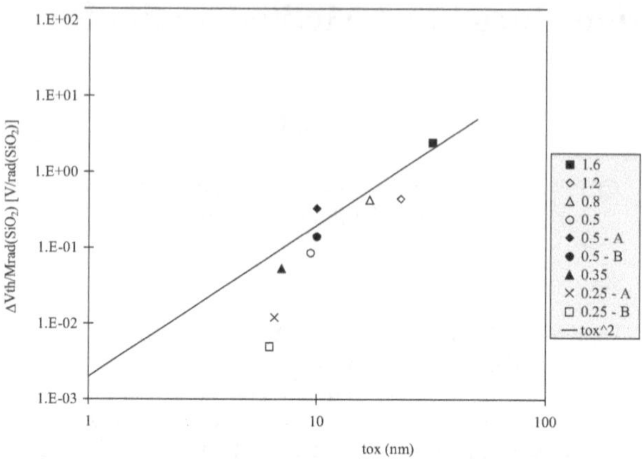

Figure 1. The threshold voltage shift of commercial-grade transistors from several technology generations, here normalized to 1Mrd of dose, follows the trend highlighted by studies on laboratory oxides in the 80s.

failure occurs at some point typically in the 10-100 krd TID range. The use of HBD methodologies can successfully eliminate all sources of radiation-induced leakage currents, therefore pushing the TID tolerance of ASICs up to the multi-Mrd levels allowed by the intrinsic properties of the thin gate oxides. These techniques will be described in detail in section 2, where examples of their large-scale utilization will be given.

Once the TID hardness has been addressed, the other fundamental problem that remains to be solved to ensure reliable operation of the ASIC in a radiation environment is linked to its sensitivity to SEEs. If a very significant decrease in the sensitivity to SEL can be achieved with the same methodology used to harden the circuit to TID effects, Single Event Upset (SEU) sensitivity can be decreased with a combination of HBD techniques, relying on either the use of modified cells or on redundancy. These techniques will be described in section 3.

2. Hardening against TID effects

TID affects the electrical characteristics of a transistor because radiation-induced charges are trapped either in the oxides (gate oxide, STI oxide) or at its interface with silicon. As already mentioned in section 1, the less than 5 nm think gate oxides of modern deep submicron processes are almost free from the effects of total dose and do not constitute a limitation to the use of the transistors even in a multi-Mrd radiation environment. The limitation to the use of commercial technologies in radiation environments comes rather from the large density of holes trapped in the thick STI oxide. This layer of positive charges builds up an electric field that ultimately inverts the p-doped silicon underneath the STI, opening conductive channels where charges can flow between n+ doped regions at different potential (source and drains of NMOS transistors, n-wells). For this reason, the typical TID-induced failure mechanism in commercial integrated circuits starts with an increase in the current consumption of the circuit, which can tolerate such increase up to a point where the leakage is so large that basic circuit functionality is lost.

Since the creation of a layer of trapped holes in the oxides is strongly dependent on the electric field in the oxide during irradiation, and since the electric field also determines the accumulation or inversion condition in the p-doped silicon regions, the leakage problem is worst at the edges of NMOS transistors. At the transistor edges, the polysilicon gate extends beyond the transistor channel and overlaps the lateral STI as schematically shown in Fig. 2. As a consequence, when the NMOS transistor is turned on an electric field is also present across the STI at the edge of the transistor: this field is not sufficient for the inversion of the p-doped silicon at the edges of the transistor. The accumulation of trapped holes in the STI in fact reinforces the electric field in this region, and eventually inversion at the transistor edges is reached even when the main transistor is turned off, and leakage current can flow in the inversion layer [8].

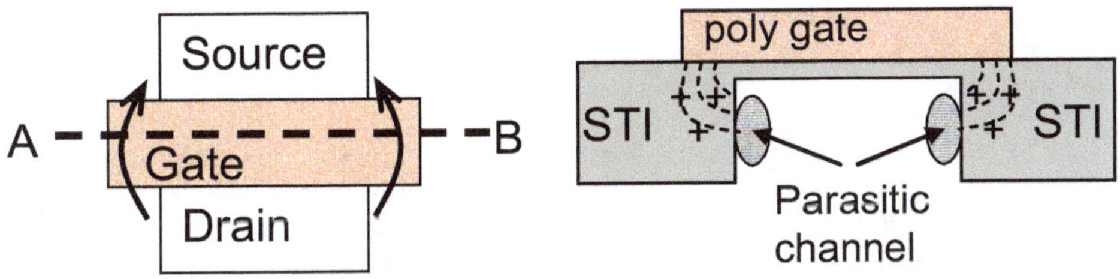

Figure 2. Top view of an open-layout NMOS transistor (left), and along its A-B line (right, view from the source or the drain electrode to the transistor channel).The dashed lines across the STI oxide represent schematically the electric field at the edge of the transistor, where the polysilicon gate overlaps the STI. Radiation-induced positive trapped charge in the STI, represented with the + symbol, enhances the electric field until eventually inversion of the p-doped substrate is achieved at the edges. This opens two parasitic channels where leakage current can flow between source and drain.

The obvious design-based solution to eliminate the above described problems is to avoid the contact between the STI oxide and any p-doped region where leakage current could possibly flow. For the NMOS transistor's edges, this is possible by completely surrounding one of the two n+ diffusions (source or drain) with the thin gate oxide [9]. Several transistor layouts are possible in this respect, some of which are shown in Fig. 3. If ringed layouts (ringed source, ringed inter-digitated) have the advantage of being more compact and allowing for any transistor size, they often require violation of design rules and have sometimes been shown to still exhibit some TID effects [10]. Despite some disadvantages that will be detailed later, the most commonly used layout is the Enclosed Layout Transistor (ELT) where one of the diffusions is entirely surrounded by the other, hence avoiding the violation of any design rule [11].

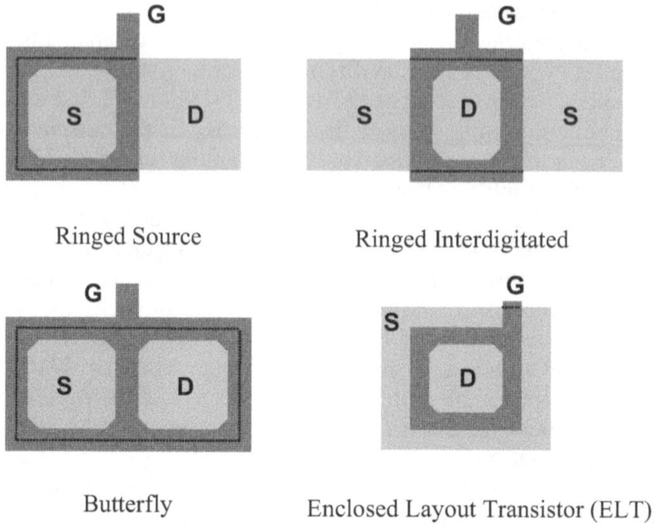

Figure 3. Transistor layout view for some of the possible NMOS designs eliminating the radiation-induced leakage current between source and drain. The solid line in each design evidences the end of the active area, or the beginning of the STI oxide. The active area under the gate does not get n+ doping, but it is covered by the radiation-tolerant thin gate oxide that surrounds either the source or the drain, or both.

Radiation-induced charge trapping in the isolation oxide can open leakage current paths between n+ diffusions at different potential (transistor to transistor or n-well to transistor leakage), as shown in Fig. 4. To prevent this to happen, an effective technique is the introduction of p+ "guardrings" between the n+ diffusions. This guardring is designed as a minimum-width p+ diffusion which is so heavily doped that it can practically never be inverted by radiation-induced trapped holes in the oxide above it. Care must be taken not to draw polysilicon lines above the guardring, because in CMOS processes this would automatically prevent the p+ doping under the polysilicon, hence introducing a weakness in the guardring: the region below the polysilicon would in that case be lightly p-type doped and would be prone to inversion.

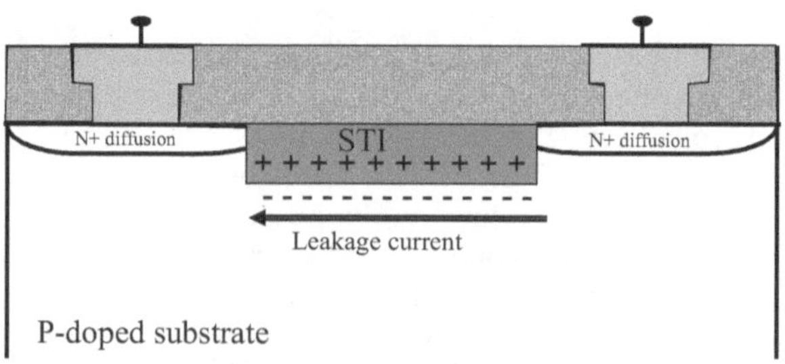

Figure 4. Radiation-induced positive charge trapping in the STI oxide can create an inversion layer in the p-doped substrate. In this layer, leakage current can flow between n+ diffusions at different potentials (the diffusions can either be sources/drains of NMOS transistors or n-wells).

The combination of systematic use of ELT NMOS transistors and guardrings, often referred to as "radiation tolerant layout techniques", has been demonstrated to be very effective in CMOS processes of different technology nodes [4],[12],[13]. Since it only relies on the natural TID tolerance of modern thin gate oxide, a physical parameter that has virtually no dependence on the particular manufacturing process, it is successfully applicable in all technologies and has no sensitivity on process changes. Compared to the use of dedicated radiation-hard technologies, it has the advantage of allowing the designer to use state-of-the-art CMOS technologies with their inherent advantages in terms of low power, high performance, high yield, short turnaround times, and last but not least low cost.

To date, the largest field application of such HBD approach, hence the best documented, has been in High Energy Physics (HEP) experiments at the CERN Large Hadron Collider (LHC) where a commercial 0.25 μm CMOS technology has been used [6]. In the following sub-sections, difficulties and practical considerations from the large experience accumulated at CERN in this domain are summarized.

2.1 Difficulties in the application of radiation tolerant layout techniques

The difficulties in the application of this approach are linked to some peculiarities of the ELT transistor itself, to the lack of availability of a commercial library systematically using radiation tolerant layout techniques, to the loss in density and to concerns in yield and reliability of the designs [14].

Peculiarities of the ELT transistor. The use of ELT transistors introduces three main topics: the need for a good model to compute their aspect ratio (effective W and L), the limitation in the W/L ratio that can be achieved, and the lack of symmetry in the device. Additionally, when designing with ELTs, one should not forget that for a given W/L ratio the ELT transistor has larger gate capacitance than a standard transistor, and correction for that should be introduced at the time of simulating the circuits.

Modeling the ELT transistor. In the first place, there are many different possible shapes for an ELT transistor, for instance square, octagonal, square with the corners cut at 45°. Each shape needing to be modeled in a different way [15],[16], it is wiser to use only one shape. Our choice fell on the square shape with the corners cut at 45°, a shape that is compatible with the ground design rules of most technologies and that avoids too sharp corners where the electric field could be excessively large and affect reliability. The details of the chosen shape – for instance the length of the cut in the corners and the size of the inner diffusion – have been decided once and for all transistors, and from that choice we elaborated a formula for the computation of the W/L ratio [14]. The good agreement between the formula and the size extracted from measurements on transistors has been verified for different channel lengths, giving us confidence in the model.

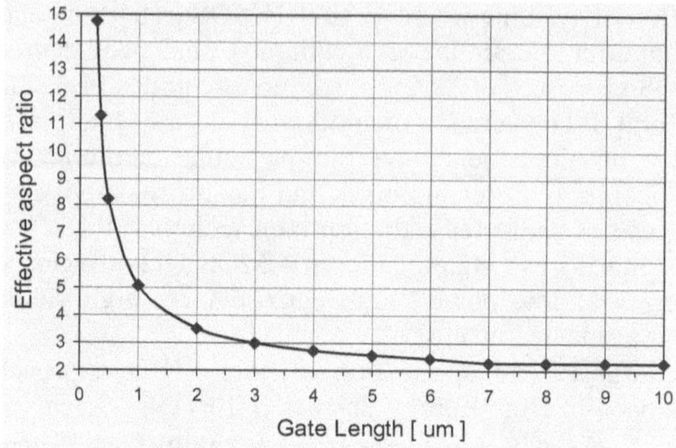

Figure 5. Minimum achievable aspect ratio for ELTs in the quarter micron process. The size is given for transistors with the inner diffusion at minimum size.

Limitation in the W/L ratio of ELTs. Since there is a direct relationship between the chosen gate length and the minimum gate width of the enclosed transistor, it is not possible to design transistors with aspect ratios lower than a certain value, as shown in Fig. 5. To obtain high W/L values it is sufficient to stretch the base shape in one or two dimensions, without modifying the corners; the calculation of the obtained W/L is straightforward. To have low aspect ratios the only way is to increase the channel length L keeping the minimum size for the inner diffusion: in the quarter micron process this is only effective up to an L of about 7μm, resulting in a minimum W/L ratio of about 2.26. Values close to this also imply a considerable waste of area and should be avoided using different circuit topologies.

ELT asymmetry. The evident lack of symmetry in the ELT layout translates in asymmetries in some of the transistor electrical characteristics. In particular, we have observed an asymmetry in the output conductance and in the matching of transistor pairs [6,4]. Since the gate in ELT transistors is annular, the source and drain contacts can be chosen inside and outside the ring of the gate, or the other way round. The measurement of the output conductance evidences larger values when the drain is chosen to be inside the ring: the choice of an outside drain lowers the output conductance by 20% (for the shorter gate length) to 75% (for a gate length of 5μm) [4]. In comparison, a transistor with standard linear layout has an output conductance that is close to the mean of the two values measured for the ELT.

Also the matching of ELT transistor pairs reveals the peculiarities of this type of transistor [6],[17]. In the first place, it appears that ELTs have an additional mismatch source, independent of the gate area and never observed before in standard transistors. The mismatch contribution of this additional source depends on the size and shape of the drain electrode, hence the mismatch properties are different for the same transistor when the drain is chosen to be the inner or outer diffusion, another evidence of the lack of symmetry of the transistor.

HBD library for digital design. No commercial library for digital design has been developed with radiation hardness in mind, therefore no library is commercially available where ELT NMOS transistors and guardrings are systematically used. Different design groups involved in ASIC development for applications in radiation environments have been confronted to this problem, and as a result a few "radiation-tolerant" libraries exist today. The work involved in the development of a small library of digital cells, enabling the design of small digital ASICs, should not be under-estimated. To support this statement, we can compare the number of logic cells included in the libraries: where commercial-grade libraries aimed at multi-million logic gate designs contain thousands of cells, each of the small number of existing HDB libraries does not exceed the 100-200 cells. Moreover, with the exception of the IMEC library [18] focused on a 0.18 μm technology that is freely available to European Institutes/Industries, these libraries are not easily available outside the communities that have produced them. Other than the IMEC library above, a CERN "radtol" library for a quarter micron technology and a Mission Research Corporation (MRC) library for a 0.18 μm technology are used for circuit design [19],[20].

Loss in density. As already pointed out, for an identical effective W/L ratio the ELT layout requires a larger area with respect to a standard layout. Moreover, the systematic insertion of guardrings is an area-consuming practice. As a result, designing with HDB techniques in a given technology implies an important loss of density, in terms of gates per mm^2. In the case of the quarter micron technology extensively used by CERN, the area overhead introduced by HDB was estimated for a number of digital cells in the radiation-tolerant library. If very simple cells such as inverters impose a 35% penalty only, more complex DFF cells required up to 75% more area. Overall, the penalty was estimated to about 70%. It is nevertheless important to consider that this comparison is made between a radiation-hard design and a commercial design, the area overhead being the price to pay for radiation tolerance.

When radiation tolerance is a design requirement, it looks more interesting to compare the density that can be achieved with different approaches at any moment in time. For instance, when CERN started the development of the quarter micron radiation tolerant library, the alternative to the HBD approach was the use of a dedicated radiation-hard technology. At that time, the most advanced technology satisfying the TID requirements was a 0.8μm BiCMOS technology capable of a density 8 times smaller than the quarter micron with HBD. Since the typical delay of radiation-hard processes with respect to the state-of-the-art commercial CMOS is 2-3 generations, it is reasonable to assume that the comparison in terms of density will always be largely in favor of the approach where HDB techniques are used in a commercial technology.

Yield and reliability concerns. Due to the lack of industrial experience with designs using ELT transistors, a justified concern existed until recently on the possible impact of this layout on the yield and reliability of the ASICs. Yield data can be obtained from the large scale application of HBD in the CERN LHC experiments. In this case, about 100 different ASICs have been designed, with a number of chips needed per circuit type ranging from 120,000 to 100. About 14 ASICs were needed in more than 20,000 pieces, and another 31 in a number between 2,000 and 20,000. Overall, the production volume for this application exceeded 2000 8-inches wafers. These ASICs included very different functions: readout for particle detectors, A/D converters, digital and analog memories, system control functions, time-to-digital conversion, transmission of digital and analog data via optical fiber and clock recovery [21],[22],[23],[24],[25],[26]. The measured yield of all these ASICs was comparable to the yield of other standard products run in the same line, demonstrating the absence in yield penalties associated to the use of ELT transistors.

Reliability concerns arise because of the corners used in the ELT transistor design: in those regions the electric field could be more intense and give origin to a larger vulnerability to hot carrier damage. A detailed study has been carried out in this respect [27] and demonstrated that in fact the hot carrier lifetime in ELT transistors depends – as it was for the output conductance and the matching – on the choice of the position of the source and drain diffusions. Transistors with the inner diffusion as drain have about 3 times lower hot carrier lifetime than standard transistors, which in turn have about 3 times lower lifetime than ELTs with outer diffusion as drain. Since the best performance in terms of power consumption and speed can usually be achieved by minimizing the drain capacitance, the typical design will use the inner diffusion of the ELT transistors for the drain. This will hence result in a decrease in reliability. Clearly, the circuit designer looking primarily to reliability could trade off performance for that, and use the outer diffusion as drain. In some circuit configuration, such as in gain nodes of amplifier circuits, this choice will be motivated also by performance since gain increases when the output conductance decreases, which can be obtained with an outer drain diffusion.

2.2 Trends in state-of-the-art CMOS technologies

The HBD work described above concerned CMOS technologies in the 0.25 and 0.18 μm nodes. Standard layout transistors in these technologies could typically stand TID levels on the order of some tens of krd before a strong increase in the leakage current. With the semiconductor industry racing towards ever smaller feature sizes, the 90 nm node is in use today for top-end digital products, and the 130nm is widely available for ASIC design. These advanced processes requiring increasingly complex sequences of processing steps, their natural radiation tolerance could well differ from older technologies and even between different foundries. In particular, STI-related steps (trench etching, thermal and High Density Plasma oxidation) that need to be tuned for optimal source-drain isolation and to reduce narrow channel effects could influence the radiation-induced leakage current at the transistor edges and between adjacent n+ diffusions. Due to the increasing popularity of the HBD approach, several works have been carried to study the natural radiation tolerance of the 130 nm technologies, and a similar study is now starting also in the 90 nm node. The first published results on the 130 nm node show a superior natural radiation tolerance with respect to older nodes [28]. If as expected the about 2 nm thick gate oxide does not show any significant sign of TID-induced effects up to doses larger than 100 Mrd, also in terms of leakage current these new technologies appear surprisingly robust. For instance, available results on 2 commercial technologies point out that the leakage between adjacent n+ junctions gets so small that practically most often the use of a p+ guardring is not necessary. Effects that have not been observed previously also appear at the transistor edges: the radiation-induced source-drain leakage for NMOS transistors peaks at a dose of 1-6 Mrd to decrease when the device is further irradiated to larger doses, and a Radiation-Induced Narrow Channel Effect (RINCE) has been reported to influence the threshold voltage of narrow channel NMOS and PMOS transistors. Overall, it looks that applications requiring TID tolerance up to a few hundreds krd would not even require HBD techniques to be employed, whilst for larger doses the tolerance could be strongly design dependent, the safest approach still being the systematic use of HBD. Additionally, it has to be pointed out that the first available results seem to indicate that the better radiation performance is not specific of one only foundry, but rather common to the 130 nm node.

3. Hardening against SEEs

Section 2 illustrated how the downscaling of commercial CMOS technologies can help to improve the TID tolerance of ASICs: what is the impact on vulnerability to SEEs?

For what concerns sensitivity to SEUs it has been predicted for years that, owing to the decrease in power supply voltage and in node capacitance, the sensitivity in modern technologies should be significantly larger [29]. Further supporting this picture, other works in the 90s also pointed out the existence of new mechanisms contributing to an increase in the upset sensitivity [30]. Extensive data has been collected by semiconductor industries to strictly monitor the SEU sensitivity of memories and CPUs, especially because of growing concern for large neutron-induced error rates in terrestrial large-scale applications [31]. Such works demonstrated that the sensitivity has instead been scaling down for DRAMs, whilst the picture is somewhat less clear for SRAMs: the sensitivity per bit has been decreasing, but due to the larger on-chip memory volume the sensitivity per chip has increased. The bottom line of such studies is that not only the node capacitance and voltage supply should be considered: the decrease of the sensitive area and of the charge collection efficiency with down-scaling also plays an important role in the overall SEU response.

Another upset mechanism that has been feared since a decade is finally starting to be observed in the field: hits in the combinatorial logic can now propagate through a long series of gates and finally reach the input of a register where, if they hazardously happen to be synchronous to the proper clock transition, they can be latched [32]. This phenomenon, due to the increased speed of the new technologies (the gate delay is shorter than the duration of the perturbation introduced by the strike of the ionizing particle), has been called "Digital" Single Event Transient (DSET). Since the wrong value can only be latched during a clock transition, the DSET error rate depends linearly on the clock frequency.

For what concerns SEL, it is clear that the introduction of STI and retrograde wells together with the reduction of the voltage supply have been very beneficial in reducing the general sensitivity [33]. This alone is not a guarantee that latchup is not occurring, since SEL is extremely design dependent and very sensitive components can still be found. But clearly, for ASIC design, with careful layout the designer can benefit from these favorable technological features.

This short review of the impact of the down-scaling of CMOS technologies on SEEs sensitivities evidences how such effects can still be responsible for circuit failure in a radiation environment, and that HBD techniques have to be used to prevent this to happen. The following sub-sections indicate some possible solutions.

3.1 Hardening against SEU

Memory cells and latches used in registers are typically the circuit elements most vulnerable to SEUs, therefore one possible solution to harden by design is to modify the cell layout or architecture to decrease its sensitivity. This approach requires the identification of a cell design that can satisfy the immunity criteria defined for the application, which differs with different radiation environments and with different functions for the cell. For instance, hardening a circuit for applications in a particle environment determined only by charged and neutral hadrons (protons, pions, neutrons) such as the one typical of High Energy Physics experiments is very different than hardening it for an environment where heavy ions are present such as for satellite applications or deep space missions. Hardening memory cells also drives the design differently than for flip-flops or latches.

Different SEU-hardened cells. A conceptually simple method of hardening a memory or latch cell is to increase the charge needed to upset it (this quantity is commonly called "critical charge"). This can be achieved with the addition of some capacitance to the sensitive nodes, and it is very effective for hadron radiation environments. Applications of such technique can be found in fact for high energy physics experiments and for terrestrial applications with high reliability requirements. In the former case, the larger capacitance was integrated either by increasing the size of some transistors, which had also the beneficial side effect of increasing their current drive, or by adding metal-metal capacitors on top of the cells [34]. A reduction of the error rate by a factor of 10 could be easily reached without directly loosing area, but the price to pay was larger power consumption and the loss of 2 metal layers for routing on top of the cells. An extension to this concept is used by industry to decrease the SRAM sensitivity in terrestrial and avionic applications by more than 2 orders of magnitude: in that case the additional capacitance is integrated with the addition of a DRAM-like capacitor on top of the SRAM [35].

Another approach relies on the development of a modified cell architecture rendering it less sensitive or – in some cases – insensitive to the charge deposited by the particle strike. This is a somewhat popular research subject, therefore a relatively large number of different cells have been proposed, out of which in reality only a few have actually been used. The problem is that often the cell can be hardened at the cost of introducing strict requirements at the circuit level, such as the distribution of 2 clocks with a precise skew, that are extremely challenging to reliably meet over a full circuit.

Probably the first solution proposed and extensively used to harden SRAM cells in old technologies is the addition of two large resistors to the cell loop [36]. The cell being formed by two cross-coupled inverters, the function of the resistors inserted between the output of each inverter and the input of the other is to delay the propagation of the signal across the loop. The charge deposited by a striking particle in any sensitive node can be removed by the conductive transistor of the stroke inverter before the induced voltage glitch travels across the loop to reinforce the change of state. This solution was effectively used in the past, but the speed penalty it introduces is incompatible with the performance of advanced SRAM circuits.

Another cell that has been used in the past for SRAM designs is the Whitaker cell [37]. This cell uses the knowledge that particle strikes in n+ diffusion can only induce a change of state when the stored value is a logic 1, whilst the opposite is true for p+ diffusions. A clever duplication of the nodes where the information is stored creates a cell with 4 storing nodes, 2 of which only have n+ diffusions and the other 2 only p+ diffusions, the two nodes of the same type storing opposite information. In this way, there are always 2 nodes that are not vulnerable to upset, and a careful connection of the transistors to form the memory loop ensures that in no condition a transient in one of the vulnerable nodes can propagate.

Other dedicated architectures are the HIT (Heavy Ion Tolerant) [38] and DICE (Dual Interlock Cell) [39], both of which have been used in real applications. These cells also use the concept of duplicating the nodes storing the information, and in particular the DICE is very attractive and extensively used because of its property of being compact, simple and hence compatible with the design of high performance circuits in advanced CMOS processes. A schematic of the DICE cell used as a latch is shown in Fig. 6, evidencing how the input of each of the 4 inverters constituting the memory loop is split in two: each of the 2 transistors of the inverter is connected to a different node. In fact the output of each of the inverters, that are connected as a closed loop, controls the gate of one only transistor (of one polarity) in the following inverter in the loop, and one only transistor (of the other polarity) in the previous inverter in the loop. A particle-induced glitch at the output of the inverter will propagate hence in two directions in the loop, but neither the previous nor the following inverter will in fact change their output and the initial condition will soon be re-established: the error is not latched in the cell. To write in the cell, or to change its condition, it is in fact necessary to write at the same time in two nodes, namely either in the odd or in the even inverters in the loop (nodes B and D in Fig. 6). The DICE architecture is often used for DFF and register cells, although care must be used in the layout of the cell: the sensitive nodes of the 2 even and odd inverters have to be spaced sufficiently to avoid a single particle strike to induce a glitch in both, because this would be equivalent to a good writing sequence and would induce the error to be latched. This characteristic is about to get a limitation of the DICE cell for its use in 130 nm CMOS technologies and beyond, where the transistor density is such that all sensitive nodes of the cell are within the reach of a single ionizing particle, unless the cell is "stretched" to dimensions that are not compatible with a high-density design. A study performed on the integration of the DICE cell in the 90 nm technology has shown an only tenfold improvement in the error rate with respect to a standard cell. Another inconvenience of the cell is that, although the strike on one node does not latch the wrong data in the cell, the output of the cell can temporarily be wrong during the strike and until the good condition is restored at all nodes. This temporary error at the cell output can propagate to the next cell in some conditions, and eventually be latched somewhere else in the circuit.

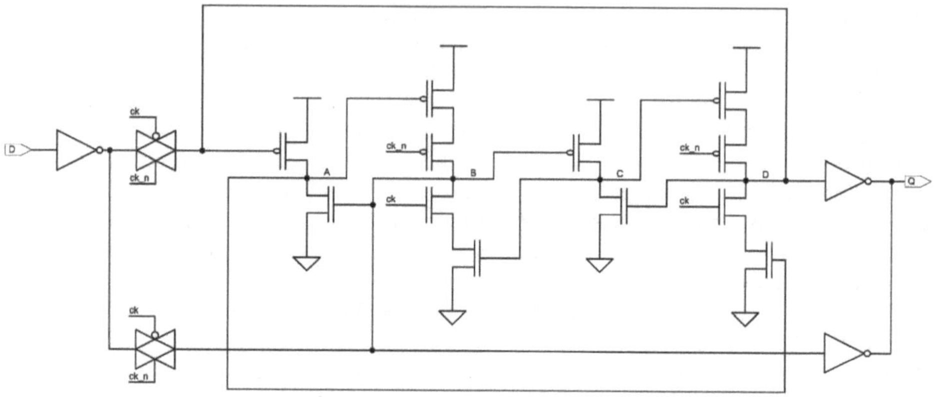

Figure 6. Schematic representation of the DICE cell used here as a latch.

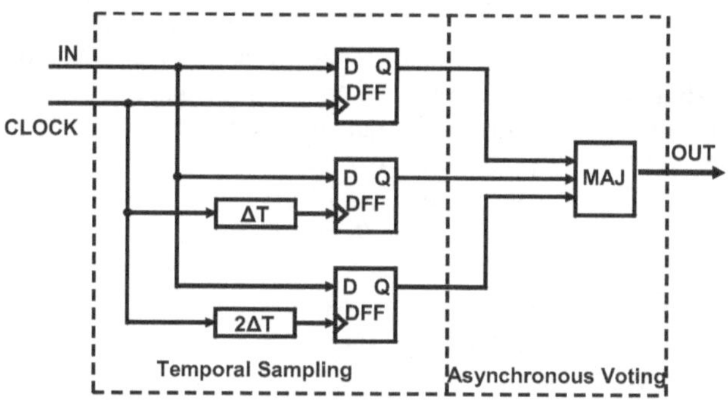

Figure 7. Architecture of a cell implementing the concept of temporal redundancy.

A modified version of the DICE cell has been also proposed, and demonstrated to be very effective in protecting the content of data in storage cells [40]. This cell being sensitive to SEU during a particular phase of the writing cycle, its use is discouraged whenever it is necessary to change the stored data often.

A somewhat different approach that has recently been proposed to protect registers against SEU but also against DSET is based on the concept of temporal sampling [41]. This approach uses the concept of redundancy and builds it inside the basic hardened cell, as shown in Fig. 7. The DFF storing the data is triplicated inside the basic cell, and the output of the 3 FFs is compared by a voter. In this way, if the content of one of the FFs is upset, the output of the cell is still correct since the voter will output only the value given by the majority of the FFs. The temporal sampling idea has been introduced in the cell to protect it from DSET. If the cell is getting its input from a sequence of combinatorial logic gates, a transient in one of the gates can propagate along the logic chain and reach the input of the cell. A standard DFF would latch the logic value at its input at the moment of the clock transition: if this transition happens in coincidence with the arrival of the transient at the input, the wrong data is latched. To prevent this to happen, the 3 DFFs in the hardened cell sample the status of the input at 3 different times, which is implemented by delaying the input by ΔT for one of the FFs and by $2\Delta T$ for another FF. To avoid the transient to be latched, it is sufficient to ensure that ΔT is longer than its duration.

SEU hardening with redundancy. Another approach to protect the circuit from SEU is to actually add redundancy to the stored information. This can be done in two ways, either by triplicating the cells storing the information (Triple Modular Redundancy, TMR) or by encoding the data and employing Error Detection And Correction (EDAC) techniques.

As already presented for the last cell of the previous sub-section, triplication of the storing cell is a valid way of protecting the content, although very space and power consuming. A single voter circuit can be used to compare the output of the triplicated cell, but in this case the final output could be affected by an error induced in the voter itself. A safer technique is to triplicate the voter as well as shown in Fig. 8. In this case, the final output is protected by errors in both the latches and the voters. Although a compact design for the voter has been proposed (G.Cervelli, CERN), this approach is very area hungry since it requires the integration of the voter on top of the triplication of the basic logic cells. The triplication approach is nevertheless widely used, for instance in High Energy Physics applications where it is often acceptable to pay an important area penalty.

Contrary to the TMR technique where each bit of information is triplicated, EDAC techniques require a much smaller information redundancy [42]. EDAC is used very extensively for data transmission and for semiconductor memories, but also whenever data need to be stored reliably (for instance, CD and DVD extensively use it). The information to be stored is encoded by a complex logic block, and some redundant information is added in this process: the larger the number of redundant bits, the safer the original data is from any source of error. At the time of reading the stored information, another complex set of logic operations is needed for decoding. Several codes can be used for EDAC, such as Hamming, Reed-Solomon and BCH [43], and each of them has different detection and correction capabilities, and different complexity.

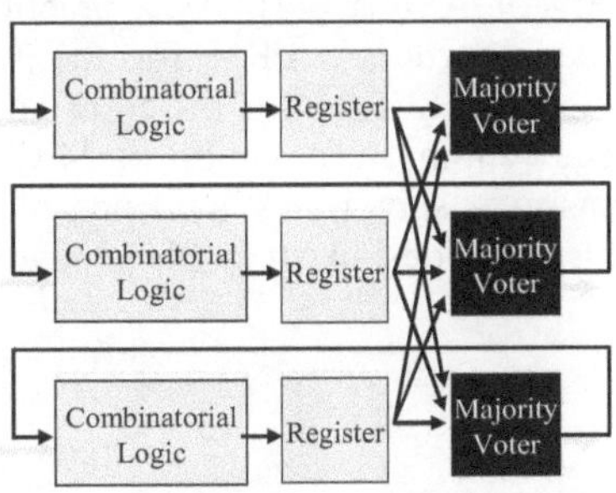

Figure 8. SEU hardening with redundancy: TMR is implemented all along the information path, and also in the voters.

To understand a little deeper the idea behind EDAC, let us consider the easier case of Hamming encoding originally proposed in the 50s, and let us assume that the information to protect is a word of 8 bits. A word of 8 bits can code 2^8 different combinations of bits, each of which is a valid word in the code since each of them could well be the information that has to be stored. Adding some redundant bit, it is possible to increase the different possible combinations of bits, out of which still only 2^8 are valid words. With Hamming encoding, the number of redundant bits added is such that every valid word differ by at least 3 bits from any other. In this way, an error affecting only one bit can not only be detected, but also corrected during the decoding since there is only one valid code differing by one bit from the word that has been read. In the case of errors affecting two bits, it is still possible to detect the presence of the error but it is not possible to guess which of the two valid codes differing by 2 bits from the read word was originally stored. The above is expressed saying that Hamming encoding allows for single error detection and correction and double error detection. Other more powerful and complex codes allow for multiple error correction as well.

3.2 Hardening against SEL

In addition to the beneficial effects of retrograde wells and shallow trench isolation typical of modern CMOS technologies, ASIC designers can effectively decrease the circuit sensitivity to SEL with the use of some design practices. This is better understood when looking at the parasitic thyristor structure responsible for latchup in Fig. 8. When some current starts to flow in the loop formed by the two parasitic bipolar transistors, and due to the presence of the parasitic resistors, a voltage builds between the base and emitter of the bipolar transistors. As a consequence, the transistors inject more current in the loop, and this condition continues until eventually a large current can flow between Vdd and Vss. Intuitive methods to prevent this situation are the reduction of the resistances and of the bipolar transistors' gain. This can be achieved for instance by increasing systematically the distance between the two parasitic complementary transistors, or by using a large number of Vdd and Vss contacts all over the circuit. The latter can be done very effectively by surrounding all regions containing NMOS transistors with p+ guardrings, which are nothing but Vss contacts. Therefore, the same layout technique introduced in section 2 to protect the circuit from TID-induced leakage currents is also very effective to reduce the SEL sensitivity of ASICs [44]. As a practical demonstration of this, a number of the ASICs developed at CERN for the LHC in the quarter micron technology has been tested for SEEs with different sources: heavy ions up to an equivalent LET of 120 Mev/cm^{2}mg, protons up to 24 GeV/c, pions up to 300 MeV. None of these circuits has ever shown any evidence of SEL, despite their different functionality and their different design style (they have been designed by different groups).

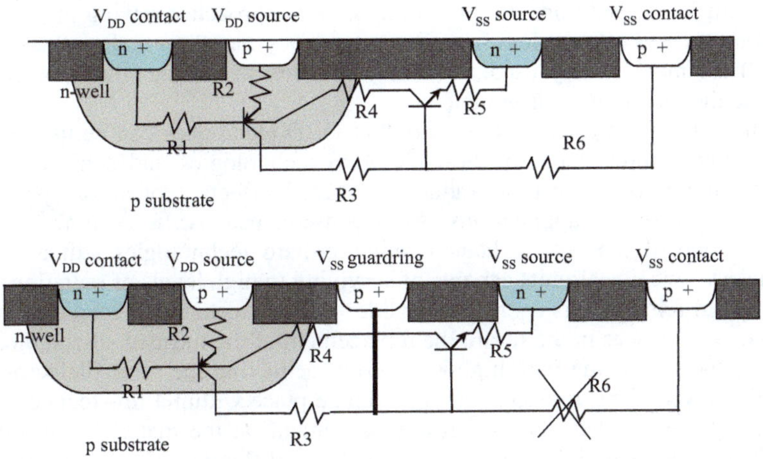

Figure 9. Parasitic thyristor responsible for SEL (top), formed by two parasitic bipolar transistors and several resistors. The introduction of a p+ guardring around all NMOS transistors (bottom), also eliminating TID-induced leakage currents, effectively reduces the gain of the lateral npn transistor in the loop and ensures its base is kept at Vss potential. As a result, the structure is considerably less sensitive to SEL.

4. Conclusion

The decrease in the gate oxide thickness accompanying the downscaling of CMOS technologies towards smaller feature sizes is rendering it increasingly less sensitive to TID effects. Taking advantage of this intrinsic property of the transistors, it is possible to eliminate all remaining sources of TID-induced failure with the use of HBD techniques and achieve ASICs able to stand even multi-Mrd radiation environments without any noticeable performance degradation. This approach relying only on a well verified physical property of the silicon dioxide, it is not dependent on specific details in the processing sequence and hence can be safely used in all commercial CMOS technologies. This characteristic explains why HBD is gaining in popularity, since it allows for decreasing considerably the cost of the ASICs that can be produced in a commercial-grade technology, and at the same time for an increase in the performance.

The same HBD techniques used for TID can also be effective in decreasing the design susceptibility to radiation-induced SEL. The addition of a p+ guardring around the NMOS transistors serves in fact the double purpose of cutting inter-transistor leakage current paths and decreasing the gain of one of the parasitic bipolar transistors forming the thyristor structure responsible for latch-up.

Decreased sensitivity or even immunity to SEUs, including those eventually latched as a consequence of radiation-induced transient propagation across combinatorial logic (DSET), can also be achieved with a multiplicity of HBD techniques. The basic memory or latch cell can be modified for that purpose, as proposed in several works, although this often implies complications in the routing of the circuit and imposes difficult conditions on the timing. Other approaches relying in triplication of the information (TMR) or less invasive redundancy based on coding the information (EDAC) are more widely used, including in fields where the source of error is not related to the radiation environment.

At the end of this review, it appears that HBD techniques can be used to develop ASICs in commercial state-of-the-art CMOS technologies and achieve the desired level of immunity against all radiation-induced effects typical of space, nuclear industry and research applications. As a consequence, ASIC designers not having access to one of the few dedicated radiation-hard technologies still available can nonetheless conceive circuits capable of surviving to high levels of radiation.

As a final remark, it is useful to consider that the technology roadmap for CMOS technologies foresees in the future the replacement of the nitrided silicon dioxide used today for the gate oxide with high-k materials achieving the same dielectric constant on thicker layers. This replacement should take place within a few technology nodes, although at present there is not yet any agreement on the material that will replace SiO_2. This new material will most likely have different radiation properties than silicon dioxide, therefore it is impossible at this time to make any forecast on its tolerance to TID. Should the new material be very sensitive to TID, the limitation in the radiation tolerance of the circuits will again be intrinsic (as it was in the times of large gate oxide thickness) and no HBD technique will enable the overcome of this limitation.

References

[1] N.S. Saks, M.G.Ancona and J.A.Modolo, "Radiation effects in MOS capacitors with very thin oxides at 80°K", IEEE Trans. Nucl. Science, Vol.31, pp.1249-1255, December 1984.

[2] N.S. Saks, M.G.Ancona and J.A.Modolo, "Generation of interface states by ionizing radiation in very thin MOS oxides", IEEE Trans. Nucl. Science, Vol. 33, pp. 1185-1190, December 1986.

[3] G.E. Moore, "Cramming More Components onto Integrated Circuits", Electronics, vol. 38, no. 8, 1965.

[4] G. Anelli et al., "Total Dose behavior of submicron and deep submicron CMOS technologies", in the proceedings of the Third Workshop on Electronics for the LHC Experiments, London, September 22-29, 1997, pp. 139-143 (CERN/LHCC/97-60, 21 October 1997).

[5] W. Snoeys et al., "Layout techniques to enhance the radiation tolerance of standard CMOS technologies demonstrated on a pixel detector readout chip", Nuclear Instruments and Methods in Physics Research A 439 (2000) 349-360.

[6] F. Faccio, "Radiation issues in the new generation of high energy physics experiments", International Journal of High Speed Electronics and Systems, Vol. 14, No. 2 (2004), 379-399.

[7] M.R. Shaneyfelt et al., "Challenges in Hardening Technologies Using Shallow-Trench Isolation", IEEE Trans. Nucl. Science, Vol. 45, No. 6, pp. 2584-2592, December 1998.

[8] T.R. Oldham et al., "Post-Irradiation effects in field-oxide isolation structures", IEEE Trans. Nucl. Science, Vol. 34, No. 6, pp. 1184-1189, December 1987.

[9] D.R. Alexander, "Design issues for radiation tolerant microcircuits for space", Short Course of the Nuclear and Space Radiation Effects Conference (NSREC), July 1996.

[10] R.N. Nowlin, S.R. McEndree, A.L. Wilson, D.R. Alexander, "A New Total-Dose Effect in Enclosed-Geometry Transistors", presented at the 42nd NSREC conference in Seattle, July 2005, to be published in the IEEE TNS, Vol. 52, No. 6, December 2005.

[11] W. Snoeys et al., "Layout techniques to enhance the radiation tolerance of standard CMOS technologies demonstrated on a pixel detector readout chip", Nuclear Instruments and Methods in Physics Research A 439 (2000) 349-360.

[12] F. Faccio et al., "Total Dose and Single Event Effects (SEE) in a 0.25μm CMOS Technology", in the proceedings of the Fourth Workshop on Electronics for LHC Experiments, Rome, September 21-25, 1998, pp. 105-113 (CERN/LHCC/98-36, 30 October 1998).

[13] N. Nowlin, K. Bailey, T. Turfler, D. Alexander, "A total-dose hardening-by-design approach for high-speed mixed-signal CMOS integrated circuits", International Journal of High Speed Electronics and Systems, Vol. 14, No. 2 (2004), 367-378.

[14] G. Anelli et al., "Radiation Tolerant VLSI Circuits in Standard Deep Submicron CMOS Technologies for the LHC Experiments: Practical Design Aspects", IEEE Trans. Nucl. Science, Vol. 46, No. 6, pp.1690-1696, December 1999.

[15] A. Giraldo, A. Paccagnella and A. Minzoni, "Aspect ratio calculation in n-channel MOSFETs with a gate-enclosed layout", Solid-State Electronics, Vol.44, 1st June 2000, pp. 981-989.

[16] A.Giraldo, "Evaluation of Deep Submicron Technologies with Radiation Tolerant Layout for Electronics in LHC Environments", PhD. Thesis at the University of Padova, Italy, December 1998 (http://wwwcdf.pd.infn.it/cdf/sirad/giraldo/tesigiraldo.html).

[17] G. Anelli, "Conception et caractérisation de circuits intégrés résistants aux radiations pour les détecteurs de particules du LHC en technologies CMOS submicroniques profondes", Ph.D. Thesis at the Politechnic School of Grenoble (INPG), France, December 2000, availeble on the web at the URL: http://rd49.web.cern.ch/RD49/RD49Docs/anelli/these.html.

[18] S. Redant et al., "The design against radiation effects (DARE) library", presented at the RADECS2004 Workshop, Madrid, Spain, 22-24 September 2004.

[19] K. Kloukinas, F. Faccio, A. Marchioro and P. Moreira, "Development of a radiation tolerant 2.0V standard cell library using a commercial deep submicron CMOS technology for the LHC experiments", in the proceedings of the Fourth Workshop on Electronics for LHC Experiments, Rome, September 21-25, 1998, pp. 574-580 (CERN/LHCC/98-36, 30 October 1998).

[20] D.Mavis, "Microcircuits design approaches for radiation environments", presented at the 1st European Workshop on Radiation Hardened Electronics, Villard de Lans, France, 30th March – 1st April 2004.

[21] M. Campbell et al., "A Pixel Readout Chip for 10-30 Mrad in Standard 0.25μm CMOS", IEEE Trans. Nucl. Science, Vol.46, No.3, pp. 156-160, June 1999.

[22] P. Moreira et al., "G-Link and Gigabit Ethernet Compliant Serializer for LHC Data Transmission", 2000 IEEE Nuclear Science Symposium Conference Record, pp.96-99. Lyon, October 15-20, 2000.

[23] F. Faccio, P. Moreira and A. Marchioro, "An 80Mbit/s radiation tolerant Optical Receiver for the CMS optical digital link", in the proceedings of SPIE 4134, San Diego, July 2000.

[24] G. Anelli et al., "A Large Dynamic Range Radiation-Tolerant Analog Memory in a Quarter-Micron CMOS Technology", IEEE Transactions on Nuclear Science, vol. 48, no. 3, pp. 435-439, June 2001.

[25] Rivetti et al., "A Low-Power 10-bit ADC in a 0.25-mm CMOS: Design Considerations and Test Results", IEEE Transactions on Nuclear Science, vol. 48, no. 4, pp. 1225-1228, August 2001.

[26] W. Snoeys et al., "Integrated Circuits for Particle Physics Experiments", IEEE Journal of Solid-State Circuits, Vol. 35, No. 12, pp. 2018-2030, December 2000.

[27] D.C. Mayer et al., "Reliability Enhancement in High-Performance MOSFETs by Annular Transistor Design", IEEE Trans. Nucl. Science, Vol. 51, No. 6, pp. 3615-3620, December 2004.
[28] F. Faccio, G. Cervelli, "Radiation-induced edge effects in deep submicron CMOS transistors", presented at the 42nd NSREC conference in Seattle, July 2005, to be published in the IEEE TNS, Vol. 52, No. 6, December 2005.
[29] P.E. Dodd et al., "Impact of Technolgy Trends on SEU in CMOS SRAMs", IEEE Trans. Nucl. Science, Vol. 43, No. 6, p. 2797, December 1996.
[30] C. Detcheverry et al., "SEU Critical Charge And Sensitive Area In A Submicron CMOS Technology", IEEE Trans. Nucl. Science, Vol. 44, No. 6, pp. 2266-2273, December 1997.
[31] R. Baumann, "Single-Event Effects in Advanced CMOS Technology", Short Course of the Nuclear and Space Radiation Effects Conference (NSREC), July 2005.
[32] P.E. Dodd, M.R. Shaneyfelt, J.A. Felix, J.R. Schwank, "Production and Propagation of Single-Event Transients in High-Speed Digital Logic ICs", IEEE Trans. Nucl. Science, Vol. 51, No. 6, pp. 3278-3284, December 2004.
[33] A.H. Johnston, "The Influence of VLSI Technology Evolution on Radiation-Induced Latchup in Space Systems", IEEE Trans. Nucl. Science, Vol.43, No.2, p.505, April 1996.
[34] F. Faccio et al., "SEU effects in registers and in a Dual-Ported Static RAM designed in a 0.25μm CMOS technology for applications in the LHC", in the proceedings of the Fifth Workshop on Electronics for LHC Experiments, Snowmass, September 20-24, 1999, pp. 571-575 (CERN 99-09, CERN/LHCC/99-33, 29 October 1999).
[35] P. Roche, F. Jacquet, C. Caillat, J.P. Schoellkopf, "An Alpha Immune and Ultra Low Neutron SER High Density SRAM", proceedings of *IRPS 2004*, pp. 671-672, April 2004.
[36] J. Canaris, S. Whitaker, "Circuit techniques for the radiation environment of Space", IEEE 1995 Custom Integrated Circuits Conference, p. 77.
[37] M.N. Liu, S. Whitaker, "Low power SEU immune CMOS memory circuits", IEEE Trans. Nucl. Science, Vol. 39, No. 6, pp. 1679-1684, December 1992.
[38] R. Velazco et al., "2 CMOS Memory Cells Suitable for the Design of SEU-Tolerant VLSI Circuits", IEEE Trans. Nucl. Science, Vol. 41, No. 6, p. 2229, December 1994.
[39] T. Calin, M. Nicolaidis, R. Velazco, "Upset Hardened Memory Design for Submicron CMOS Technology", IEEE Trans. Nucl. Science, Vol. 43, No. 6, p. 2874, December 1996.
[40] F. Faccio et al., "Single Event Effects in Static and Dynamic Registers in a 0.25μm CMOS Technology", IEEE Trans. Nucl. Science, Vol.46, No.6, pp.1434-1439, December 1999.
[41] P. Eaton, D. Mavis et al., "Single Event Transient Pulsewidth Measurements Using a Variable Temporal Latch Technique", IEEE Trans. Nucl. Science, Vol. 51, no. 6, p.3365, December 2004.
[42] S. Niranjan, J.F. Frenzel, "A comparison of Fault-Tolerant State Machine Architectures for Space-Borne Electronics", IEEE Trans. On Reliability, Vol. 45, No. 1, p. 109, March 1996.
[43] S. Lin, D.J. Costello Jr., "Error Control Coding", Second edition, Pearson Prentice Hall, 2004, ISBN 0-13-017973-6.
[44] T. Aoki, "Dynamics of heavy-ion-induced latchup in CMOS structures", IEEE Trans. El. Devices, Vol. 35, No. 11, p. 1885, November 1988.

Fault Tolerance in Programmable Circuits

Fernanda Lima Kastensmidt and Ricardo Reis

Instituto de Informática
Universidade Federal do Rio Grande do Sul
CP 15064 CEP 91501-970 Porto Alegre Brazil
fglima@inf.ufrgs.br, reis@inf.ufrgs.br

Abstract. This chapter is dedicated to the effects of radiation on programmable circuits. It is described the radiation effects on integrated circuits manufactured using CMOS process and it is explained in detail the difference between the effects of a SEU in an ASIC and in a SRAM-based FPGA architecture. It is also discussed some SEU mitigation techniques that can be applied at the FPGA architectural level. The problem of protecting SRAM-based FPGAs against radiation in the high level description is also defined.

1. Introduction

Fault-tolerance on semiconductor devices has been a meaningful matter since upsets were first experienced in space applications several years ago. Since then, the interest in studying fault-tolerant techniques in order to keep integrated circuits (ICs) operational in such hostile environment has increased, driven by all possible applications of radiation tolerant circuits, such as space missions, satellites, high-energy physics experiments and others [1]. Spacecraft systems include a large variety of analog and digital components that are potentially sensitive to radiation and must be protected or at least qualified for space operation.

Designers for space applications currently use radiation-hardened devices to cope with radiation effects. However, there is a strong drive to utilize standard commercial-off-the-shelf (COTS) and military devices in spaceflight systems to minimize cost and development time as compared to radiation-hardened devices, which brings the necessity of using fault-tolerant techniques in standard circuit to ensure reliability.

In addition, because of the continuous evolution of the fabrication technology process of semiconductor components, in terms of transistor geometry shrinking, power supply, speed, and logic density [2], the fault-tolerance starts to be a matter on circuits operating at ground level as well. As stated in [3, 4], drastic device shrinking, power supply reduction, and increasing operating speeds significantly reduce the noise margins and thus the reliability that very deep submicron (VDSM) ICs face from the various internal sources of noise. This process is now approaching a point where it will be unfeasible to produce ICs that are free from these effects. Consequently, fault-tolerance is not anymore a matter exclusively for space designers but also for designers in next generation products, which must cope with errors in the ground level due to the advanced technology.

R. Velazco et al. (eds.), Radiation Effects on Embedded Systems, 161–181.

The programmable circuits can be divided in two kinds, the ones programmed by the user and the ones programmed at the foundry using some metal layers. This chapter is dedicated to the first ones.

Field Programmable Gate Arrays (FPGAs) are very popular for design solutions because of the high flexibility and reconfigurability feature, which reduces the time to market. They are also an attractive candidate for space applications in terms of high density, high performance, low NRE (Non-Recurring Engineering) cost and fast turnaround time. SRAM-based FPGAs can offer an additional benefit for remote missions. For instance, by allowing in-orbit design changes thanks to re-programmability, with the aim of reducing the mission cost by correcting errors or improving system performance after launch.

There are two ways to implement fault-tolerant circuits in SRAM-based FPGAs. The first possibility is to design a new FPGA matrix composed of fault-tolerant elements. These new elements can replace the old ones in the same architecture topology or a new architecture can be developed in order to improve robustness. The cost of these two approaches is high and it can differ according to the development time, number of engineers required to perform the task and the foundry technology used. Another possibility is to protect the high-level description by using some sort of redundancy, targeting the FPGA architecture. In this way, it is possible to use a commercial FPGA part to implement the design and the SEU mitigation technique is applied to the design description before being synthesized in the FPGA. The cost of this approach is inferior to the previous one because in this case the user is responsible for protecting the own design and it does not require new chip development and fabrication. In this way, the user has the flexibility of choosing the fault-tolerant technique and consequently the overheads in terms of area, performance and power dissipation.

All of them have their own space in the market, as each application requires different constraints. But because the semiconductor industry trends to emphasize time-to-market and low-cost production, the implementations based on high-level design look more interesting. In this work, both architectural and the high-level methods are presented and discussed, but because of the high cost of the implementations of the architecture based one, only implementations based on high-level were designed and tested in details.

Figure 1 shows a SEU and SET mitigation technique design flow for programmable logic.

This chapter is organized as follows. Section 2 describes the radiation effects on integrated circuits manufactured using CMOS process and it explains in detail the difference between the effects of a SEU in ASIC and in SRAM-based FPGA architectures. Section 3 discusses some SEU mitigation techniques that can be applied at the FPGA architectural level. Section 4 defines the problem of protecting SRAM-based FPGAs against radiation in the high level description. Section 5 discusses conclusions and future work, followed by the references.

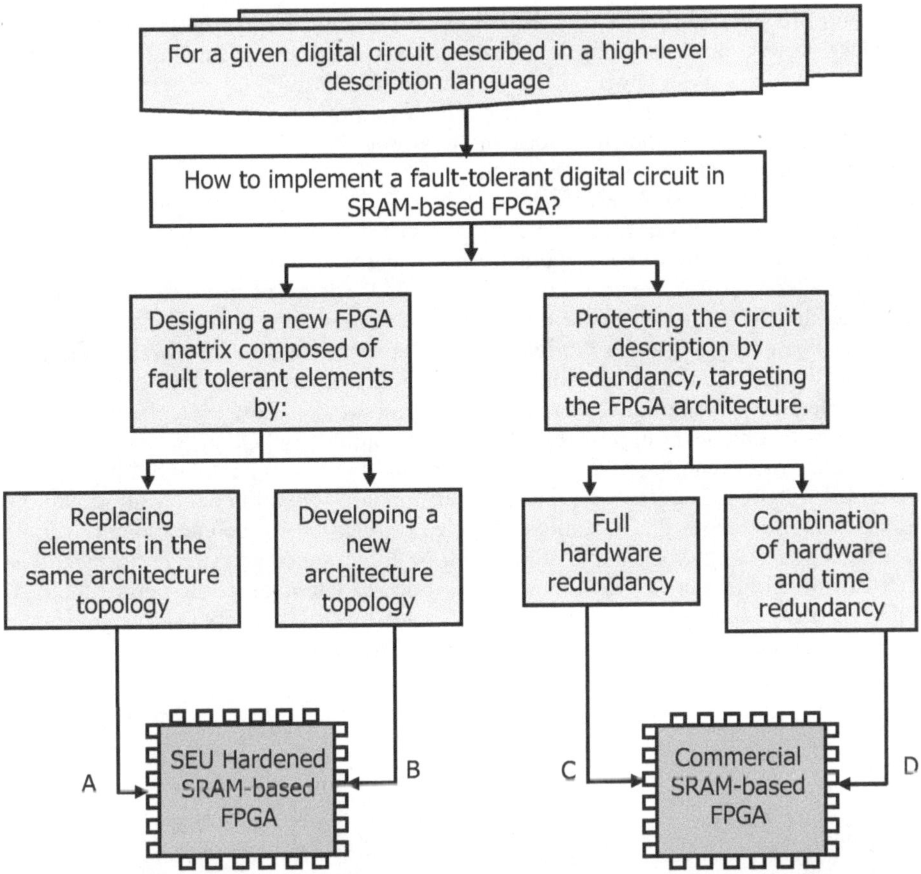

Figure 1. SEU and SET Mitigation Technique Design Flow for Programmable Logic.

2. Radiation effects on SRAM-based FPGAs

The radiation environment is composed of various particles generated by sun activity [5]. The particles can be classified as two major types: (1) charged particles such as electrons, protons and heavy ions, and (2) electromagnetic radiation (photons), which can be x-ray, gamma ray, or ultraviolet light. The main sources of charged particles that contribute to radiation effects are protons and electrons trapped in the Van Allen belts, heavy ions trapped in the magnetosphere, galactic cosmic rays and solar flares. The charged particles interact with the silicon atoms causing excitation and ionization of atomic electrons.

When a single heavy ion strikes the silicon, it loses its energy via the production of free electron hole pairs resulting in a dense ionized track in the local region. Protons and neutrons can cause nuclear reaction when passing through the material. The recoil also produces ionization. The ionization generates a transient current pulse that can be interpreted as a signal in the circuit causing an upset.

At the ground level, the neutrons are the most frequent cause of upset [6]. Neutrons are created by cosmic ion interactions with the oxygen and nitrogen in the upper atmosphere. The neutron flux is strongly dependent on key parameters such as altitude, latitude and longitude. There are high-energy neutrons that interact with the material generating free electron hole pairs and low energy neutrons. Those neutrons interact with a certain type of Boron present in semiconductor material creating others particles. The energized alpha particles are the greatest concern in this case and they are addressable through processing and packaging material. In principle, a very careful selection of materials can minimize alpha particles. However, this solution is very expensive and never eliminates the problem completely [7].

A single particle can hit either the combinational logic or the sequential logic in the silicon [8]. Figure 2 illustrates a typical circuit topology found in nearly all sequential circuits. The data from the first latch is typically released to the combinatorial logic on a falling or rising clock edge, at which time logic operations are performed. The output of the combinatorial logic reaches the second latch sometime before the next falling or rising clock edge. At this clock edge, whatever data happens to be present at its input (and meeting the setup and hold times) is stored within the latch.

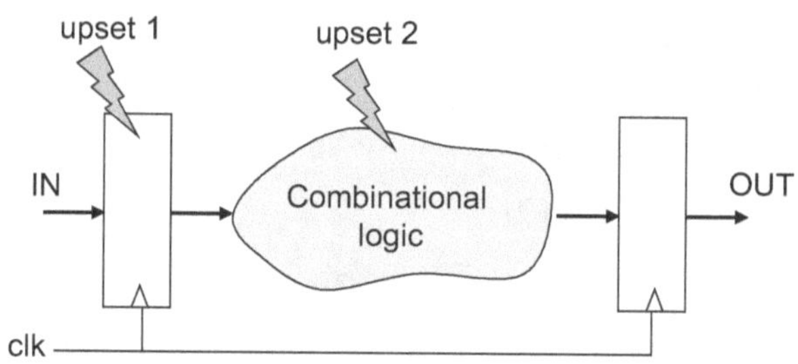

Figure 2. Upsets hitting combination and sequential logic.

When a charged particle strikes one of the sensitive nodes of a memory cell, such as a drain in an off state transistor, it generates a transient current pulse that can turn on the gate of the opposite transistor. The effect can produce an inversion in the stored value, in other words, a bit flip in the memory cell. Memory cells have two stable states, one that represents a stored '0' and one that represents a stored '1.' In each state, two transistors are turned on and two are turned off (SEU target drains). A bit-flip in the memory element occurs when an energetic particle causes the state of the transistors in the circuit to reverse. This effect is called Single Event Upset (SEU) or soft error and it is one the major concerns in digital circuits.

When a charged particle hits the combinational logic block, it also generates a transient current pulse. This phenomenon is called single transient effect (SET) [9]. If the logic is fast enough to propagate the induced transient pulse, then the SET will eventually appear at the input of a latch, where it may be interpreted as a valid signal. Whether or not the SET gets stored as real data depends on the temporal relationship between its arrival time and the falling or rising edge of the clock. In [10], the probability of a SET becoming a SEU is discussed. The analysis of SET is very complex in large circuits composed of many paths. Techniques such as timing analysis [11] could be applied to analyze the probability of a SEU in the combinational logic being stored by a memory cell or resulting in an error in the design operation. Additional invalid transients can occur at the combinatorial logic outputs as a result of SETs generated within global signal lines that control the function of the logic. An example of this would be SETs generated in the instruction lines to an ALU (Arithmetic Logic Unit).

In FPGAs, the upset has a peculiar effect when hit the combination and sequential logic mapped into the programmable architecture. Let's take for example the SRAM-based FPGAs such as the Virtex® family from Xilinx that is one of the most popular programmable devices used in the market nowadays.

Virtex® devices [12] consist of a flexible and regular architecture composed of an array of configurable logic blocks (CLBs) surrounded by programmable input/output blocks (IOBs), all interconnected by a hierarchy of fast and versatile routing resources. The CLBs provide the functional elements for constructing logic while the IOBs provide the interface between the package pins and the CLBs. The CLBs are interconnected through a general routing matrix (GRM) that comprises an array of routing switches located at the intersections of horizontal and vertical routing channel. The Virtex matrix also has dedicated memory blocks called Block SelectRAMs of 4096 bits each, clock DLLs for clock-distribution delay compensation and clock domain control, and two 3-State buffers (BUFTs) associated with each CLB.

Virtex devices are quickly programmed by loading a configuration bitstream (collection of configuration bits) into the device. The device functionality can be changed at anytime by loading in a new bitstream. The bitstream is divided into frames and it contains all the information to configure the programmable storage elements in the matrix located in the Look-up tables (LUT) and flip-flops, CLBs configuration cells and interconnections, Figure 3. All these configuration bits are potentially sensitive to SEU and consequently they were our investigation targets.

In an ASIC, the effect of a particle hitting either the combinational or the sequential logic is transient; the only variation is the time duration of the fault. A fault in the combinational logic is a transient logic pulse in a node that can disappear according to the logic delay and topology. In other words, this means that a transient fault in the combinational logic may or may not be latched by a storage cell. Faults in the sequential logic manifest themselves as bit flips, which will remain in the storage cell until the next load.

On the other hand, in a SRAM-based FPGA, both the user's combinational and sequential logic are implemented by customizable logic memory cells, in other words, SRAM cells, as represented in Figure 3. When an upset occurs in the combinational logic synthesized in the FPGA, it corresponds to a bit flip in one of the LUTs cells or in the cells that control the routing. An upset in the LUT memory cell modifies the implemented combinational logic, Figure 4. It has a permanent effect and it can only be corrected at the next load of the configuration bitstream. The effect of this upset is related to a stuck-at fault (one or zero) in the combinational logic defined by that LUT. This means that an upset in the combinational logic of a FPGA will be latched by a storage cell, unless some detection technique is used. An upset in the routing can connect or disconnect a wire in the matrix, please see Figure 5. It has also a permanent effect and its effect can be mapped to an open or a short circuit in the combinational logic implemented by the FPGA. The fault will also be corrected at the next load of the configuration bitstream.

When an upset occurs in the user sequential logic synthesized in the FPGA, it has a transient effect, because the next load of the flip-flop corrects an upset in the flip-flop of the CLB. An upset in the embedded memory (BRAM) has a permanent effect and it must be corrected by fault tolerant techniques applied in the architectural or in the high-level description, as the load of the bitstream cannot change the memory state without interrupting the normal operation of the application. In [13, 14], the effects of upsets in the FPGA architecture are also discussed. Note that there is also the possibility of having single event transient (SET) in the combinational logic used to build the CLB such as input and output multiplexors used to control part of the routing.

Radiation tests performed in Xilinx FPGAs [15, 16] show the effects of SEU in the design application and prove the necessity of using fault-tolerant techniques for space applications. A fault-tolerant system designed into SRAM-based FPGAs must be able to cope with the peculiarities mentioned in this section such as transient and permanent effects of a SEU in the combinational logic, short and open circuit in the design connections and bit flips in the flip-flops and memory cells. In [17], the effect of neutrons was also analyzed in a SRAM-based FPGA from Xilinx. In that time, the FPGA presented very low susceptibility to neutrons, but the vulnerability is increasing as the technology is reaching smaller transistor size and consequently higher logic density.

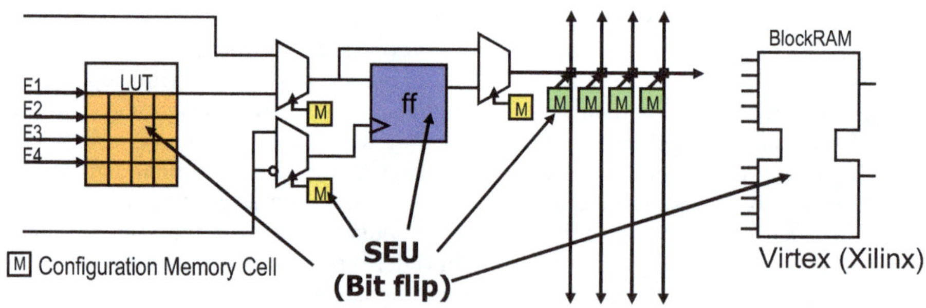

Figure 3. SEU Sensitive Bits in the CLB Tile Schematic.

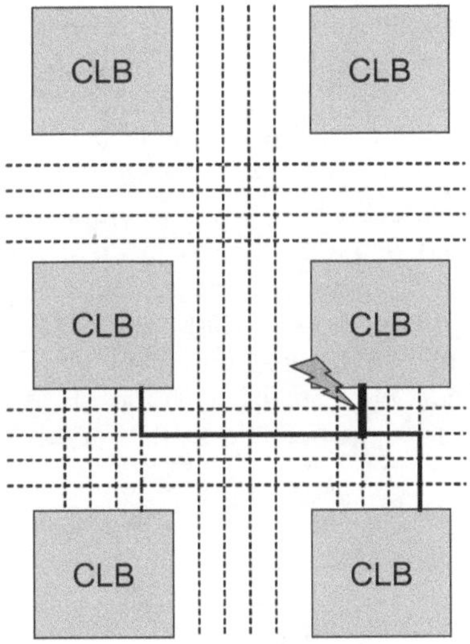

Figure 4. Examples of upsets in the SRAM-based FPGA architecture in the routing.

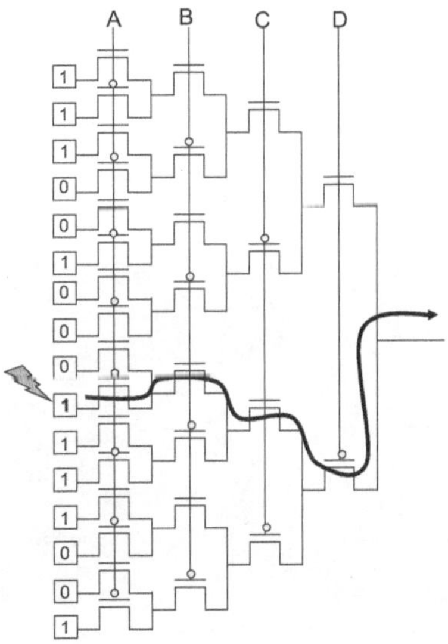

Figure 5. Examples of upsets in the SRAM-based FPGA architecture in the LUT.

2.1 Fault injection mechanisms

Fault injection techniques have been used to evaluate the effects of SEUs in the configuration memory for switch matrices. Based on the results of these techniques, the routing defects in switch matrices can be classified as either open defects or short defects. Open defect can be seen as a single line broken causing the line to be driven constantly by an unknown value. In Figure 6, a single line is broken, causing the net connected that specific single line to go to ground value. Short defect occurs when two previously unconnected lines are now connected due to an SEU as shown in Figure 7. The short can be characterized as either a wired-AND or a wire-OR model. Two special cases of short defects are stuck-at-0 and stuck-at-1 defects, where a line is shortened with the ground or power line respectively. In fact, an open defect is equivalent to a stuck-at-0 or a stuck-at-1 defect, depending on whether the net is by default pulled-up or pulled-down. A third kind of defect is the creation of a net that is driven by an existing net, and is not connected to any other net. Such a defect does not affect the logic of the circuit but can increase the power consumption. This paper primarily addresses open and short defects, as these defects can change the logic of the circuit.

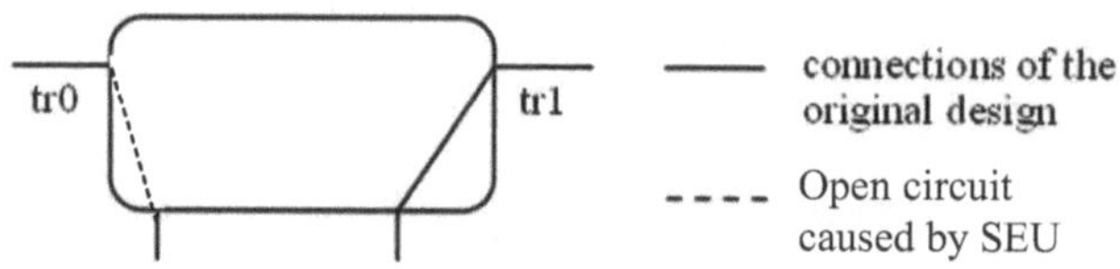

Figure 6. Example of Open circuit in a Single Line.

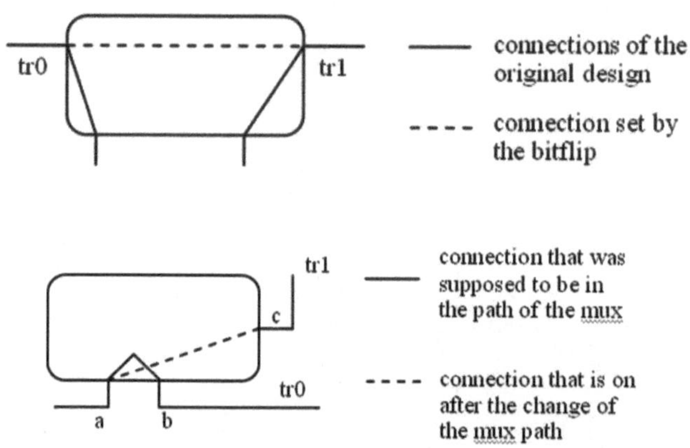

Figure 7. Example of Short circuit between Single Lines and Single Lines and Hex lines.

3. Architectural SET and SEU mitigation techniques for SRAM-based FPGAs

The first SEU mitigation solution that has been used for many years in spacecraft was shielding, which reduces the particle flux to very low levels but it does not completely eliminate it. This solution was sufficient to avoid errors caused by radiation effects for many years in the past. However, due to the continuous evolution of the fabrication technology process, as explained previously, the electronic circuits are becoming more and more sensitive to radiation particles, and the charged particles that once were negligible are now able to cause errors in the electronic design. Consequently, extra techniques must be applied to avoid radiation effects.

Several SEU mitigation techniques have been proposed in the last few years in order to avoid faults in digital circuits, including those implemented in programmable logic. They can be classified as: fabrication process-based techniques, such as: epitaxial CMOS processes and advanced process such as silicon-on-insulator (SOI); design-based techniques for mitigation, such as: triple modular redundancy (TMR), time redundancy, multiple redundancy with voting, EDAC (error detection and correction coding) and hardened memory cell level; and recovery techniques (applied to programmable logic only), such as: reconfiguration, partial configuration, rerouting design.

However, in the case of SRAM based FPGAs, the problem of finding an efficient technique in terms of area, performance and power is very challenging, because of the high complexity of the architecture. As previously mentioned, when an upset occurs in the user's combinational logic implemented in an FPGA, it provokes a very peculiar effect not commonly seen in ASICs. The SEU behavior is characterized as a transient effect, followed by a permanent effect. The upset can affect either the combinational logic or the routing. The consequences of this type of effect, a transient followed by a permanent fault, cannot be handled by the standard fault tolerant solutions used in ASICs, such as Error Detection and Correction Codes (EDAC), hamming code, or the standard TMR with a single voter, because a fault in the encoder or decoder logic or in the voter would invalidate the technique. The problem of protecting SRAM-based FPGAs against SEU is not yet solved and more studies are required to reduce the limitation of the methods currently used.

In the architectural level, the previous solutions leave open at least two problems to be solved:

- how to cope with SETs in the CLB logic to avoid upsets being stored in the flip-flop,
- how to cope with multiple bit upsets in the LUTs, routing and especially in the embedded memory.

In this section, we propose the investigation and development of SEU mitigation techniques for SRAM-based FPGAs that can be applied to FPGAs with or without embedded processors that can cope with the two problems still not solved. The SRAM based FPGAs were chosen because of their high applicability in space. Different than FPGAs programmed by anti-fuse that can be programmed just once, SRAM based FPGAs can be reprogrammed by the user as many times as necessary in a very short period. So, applications can be updated and corrected after launch. This feature is very valuable for space applications because it can reduce the cost in update missions or even save missions that were launched with design problems.

First, it is necessary to analyze the amount of the sensitive area in the programmable matrix and the characteristics of them to propose improvements in the SEU mitigation techniques for SRAM-based FPGAs. Figure 2 had shown the set of configuration cells in a CLB tile of the Virtex family. There are 864 memory bits responsible for the customization of the logic. Analyzing the percentage of each type of SRAM cell in the whole set of memory elements in the CLBs, the LUTs represent 7.4%, the flip-flops represent 0.46%, the customization bits in the CLB represent 6.36% and the general routing represents 82.9%, Figure 8.

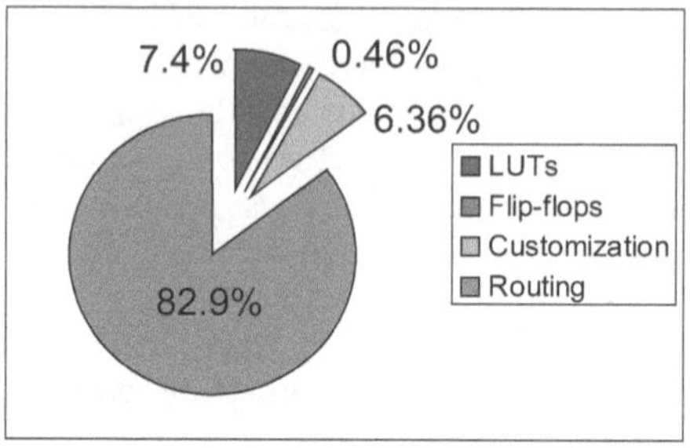

Figure 8. Percentage of sensitive SRAM bits in the CLB of Virtex FPGAs.

Based on these results, the effect of an upset in the routing configuration (customization bits of the CLB and general routing) seems to be the major concern, totaling approximately 90% of the sensitive area in each CLB. This type of fault, as mentioned previously, has a permanent effect, which represents an open or short circuit in the final connections of the logic design. A solution that can increase the area of this customization logic too much is not very attractive in final area and cost of this FPGA.

In addition to these programmable cells presented in Figure 8, there are other memory elements in FPGA devices that can also be affected by SEU: SelectMAP (Selectable Microprocessor Access Port) latches, JTAG (Joint Test Action Group - IEEE Std. 1149.1x), TAP (Test Access Port) latches and others latches of other built-in non-programmable features. The main effects of a SEU in these latches are SEFI (Single Event Functional Interrupt) such as configuration circuit upsets and JTAG circuit upsets. There are few flip-flops or latches in the POR, less than 40 latches or flip-flops, which leads to a very small cross-section. But they cannot be disregarded because an upset in one of these latches can force the chip to be reprogrammed.

Some solutions to protect the POR can be: TMR the whole block, replace the cells by SEU hardened memory cells or use extra logic to turn off the POR after the device is programmed by an external pin. In the next paragraphs, some fault-tolerant techniques will be discussed to protect the SRAM cells of the LUT, flip-flops, routing and customization cells, and the embedded block RAM. The advantages and disadvantages of each technique were analyzed based on previous work results in ASICs implementations.

The first solution that can be studied is to replace some or all of the latches in the FPGA by SEU hardened flip-flops. Many hardened memory cells were designed during the last years. However each one has different characteristics that can show more efficiency for some applications. They differ from each others in some characteristics such as: the number of transistor, the method, the SEU order effect, the ability to accumulate or not upsets and the SET immunity in combinational logic. For example, standard latches have a first order of susceptibility; in other words, they are upset by a single node strike. Some of them require multiple node strikes to upset cells such as TMR memory cells, DICE memory cell [18] and simple temporal memory cells [19]. Temporal latches built from DICE cells, for example, have a second and third order of susceptibility [20], which means that is robust to upsets in multiple memory cells and also a ion strike in multiple nodes of a single cell.

The hardened memory solution is suitable to replace the SRAM cells in the routing, general customization and lookup tables because they present a small overhead compared to logic redundancy technique and EDAC. Solutions such as IBM [21], NASA [22], DICE [18], HIT [23] and resistor [24] memory cells look interesting in the number of transistors and fault coverage. The final area will be around 2 times the original one, which is a very good result in terms of high-reliability.

For the LUT, for instance, if the cells are placed too close to each other, it is possible to use the solution of a TMR memory cell, where each cell is a DICE memory cell. In this case, this solution is robust to the 1st, 2nd and 3rd order of upsets. And because the LUT cells comprise only 7.4% of the cells, the impact in area will not be so intense. In [25], a SEU immune memory cell based on decoupling resistors was developed for FPGAs. The design is asymmetric to provide that the data cell powers-up in a know state. In the referenced paper, the size and the speed of the cell are discussed. The cells that are not in the critical path, such as the cells that control the routing, for example, do not need to be high-speed ones. In this case, the tolerance and the size are the main issue. Results show the high reliability of the cell for heavy ions strike.

The embedded memory in the FPGA must be protected in order to avoid errors. EDAC is a suitable technique to correct upsets in memory structures, as discussed previously. An example is the hamming code that can be applied to embedded FPGA memory. However, hamming code is not able to cope with multiple upsets in the same coded word. And in the case of the embedded memory, it is very important to protect the cells against multiple bit upsets (MBU) for two main reasons: new SRAM technologies (VDSM) are susceptible to MBU, the scrubbing procedure does not reconfigure (update) the internal memory, and consequently, upsets have a higher probability of accumulating in the memory.

So, a new code is needed to correct all possible double errors. The initial option would be using a Reed-Solomon code with capability to correct two different symbols. But this RS code has more than twice the area and delay overhead of the single symbol correction RS [26], which makes this solution inappropriate for hardware implementation in memory architectures. Previous work has been published on the use of RS code to protect memory [27], however it does not take into account double bit upsets in the same coded word, which is likely to occur in VDSM technologies.

An innovative solution has been developed able to correct all double bit upsets in VDSM memories [28]. This solution combines hamming code and RS code with single symbol correction capability. This technique solves the problem of how to achieve 100% of double fault correction with a low-cost RS code. The hamming code protects the bits between the RS symbols. The number of bits protected by hamming will be the same as the number of symbols protected by Reed-Solomon, so this option does not significantly increases the area overhead.

The triple modular redundancy (TMR) is another SEU mitigation technique. There are many TMR topologies. Each implementation is associated with different area penalties and fault coverage. The system requirements and the architecture must be analyzed in order to correctly choose the most convenient approach.

Now, let's take a look in the CLB flip-flops. They receive the output of the multiplexors that set up the signal path from the LUT in the CLB slice. If a transient fault (SET) occurs in one of the multiplexors, this upset must not be stored in the flip-flops. Consequently, it is not sufficiently reliable to replace the CLB flip-flop by a hardened flip-flop. It is also necessary to insert some kind of fault detection and correction in the input of this flip-flop to filter SETs. The combinational logic does not need to be changed. A possible solution is to combine the temporal latch composed of DICE memory cells, presented in [20] with the TMR approach with refreshing. The final flip-flop shows a high reliability to 1st, 2nd and 3rd order of SEU and SETs, refreshing of SEU and additionally a small impact in the final area because the flip-flops correspond to less than 1% of the total sensitive area. Figure 9 shows this hardened flip-flop topology.

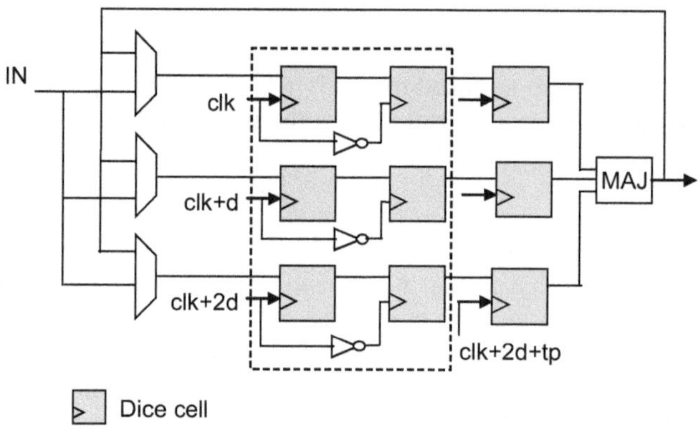

Figure 9. Proposed SEU and SET Hardened flip-flop with refreshing.

4. High-level SEU mitigation techniques for SRAM-based FPGAs based on TMR

The previous section discussed fault-tolerant techniques in the architectural level for SRAM-based FPGAs. Although these solutions can achieve a high reliability, they also present a high cost because they need investment in development, test and fabrication. So far, there are very few FPGA companies that are investing in designing fault-tolerant FPGAs as this market is still focused in only military and space application, which is very small compared to the commercial market. However, because of the technology evolution, applications at the atmosphere and at ground level have been starting to face the effect of neutrons, as mentioned in section 2. As a result, fault-tolerant techniques begin to be necessary in many commercial applications that need some level of reliability.

A less expensive solution is a high-level SEU tolerant technique that can be easily implemented by the user or by the company designers in commercial FPGAs or in parts manufactured by a technology that can avoid latch up and reduce the total ionization dose, as the Virtex® QPRO family [29]. The high-level SEU mitigation technique used nowadays to protect designs synthesized in the Virtex® architecture is mostly based on TMR combined with scrubbing [30].

Triple Modular Redundancy (TMR) is a well-known fault tolerant technique for avoiding errors in integrated circuits. The TMR scheme uses three identical logic blocks performing the same task in tandem with corresponding outputs being compared through majority voters. TMR is especially suitable for protecting designs synthesized in SRAM-based Field Programmable Gate Arrays (FPGAs) [31], due to the peculiar effect of an upset in the user's combinational logic design.

The TMR mitigation scheme uses three identical logic circuits (redundant domain 0, redundant domain 1 and redundant domain 2), synthesized in the FPGA, performing the same task in tandem, with corresponding outputs being compared through a majority vote circuit, as shown in Figure 10. It is important to notice that the use of TMR by itself it is not sufficient to avoid errors in the FPGA, it is mandatory to reload the bitstream constantly, which is called scrubbing. The scrubbing [32] allows a system to repair SEUs in the configuration memory without disrupting its operations. The scrubbing is performed through the Virtex® SelectMAP interface. When the FPGA is in this mode, an external oscillator generates the configuration clock that drives the FPGA and PROM that contains the "gold" bitstream. At each clock cycle new data are available on the PROM data pins.

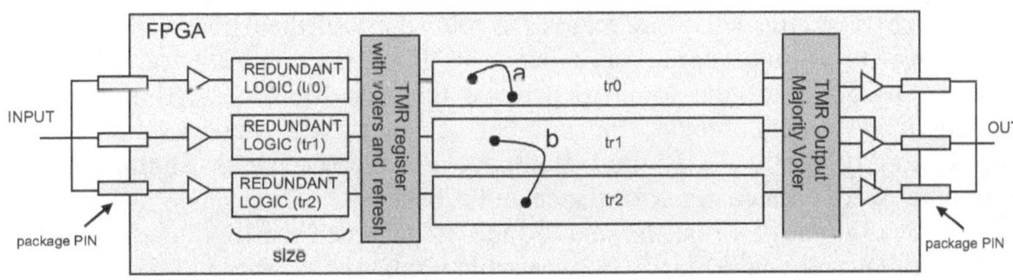

Figure 10. Triple Modular Redundancy for Xilinx FPGAs.

4.1 Solutions to improve TMR reliability

However, TMR and scrubbing do not ensure 100% of reliability in presence of upsets because there are few SEU in the routing that may affect more than one TMR domain, provoking an error in the TMR voter. The probability that an upset in the routing overcomes the TMR is related to the routing density and logic placement. In Figure 11, there are two examples of upsets in the routing. Upset "a" connects two signals from the same redundant domain, which does not generate an error in the TMR output, because the outermost voters will vote the upset effect. However, upset "b" may provoke an error in the TMR output, because it connects two signals from distinct redundant logic blocks affecting two out of three redundant domains of the TMR.

4.2 Solutions based on placement and routing

Dedicated floorplanning for each redundant part of the TMR can reduce the probability of upsets in the routing affecting two or more logic modules, but it may not be sufficient, since placement can be too complex in some cases. Remember that each time it is necessary to include voters, there are connections between the redundant parts (see Figure 10), which make impossible to place the redundant logic parts very far away from each other with no connections at all. One solution is the Reliability-Oriented Place and Route algorithm (RoRA), which is a place and route algorithm for SRAM-based FPGAs able to enforce particular technique in order to enforce every circuit mapped on SRAM-FPGAs against SEUs in their configuration memory cells [33]. Routing duplication can also be a solution to improve reliability in TMR. In [34], it is proposed a method to duplicate the routing locally inside the CLB to avoid problems with open and short circuits provoked by SEUs in the routing.

4.3 Solutions based on voting adjustments

The first voting adjustment was proposed in [35]. In this paper, it is proposed a logic partition in order to add more voter stages in the circuit. If the redundant logic parts tr0, tr1 and tr2 (represented in Fig. 11 after the TMR register with voters and refresh) are partitioned in smaller logic blocks with voters, a connection between signals from distinct redundant parts could be voted by different voters. This logic partition by voters is represented in Figure 3. Notice that now the upset "b" can not provoke an error in the TMR output, which increases the robustness of the TMR in the presence of routing upsets without being of concern to floorplanning. The problem is to evaluate the best size of the logic to achieve the best robustness. If the logic is partitioned in very small blocks, the number of voters will increase dramatically, causing an overly costly TMR implementation. The objective is finding the best partition in terms of area cost, performance and robustness.

The results [35] suggest that there is a trade off between the logic partition of the throughput logic (and consequently between the number of voters) and the number of routing upsets that could provoke an error in the TMR. In contrary to what was expected, a large number of voters does not always mean larger protection against upsets. There is an optimal logic partition for each circuit that can reduce the propagation of the upset effect in the routing. For the case study circuit, the best partition is the medium partition (TMR_p2). This version of the TMR design presents a small sensitivity to routing upsets (0.98%, a four times improvement over normal TMR) and small performance degradation (about 10%) compared to the standard version (not protected).

Another solution is based on changing the logic of the voter when a Selective Triple Modular Redundancy (STMR) approach is used [36]. Figure 12 illustrated the modified majority voter. The Cross-check & Reconfiguration module verifies the accordance in the behavior of the replicated functional modules (Cross-check module), controls the execution of the diagnosis and reconfiguration procedures (Reconfiguration module) for the faulty device.

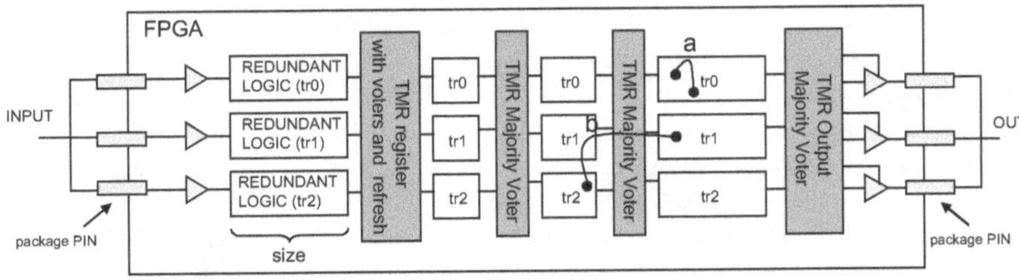

Figure 11. Triple Modular Redundancy (TMR) Scheme with Logic Partition in the FPGA.

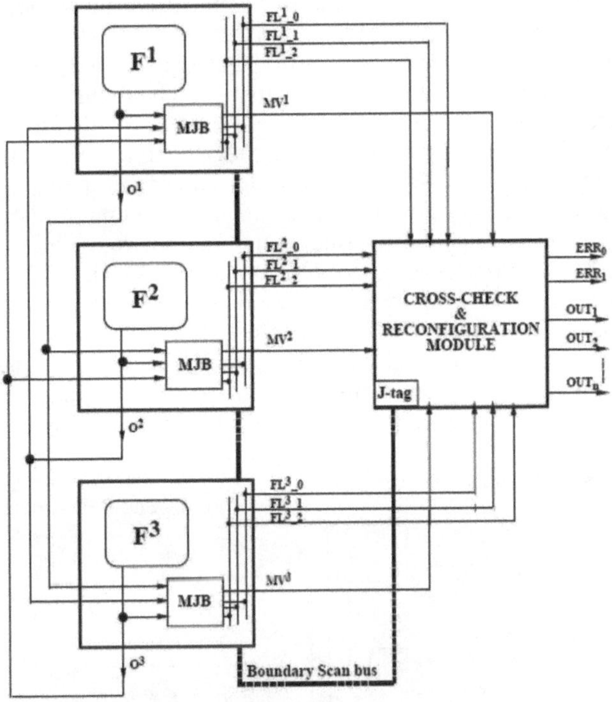

Figure 12. Modified Majority Voter [36].

4.4 Solutions to reduce TMR overhead

The TMR technique is a suitable solution for FPGAs because it provides a full hardware redundancy, including the user's combinational and sequential logic, the routing, and the I/O pads. However, it comes with some penalties because of its full hardware redundancy, such as area, I/O pad limitations and power dissipation. Many applications can accept the limitations of the TMR approach but some cannot. Aiming to reduce the number of pins overhead of a full hardware redundancy implementation (TMR), and at the same time coping with permanent upset effects, we present a new technique based on time and hardware redundancy to protect the user's combinational logic, where the double modular redundancy with comparison (DWC) is combined with a time redundancy upset detection machine, able to detect upsets and to identify the correct value to allow continuous operation. The sequential logic continues to be protected by TMR to avoid accumulation of faults as previously described, since the scrubbing does not change the content of a user's memory cell. However, experiments based on using duplication for both combinational and sequential logic for digital filters have shown high reliability, as it will be discussed further in this paper.

The reliability and the safety of TMR scheme compared to self-checking-based fault-tolerant schemes were discussed in [37]. The experimental results presented that the higher the complexity of the module, the greater the difference in reliability between self-checking and TMR. In summary, the self-checking fault-tolerant scheme can achieve a higher reliability in comparison to the TMR if the self-checking overhead bound of 73% is not exceeded. The idea of using self-checking fault-tolerant scheme can be extended for FPGAs by using the duplication with comparison (DWC) method combined with concurrent error detection (CED) technique that it works as a self-checking. Figure 13 presents the scheme, called hot backup DWC-CED. The CED is able to detect which module is faulty in the presence of an upset, and consequently, there is always a correct value in the output of the scheme, because the mechanism is able to select the correct output out of two.

In the case of SEU detection in SRAM-based FPGAs, the CED must be able to identify permanent faults in the redundant modules. The CED works by finding the property of the analyzed logic block that can help to identify an error in the output in the presence of a permanent fault. There are many methods to implement logic to detect permanent faults, most solutions are based on time or hardware redundancy and they manifest a property of the logic block that is being analyzed.

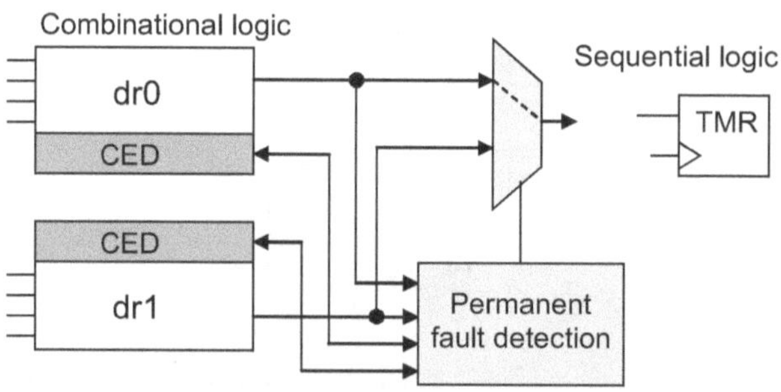

Figure 13. DWC combined with CED scheme.

The CED scheme based on time redundancy recomputes the input operands in two different ways to detect permanent faults. During the first computation at time t0, the operands are used directly in the combinational block and the result is stored for further comparison. During the second computation at time t0+d, the operands are modified, prior to use, in such a way that errors resulting from permanent faults in the combinational logic are different in the first calculation than in the second and can be detected when results are compared. These modifications are seen as encode and decode processes and they depend on the characteristics of the logic block. The scheme is presented in Figure 14.

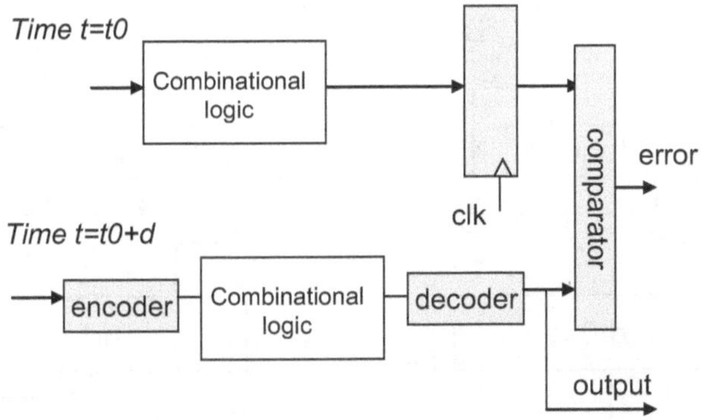

Figure 14. Time redundancy for permanent fault detection.

Figure 15 shows the scheme proposed for an arithmetic module, in the present case study: a multiplier. There are two multiplier modules: mult_dr0 and mult_dr1. There are multiplexors at the output able to provide normal or shifted operands. The output computed from the normal operands is always stored in a sample register, one for each module. Each output goes directly to the input of the user's TMR register. Module dr0 connects to register tr0 and module dr1 connects to register tr1. Register tr2 will receive the module that does not have any fault. By default, the circuit starts passing the module dr0. A comparator at the output of register dr0 and dr1 indicates an output mismatch (Hc). If Hc=0, no error is found and the circuit will continue to operate normally. If Hc=1, an error is characterized and the operands need to be re-computed using the RESO method to detect which module has the permanent fault. The detection takes one clock cycle.

In the case of a registered output, each output goes directly to the input of the user's TMR register. Figure 16 illustrates the logic scheme. Module dr0 connects to register tr0 and module dr1 connects to register tr1. While the circuit performs the detection, the user's TMR register holds its previous value. While the circuit performs the detection, the TMR register holds its previous value. When the faulty free module is found, register tr2 receives the output of this module and it will continue to receive this output until the next chip reconfiguration (fault correction). By default, the circuit starts passing the module dr0. In the case of a non-registered output, the signals can be driven directly to the next combinational module or to the I/O pads, as shown in Figure 17.

Let us consider two different fault situations when the output is saved in a TMR register. In one, the fault occurs in module dr0 (Mult_dr0). Hc indicates that there is an output mismatch; Tc0 indicates that module dr0 is faulty and Tc1 indicates that dr1 is fault free. This analysis takes one clock cycle. Consequently, the permanent fault detection block selects dr1 for the tr2 input. Note that the value stored in the user's TMR register is held for one cycle while the scheme identifies the faulty free module. In the second case, a fault occurs in the module dr1 (Mult_dr1), similar to the previous example, Hc indicates that there is an output mismatch; Tc0 indicates that module dr0 is fault free and Tc1 indicates that dr1 is faulty. The permanent fault detection block selects dr0 for the tr2 input.

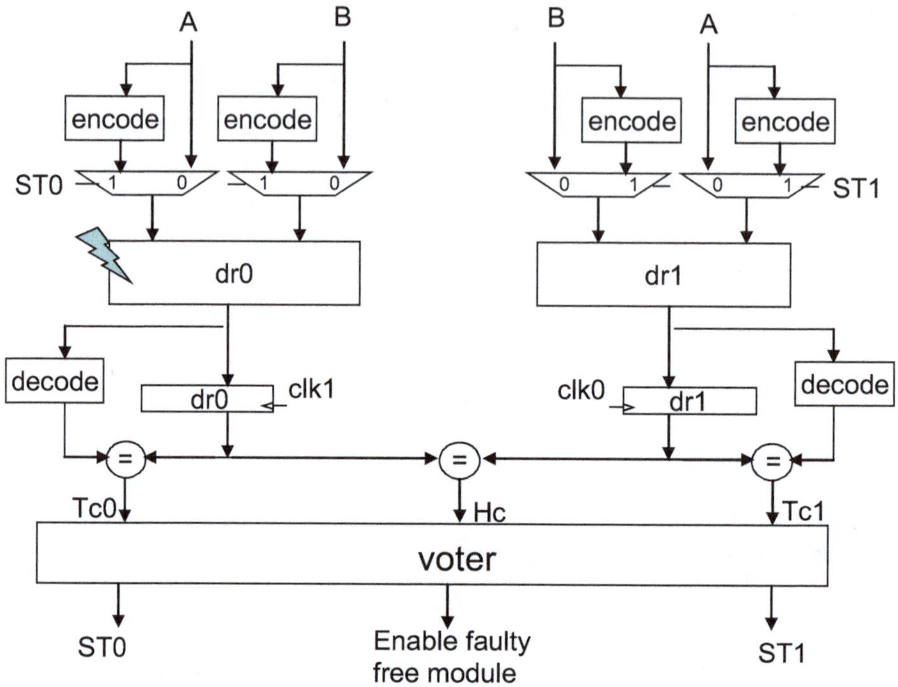

Figure 15. Fault tolerant technique based on DWC combined with CED for SRAM-based FPGAs.

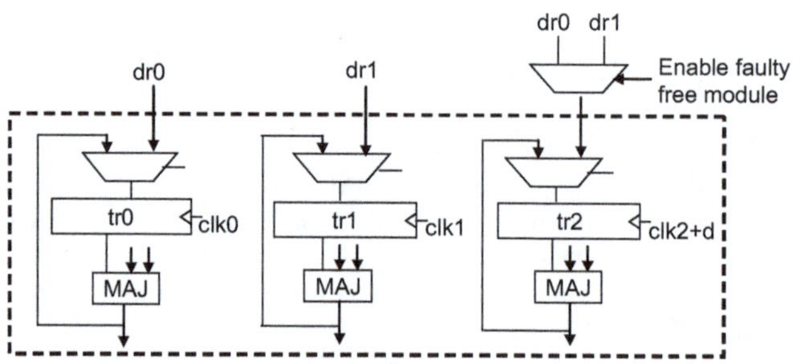

Figure 16. Examples of implementations with the combinational output registered.

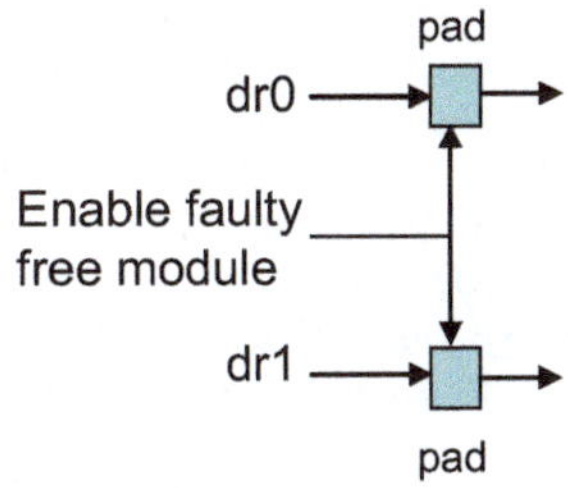

Figure 17. Examples of implementations with the combinational at the pads.

According to the user's application requirements, the designer will be able to choose between a full hardware redundancy implementation (TMR) or a mixed solution where duplication with comparison is combined to concurrent error detection to reduce pins and power dissipation in the interface, as well as area, as shown in previous examples. Figure 18 shows some implementations combining TMR and DWC with time redundancy. It is possible to use this new technique only in the interface of the FPGA, in this way reducing pins, or along the design to reduce the number of I/O pads and also area for large combinational circuits,

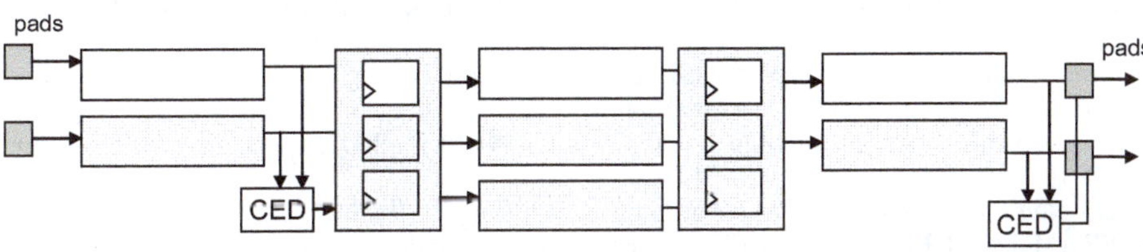

Figure 18. Implementation of the fault tolerant technique based on DWC combined with CED and TMR for SRAM-based FPGAs.

References

[1] Nasa. Radiation Effects on Digital Systems. USA, 2002. Available at: <radhome. gsfc.nasa.gov/top.htm>. Visited on January, 2006.

[2] Sia Semiconductor Industry Association. The National Technology Roadmap for Semiconductors. USA, 1994.

[3] Johnston, A. Scaling and Technology Issues for Soft Error Rates. In Research Conference On Reliability, 4., 2000. Proceedings... Palo Alto: Stanford University, 2000.

[4] O'Bryan, M. et al. Current single event effects and radiation damage results for candidate spacecraft electronics. In IEEE Radiation Effects Data Workshop, 2002. Proceedings... [S.l.]: IEEE Computer Society, 2002. p. 82-105.

[5] Barth, J. Applying Computer Simulation Tools to Radiation Effects Problems. In IEEE Nuclear Space Radiation Effects Conference, NSREC, 1997. Proceedings… [S.l.]: IEEE Computer Society, 1997. p. 1-83.

[6] Normand, E. Single event upset at ground level. IEEE Transactions on Nuclear Science, New York, v.43, n.6, p. 2742 -2750, Dec. 1996.

[7] Dupont, E.; Nicolaidis, M.; Rohr, P. Embedded robustness IPs for transient-error-free ICs. IEEE Design & Test of Computers, New York, v.19, n.3, p. 54-68, May-June 2002.
[8] Alexandrescu, D.; Anghel, L.; Nicolaidis, M. New methods for evaluating the impact of single event transients in VDSM ICs. In: IEEE International Symposium On Defect And Fault Tolerance in VLSI Systems Workshop, DFT, 17., 2002. Proceedings… [S.l.]: IEEE Computer Society, 2002. p. 99-107.
[9] Leavy, J. et al. Upset due to a single particle caused propagated transient in a bulk CMOS microprocessor. IEEE Transactions on Nuclear Science, New York, v.38, n.6, p. 1493-1499, Dec. 1991.
[10] Hass, J. Probabilistic Estimates of Upset Caused by Single Event Transients. In Nasa Symposium on VLSI Design, 8., 1999. Proceedings... [S.l.: s.n.], 1999.
[11] Guntzel, J.. Functional Timing Analysis of VLSI Circuits Containing Complex Gates, Doctoral Thesis, Instituto de Informatica, UFRGS, Porto Alegre, Brazil, 2001.
[12] Xilinx, Inc. Virtex®™ 2.5 V Field Programmable Gate Arrays: Datasheet DS003. USA, 2000.
[13] Rebaudengo, M.; Reorda, M.S.; Violante, M. Simulation-based Analysis of SEU effects of SRAM-based FPGAs. In International Workshop On Field-Programmable Logic And Applications, FPL, 2002. Proceedings... [S.l.]: IEEE Computer Society, 2002. p. 607-615.
[14] Caffrey, M.; Graham, P.; Johnson, E. Single Event Upset in SRAM FPGAs. In Military and Aerospace Applications of Programmable Logic Conference, MAPLD, 2002. Proceedings... [S.l.: s.n.], 2002.
[15] Fuller, E. et al. Radiation test results of the Virtex FPGA and ZBT SRAM for Space Based Reconfigurable Computing. In International Conference on Military and Aerospace Applications of Programmable Logic Devices, MAPLD, 2002. Proceedings... [S.l.: s.n.], 2002.
[16] Carmichael, C.; Fuller, E.; Fabula, J.; Lima, F. Proton Testing of SEU Mitigation Methods for the Virtex® FPGA. In International Conference on Military and Aerospace Applications of Programmable Logic Devices, MAPLD, 2001. Proceedings... [S.l.: s.n.], 2001.
[17] Ohlsson, M.; Dyreklev, P.; Johansson, K.; Alfke, P. Neutron Single Event Upsets in SRAM based FPGAs. In IEEE Nuclear Space Radiation Effects Conference, NSREC, 1998. Proceedings… [S.l.]: IEEE Computer Society, 1998.
[18] Canaris, J.; Whitaker, S. Circuit techniques for the radiation environment of space. In Custom Integrated Circuits Conference, 1995. Proceedings... [S.l.]: IEEE Computer Society, 1995, p. 77-80.
[19] Anghel, L.; Alexandrescu, D.; Nicolaidis, M. Evaluation of a soft error tolerance technique based on time and/or space redundancy. In Symposium on Integrated Circuits and Systems Design, SBCCI, 13., 2000. Proceedings… Los Alamitos : IEEE Computer Society, 2000. p. 237-242.
[20] Mavis, D.; Eaton, P. SEU and SET Mitigation Techniques for FPGA Circuit and Configuration Bit Storage Design. In International Conference on Military and Aerospace Applications of Programmable Logic Devices, MAPLD, 2000. Proceedings... [S.l.: s.n.], 2000.
[21] Rockett, L. R. An SEU-hardened CMOS data latch design. IEEE Transactions on Nuclear Science, New York, v.35, n.6, p. 1682-1687, Dec. 1988.
[22] Whitaker, S.; Canaris, J.; Liu, K. SEU hardened memory cells for a CCSDS Reed-Solomon encoder. IEEE Transactions on Nuclear Science, New York, v.38, n.6, p. 1471-1477, Dec. 1991.
[23] Calin, T.; Nicolaidis, M.; Velazco, R. Upset hardened memory design for submicron CMOS technology. IEEE Transactions on Nuclear Science, New York, v.43, n.6, p. 2874 -2878, Dec. 1996.
[24] Weaver, H.; et al. An SEU Tolerant Memory Cell Derived from Fundamental Studies of SEU Mechanisms in SRAM. IEEE Transactions on Nuclear Science, New York, v.34, n.6, Dec. 1987.

[25] Rockett, L. R. A design based on proven concepts of an SEU-immune CMOS configurable data cell for reprogrammable FPGAs. Microelectronics Journal, Elsevier, v.32, p. 99-111, 2001.

[26] Houghton, A. D. The Engineer's Error Coding Handbook. London: Chapman & Hall, 1997.

[27] Redinbo, G.; Napolitano, L.; Andaleon, D. Multi-bit Correcting Data Interface for Fault-Tolerant Systems. IEEE Transactions on Computers, New York, v.42, n.4, p. 433-446, Apr. 1993.

[28] Neuberger, G.; Lima, F.; Carro, L.; Reis, R. A Multiple Bit Upset Tolerant SRAM Memory. Transactions on Design Automation of Electronic Systems, TODAES, New York, v.8, n.4, Oct. 2003.

[29] Xilinx, Inc. QPRO™Virtex®™ 2.5V Radiation Hardened FPGAs: Application Notes 151. USA, 2000.

[30] Xilinx Inc. Virtex® Series Configuration Architecture User Guide: Application Notes 151. USA, 2000.

[31] Carmichael, C. Triple Module Redundancy Design Techniques for Virtex® Series FPGA: Application Notes 197. San Jose, USA: Xilinx, 2000.

[32] Lima, F.; Carmichael, C.; Fabula, J.; Padovani, R.; Reis, R. A fault injection analysis of Virtex FPGA TMR design methodology. In European Conference on Radiation and Its Effects on Components and Systems, RADECS, 2001. Proceedings... [S.l.]: IEEE Computer Society, 2001b. p. 275-282.

[33] M. Sonza Reorda, L. Sterpone, M. Violante, "Multiple errors provoked by SEUs in the FPGA configuration memory: a possible solution", 10th IEEE European Test Symposium, 2005.

[34] Kastensmidt, F. L.; Kinzel Filho, C.; Carro, Luigi . Improving Reliability of SRAM Based FPGAs by Inserting Redundant Routing. IEEE Transactions on Nuclear Science, New York, v. 53, n. 4, 2006.

[35] Kastensmidt, F. L., Carro, Luigi, Sterpone, L., Reorda, M. On the Optimal Design of Triple Modular Redundancy Logic for SRAM-based FPGAs In: Proceedings in Design Automation and Test in Europe (DATE). New York: IEEE, 2005.

[36] D'Angelo, S.; Metra, C.; Pastore, S.; Pogutz, A.; Sechi, G.R. Fault-tolerant voting mechanism and recovery scheme for TMR FPGA-based systems. In: IEEE International Symposium on Defect and Fault Tolerance in VLSI Systems, 1998. p. 233-240.

[37] Lubaszewski, M.; Courtois, B.; A reliable fail-safe system, IEEE Transactions on Computers, Volume: 47 Issue: 2, Feb. 1998, P. 236-241.

Automatic Tools for Design Hardening

Celia López-Ongil, Luis Entrena, Mario García-Valderas, Marta Portela-García

Electronic Technology Department. Carlos III University of Madrid
Avda. de la Universidad, 30. 28911 Leganés (Madrid) Spain
{celia, entrena, mgvalder, mportela}@ing.uc3m.es

Abstract. Historically, there has been a lack of CAD tools for the design of on-line testable circuits. As a consequence, the design of on-line testable circuits is currently being made manually to a large extent. In this paper we detailed an academic tool for the automatic insertion of fault-tolerant structures in designs described at Register Transfer Level (RTL). With this tool, a fault-tolerant version of the design can be automatically produced according to the user specifications. The resulting fault-tolerant design is also described at RTL and can be simulated and synthesized with commercial tools. Examples are shown to demonstrate the capabilities of this approach.

1. Introduction

Traditionally, on-line testing technology has been restricted to few application domains. The area and performance penalties introduced by on-line testing were only accepted in order to meet high reliability requirements in safety critical applications. The relatively small number of such applications did not make attractive for CAD vendors the development of tools specific to the design of on-line testable circuits. As a matter of fact, on-line testable circuits are being designed manually to a large extent. This requires large design efforts that affect negatively design productivity and time-to-market.

With the widespread use of Hardware Description Languages (HDLs), most designs are usually started at the Register Transfer Level (RTL) supported by these languages and synthesis tools. Once the design has been validated by simulation, logic synthesis and place and route are generally used to obtain the lower level models of the design in an automatic way. The increase in the abstraction level, supported by automatic CAD tools, has allowed unprecedented productivity and shorter development times in the design of ICs.

The design of an on-line testable circuit is usually accomplished by introducing fault-tolerant structures in a previously designed circuit. Within the current HDL-based methodologies, the insertion of fault-tolerant structures is usually performed by manually modifying the HDL code in order to insert hardware redundancy, information redundancy or time redundancy at critical points in the circuit. Then, the modified, fault-tolerant design obtained follows a similar design flow consisting in automatic logic synthesis and place and route.

R. Velazco et al. (eds.), Radiation Effects on Embedded Systems, 183–200.

There are several reasons why designers usually prefer performing fault-tolerant modifications at the RT level. First of all, this is a natural choice, because most designers work mainly at this level nowadays. This choice also allows the use of behavioral simulation, in order to validate the fault-tolerant design. On the other hand, the application of many fault-tolerant techniques requires a synthesis step, that cannot be accomplished by simply substituting some logic cells by their testable equivalents (as in the case of structured DFT techniques) except for some trivial cases [1] [2]. With a clever use of commercial synthesis tools, many fault-tolerant structures can be correctly synthesized, thus saving an important amount of design time.

Whenever needed, the design can be further modified at lower abstraction levels. Several self-checking designs of datapath modules such as adders, ALUs, shifters, registers, multipliers, etc. have been proposed (e.g., [3][4]). Macroblock generators for self-checking data path modules have been implemented and integrated in a CAD framework [5]. Also, synthesis tools for fault secure multilevel logic functions and FSMs that are based on parity codes or unordered codes have been proposed. Although these methods may give cost effective solutions in many cases, further investigations are still needed in order to achieve synthesis of low cost multilevel fault secure circuits [6].

In the physical domain, on-line monitoring of physical parameters such as current dissipation, temperature, radiation dose, etc. by using built-in sensors is used to detect failures. In summary, these techniques provide good solutions for some particular components. The availability of these components for reuse is a very valuable complement for the design of fault-tolerant circuits, but still a great deal of manual redesign is required at the RT level, particularly for control logic.

In this chapter, a solution for automatic insertion of fault-tolerant structures in designs described at RT level is presented. This solution, named FTI tool, provides an automatic way to substitute current manual practices in the application of hardware redundancy and information redundancy techniques in fault-tolerant designs. The automation allows to increase the design productivity and to reduce the chance of errors. The resulting fault-tolerant design is also obtained in an RTL description which can be further simulated and synthesized with commercial tools.

The FTI tool has been developed in project AMATISTA (IST #11762) along with a fault injection and simulation tool, FTV, which allows evaluating the level of fault-tolerance achieved in the design. Throughout the iteration with the insertion of fault-tolerant structures and the evaluation of the fault-tolerance achieved, a much deeper exploration of the design space is enabled early in the design cycle.

The remaining of the chapter is as follows. Section 2 describes the general schema of the fault-tolerance insertion tool. Section 3 shows the results obtained with hardware redundancy insertion. Section 4 shows the results obtained with information redundancy insertion. Section 5 presents automatic mechanisms for the insertion of error recovery actions. Finally, section 5 presents the conclusions of this work.

2. Automatic hardening of RTL designs

The main objective of FTI tool is the automatic generation of fault tolerant circuits departing from an existing design and a set of hardening specifications. Besides, FTI tool can be included in current design flow of top-down methodologies with no impact in design teams. The use of FTV tool for evaluating the fault tolerance achieved will help designers in the hardening process [7].

The structure of FTI tool is shown in Figure 1. Fault Tolerance Insertion Procedures work with two libraries, FT-library contains the set of fault tolerant structures and techniques provided together with FTI and WORK-library contains users' design to be hardened. Throughout a user interface, the user may apply fault-tolerance techniques at selected points in the design in a step-by-step fashion.

The main input to FTI tool is the original design, which is described at RT level with a hardware description language. Although FTI tool works with RTL design descriptions, it performs an elaboration for synthesis of the original design description in order to identify the hardware resources. Therefore, hardening process is always hardware-oriented.

The output produced by FTI is the hardened version of the input design, also described at RT level with an HDL. There are two main advantages of maintaining this format. First, it is possible to apply the same tests for functional validation for the original and the hardened versions of the circuit. Second, fault tolerance is evaluated in a high level of abstraction. Finally, once the hardened version of the circuit is validated, it can be processed in the next steps of a typical design flow.

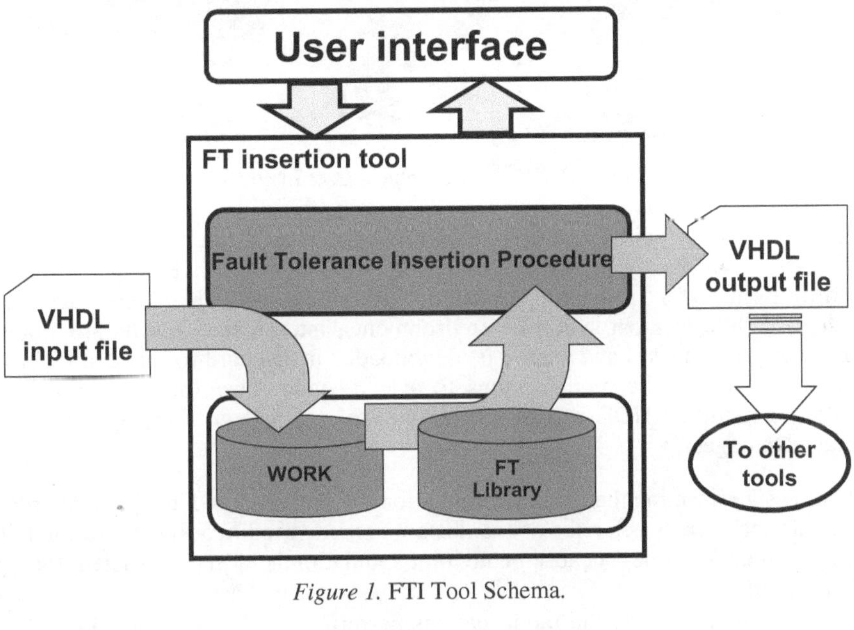

Figure 1. FTI Tool Schema.

In order to help designers in the task of hardening, the graphical user interface guides users through different steps, Figure 2. At each step a design modification is performed, users must specify the elements to be hardened and the technique to be applied. The descriptions of components needed for the various fault-tolerance techniques to be applied are available in the component library named FT-library. This FT-library is provided with the tool and could also be enhanced with user's FT components. Examples of these components are encoders, decoders, checkers, majority voters, comparators, etc. Self-checking designs of particular modules (e.g., typical data path components) can be included as well. These components are inserted in the design as required for the fault-tolerance techniques to be applied. FTI tool is intended to deal with hierarchical designs, where only critical parts are modified and, finally a complete description of the design is generated.

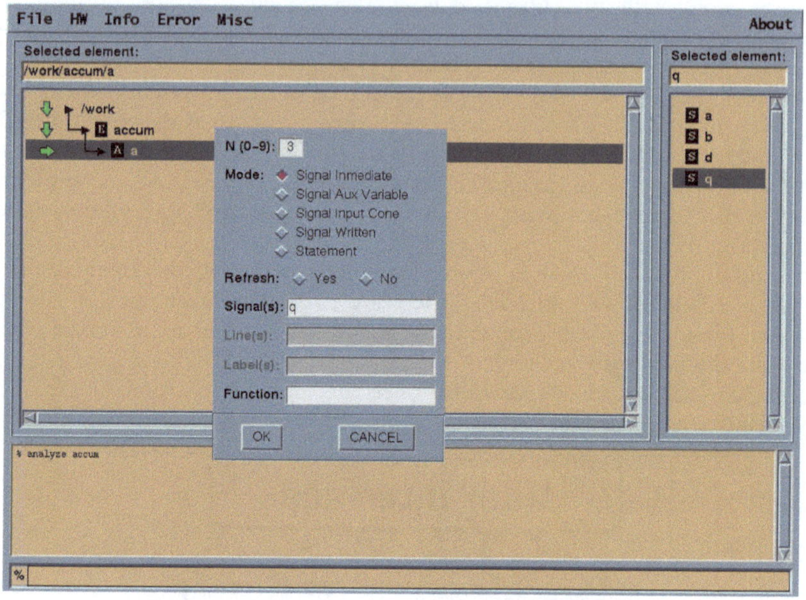

Figure 2. FTI Tool Graphical User Interface.

FTI tool works with a design representation based on a database format. Modifications and analyses are performed on this database, AIRE CE, which is a standard format for processing designs described with hardware description languages. While in most synthesis environments, intermediate results are stored in internal design databases and cannot be downloaded in standard formats, in FTI tool it is possible to revert the modifications to a HDL code. Therefore, it is possible to download a modified HDL version of the fault-tolerant design. This option provides several advantages:

- It allows performing behavioral simulation of the modified design, in order to validate the correct behavior of the design after the modifications performed. This saves simulation time, because behavioral simulation is far more efficient than gate level simulation
- The user can recognize the modifications performed in the code, providing higher user confidence in the modification process. Manual modifications are still possible, if needed. During the user learning steps, the user may compare the results obtained with the automatic and manual modification process.
- The proposed approach makes automatic insertion of fault-tolerant structures compatible with any design environment supporting an HDL input. Considering that one of the main reasons for the lack of CAD tools in the area of on-line testing has been the relatively small number of users, this is an interesting advantage as it can reach the entire design community independently of the particular design environment used.

The proposed approach allows making full use of the synthesis capabilities provided by commercial tools. Components that have been designed manually according to specific techniques can be incorporated in FT-library and inserted under the general insertion mechanisms of the tool.

The Fault-Tolerance Insertion tool (FTI) has been developed using the source code analyzer Tauri [8] and the AIRE CE format [9]. This tool provides an open analysis and elaboration database for VHDL and Verilog designs according to the standards IEEE 1076 and IEEE 1364, respectively. In the sequel, we will use VHDL for the examples. However, note that the proposed techniques can be easily extended to Verilog.

3. Automatic insertion of hardware redundancy

Hardware redundancy is the addition of extra hardware for the purpose of either detecting or tolerating faults. Hardware redundancy techniques achieve fault tolerance by replicating hardware components of the circuit. The outputs of the replicated components are then resolved to produce a single output, which may be coded to cover for possible faults in the output resolution logic. The type and characteristics of the resolution function depends on the particular hardware redundancy technique used.[1] For instance, the resolution function for N-Modular Redundancy (NMR) implements a majority voter, while for Duplication With Comparison (DWC) the resolution function implements a comparator of the two replicas and assigns one of the two replicas to the output.

Passive hardware redundancy techniques provide error-free outputs, as they are able to mask the occurrence of faults. In the FTI tool, passive technique available is n-modular redundancy (NMR).

Active hardware redundancy techniques do not provide error-free outputs, but provide an error signal, which is activated in case an error is detected. Error recovery actions are then driven by this signal. In FTI tool, active techniques available are duplication with comparison (DWC) and standby sparing. It is also possible to apply a hybrid technique which mask errors in the outputs and provides an error signal: n-modular redundancy with spares.

With respect to the testability of modified circuit, it must be taken into account that redundancy will worsen testability, but active techniques are generating an error signal that could help in the test of the circuit.

From the previous definitions, any transformation based on a hardware redundancy technique is determined by three main aspects:

- The replication target, which is defined as the piece of hardware that is to be replicated
- The particular hardware redundancy technique to be applied, which mainly defines the output resolution function
- The error recovery actions which are required to recover from a detected error

---

[1] This resolution function must not be confounded with HDL resolution functions that solve wired logic

The first two aspects will be described in the following sections. The error recovery actions are common to information redundancy techniques and will be described in section 6, after the insertion of information redundancy.

3.1 Target selection and replication

The target can be specified in terms of VHDL objects (signals, variables and ports) and statements. The associated hardware is then identified and modified in the elaborated model, which provides a RT-level view of the design.

In this model, the target may be a set of objects along with a selected portion of its input cone. Identifying the input cone of a signal can be easily performed in the elaborated model by tracing back the signal up to the primary inputs. A typical case in RTL consists in selecting the input cone of the signal up to the preceding registered signals. This case can be recursively extended to larger portions of the input cone.

Replication of the target consists in identifying the target, by marking the nodes in the selected portion of the input cone of the signal, and inserting copies of the marked nodes (the replicas) in the elaborated model, along with the new object declarations associated to the replicated nodes.

The VHDL code corresponding to the replicas is then generated by tracing back to the analyzed model. This code needs to be modified in order to cover some hardware structures that are not described explicitly in VHDL, such as default assignments to registers. In this case, auxiliary variables are inserted to model explicitly the required behavior.

3.2 Resolution function

The output resolution function must be inserted in order to obtain a single, fault-tolerant output from the outputs of the replicated components. For active hardware techniques, such as Duplication With Comparison (DWC), the resolution function also provides an error signal, which is activated in case of error.

The resolution function is inserted by means of a call to a VHDL function or procedure included in the FT-library that corresponds to the required fault-tolerant technique applied.

3.3 Example

We will illustrate the replication approach by showing the result of applying it to one example. The initial example is shown in Code 1.

Figure 3 shows the hardware inferred from this code (elaborated model) and the result after duplicating the register modeled by signal p. The duplicated signals are labeled with the suffix "_FT0" and "_FT1". A resolution function is applied afterwards to determine the value of the target signal p. In this example, the resolution function corresponds to DWC technique and is modeled by a VHDL procedure that outputs the value of **p** and an error signal. The resulting VHDL code is shown in Code 2. Modifications introduced automatically by the FTI tool are shown in bold.

Code 1: VHDL description of Example_1

```
      architecture initial of example_1 is
      ...
      begin
      ...
      P1: process(clk, reset)
      begin
         if reset = '0' then
           data1 <= "000";
           data2 <= "000";
         elsif rising_edge(clk) then
           data1 <= a;
           data2 <= b;
         end if;
      end process;

      d <= data1 or data2;

      P2: process(clk, reset)
      begin
         if reset = '0' then
           p <= "000";
         elsif rising_edge(clk) then
           if enable = '1' then
             if load = '1' then
               p <= d;
             else
               p <= p+'1';
             end if;
           end if;
         end if;
      end process;
      ...
end initial;
```

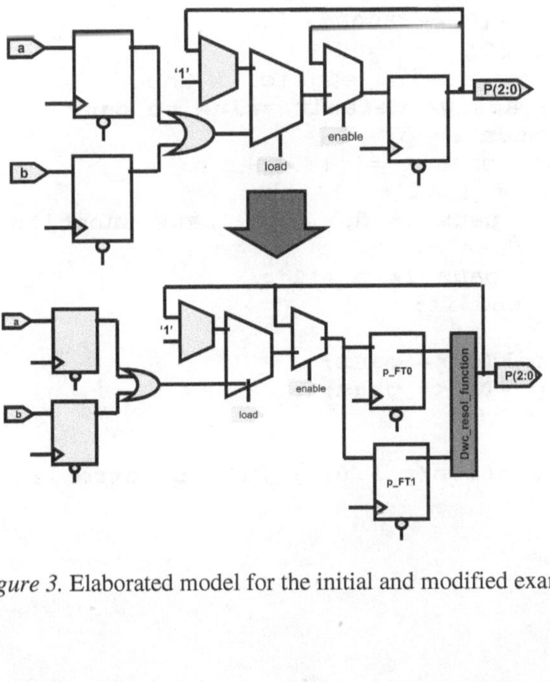

Figure 3. Elaborated model for the initial and modified example_1.

Note that the feedback input to the register is implicitly defined in VHDL. Just a replication of the statements where signal p is assigned would produce a replication of the feedback loop as well. The solution to replicate exactly the register consists in using an intermediate variable (paux) that models the input to the register and then assigning the replicas to the intermediate variable.

Code 2. Example_1 after duplicating register p

```
-- Result after duplicating register p
architecture code2 of example_1 is
...
begin
...
P1: process(clk, reset)
begin
          if reset = '0' then
           data1 <= "000";
           data2 <= "000";
          elsif rising_edge(clk) then
           data1 <= a;
           data2 <= b;
          end if;
end process;

d <= data1 or data2;

P2: process(clk, reset)
          -- create auxiliary variable
          variable paux:std_logic_vector(2 downto 0);
begin
          if reset = '0' then
           p_FT0 <= "000";
           p_FT1 <= "000";
          elsif rising_edge(clk) then
          -- assign default value to paux
           paux := p;
           if enable = '1' then
            if load = '1' then
             paux := d;     -- paux substitutes p
            else
             paux := p +'1';
            end if;
           end if;
           p_FT0 <= paux;
           p_FT1 <= paux;
          end if;
end process;
DWC_resol_function(p_FT0, p_FT1, p, error);
...
end code2;
```

Figure 4 shows the result of duplicating signal p along with its input cone up to the preceding registered signals (data1, data2). The code generated in this case is shown in Code 3

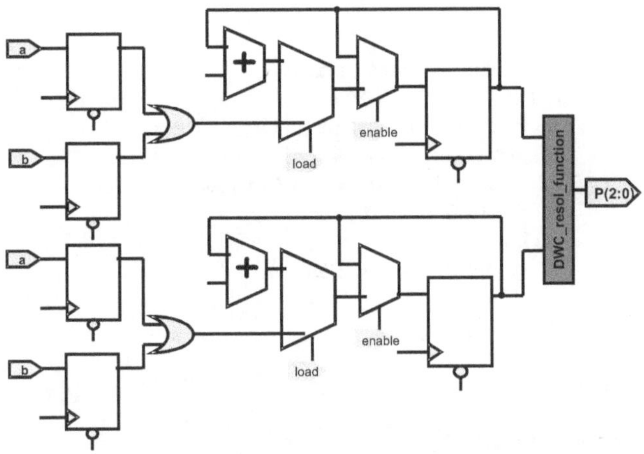

Figure 4. Elaborated model after replicating *p* and its input cone.

Code 3. Example after replicating p and its input cone

```
-- Result after duplicating register p
-- along with its input cone
architecture code3 of example is
...
begin
...
process(clk, reset)
begin
        if reset = '0' then
         data1 <= "000";
         data2 <= "000";
        elsif rising_edge(clk) then
         data1 <= a;
         data2 <= b;
        end if;
end process;

data1_FT0 <= data1;
data1_FT1 <= data1;
data2_FT0 <= data2;
data2_FT1 <= data2;

enable_FT0 <= enable;
enable_FT1 <= enable;
load_FT0 <= load;
load_FT1 <= load;
```

```
d_FT0 <= data1_FT0 or data2_FT0;
d_FT1 <= data1_FT1 or data2_FT1;

process(clk, reset)
begin
          if reset = '0' then
            p_FT0 <= "000";
          elsif rising_edge(clk) then
            if enable_FT0 = '1' then
              if load_FT0 = '1' then
                p_FT0 <= d_FT0;
              else
                p_FT0 <= p_FT0 +'1';
              end if;
            end if;
          end if;
end process;

process(clk, reset)
begin
          if reset = '0' then
            p_FT1 <= "000";
          elsif rising_edge(clk) then
            if enable_FT1 = '1' then
              if load_FT1 = '1' then
                p_FT1 <= d_FT1;
              else
                p_FT1 <= p_FT1 +'1';
              end if;
            end if;
          end if;
end process;
DWC_resol_function(p_FT0, p_FT1, p, error);
...
end code3;
```

4. Automatic insertion of information redundancy

Information redundancy techniques achieve fault-tolerance by using redundant codes for data. The result of adding information redundancy is that a portion of dataflow works with encoded data. Therefore, a set of operations must be provided in order to encode data at the beginning of the selected portion of dataflow, to decode data at the end of the selected portion of data flow, as well as to modify buses and operators for the required data code.

The encoding of a portion of dataflow can be obtained through a composition of the following basic operations:

- Insert an encoder at a selected point
- Insert a decoder/checker at a selected point
- Substitute an existing operator by the equivalent operator that operates on coded data
- Extend the size of selected data to accommodate the extra bits required by the redundant code used

The insertion of an encoder has the effect of generating a new signal (the encoded signal), which will substitute the original unencoded signal for all or some of its fan-outs. Analogously, the insertion of a decoder has the effect of generating a new signal (the decoded signal), which will substitute the original encoded signal for all or some of its fan-outs. The insertion of a checker has the effect of generating a new signal (the error signal), which will be further used to drive some error recovery actions. Note that for non-separable codes, decoding and checking operations are usually performed jointly by a decoder/checker in order to save hardware.

The insertion of a encoder, decoder or checker is implemented by inserting a new object declaration (the encoded, decoded or error signal, depending on the case) with the appropriate size, inserting the appropriate function (encoder, decoder or checker function) in between the target signal and the new signal, and substitute the target signal with the new signal for all or some of the fan-outs of the target signal.

Operator substitution consists in substituting the operator by another operator available in the FT-library. Extending the data size consists in extending the range of the declaration of the target object with the extra bits required for the particular code used.

By properly using these operations, a portion of dataflow can be made to work with a redundant code. In addition, a checker can be used in order to verify the coherence of the data after the application of these operations in order to detect possible mistakes. Such a checker validates the following properties:

- The type and range of any object must be consistent with the values assigned to the object. This property is checked during analysis in the original circuit and is part of the analyzer. The check is now extended to the modified portions of the design.
- Signals that convey encoded (unencoded) data should not be assigned to signals that convey unencoded (encoded) data, except through the appropriate element (decoder or encoder). This can be verified in the elaborated model by marking the portion of the circuit that uses encoded data.

In the proposed tool, error detecting codes available are parity codes and unordered codes (Berger and 2-out-of-n codes). Also, error correcting codes are provided, such as Hamming error codes.

4.1 Examples

Consider again the example shown in Code 1. Figure 5 shows the hardware modified after encoding just the register modeled by signal p. The code produced after the modification is shown in Code 4 (only for process P2). The encode, decode and check functions are general and can be bound to the appropriate functions in the FT-library for the particular code used. In the case of a non-separable code, the decode and check function calls will be substituted by a single procedure call to the corresponding decoder/checker.

Note that the auxiliary variable paux is again used in order to model explicitly the feedback to the register.

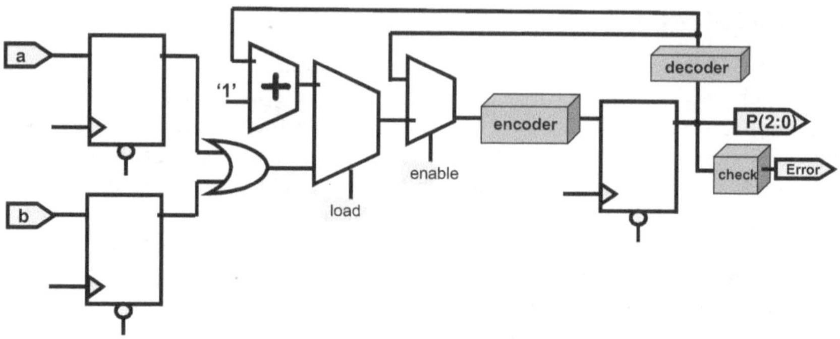

Figure 5. Elaborated model after encoding register p.

Code 4 Example_1 after encoding register p

```
architecture code4 of example
          signal p: std_logic_vector(extended_size);
          signal dec_p: std_logic_vector(2 downto 0);
begin
...
P2: process(clk, reset)
          variable paux:std_logic_vector(2 downto 0);
begin
          if reset = '0' then
           p <= "000";
          elsif rising_edge(clk) then
           paux := dec_p;
           if enable = '1' then
            if load = '1' then
             paux := d;
            else
             paux := paux +'1';
            end if;
           end if;
           p <= encode(paux);
          end if;
end process;
error <= check(p);
decoded_p <= decode(p)
end code4;
```

Bla, bla, bla

Code 5. VHDL description of Example_2: ACCUM

```
library IEEE;
use IEEE.std_logic_1164.all;
use IEEE.std_logic_unsigned.all;

entity ACCUM is
  port(
    signal Clk       : in  std_logic;
    signal Reset     : in  std_logic;
    signal En        : in  std_logic;
    signal Sel       : in  std_logic;
    signal AD        : in  std_logic_vector(7 downto 0);
    signal BD        : in  std_logic_vector(7 downto 0);
    signal S         : out std_logic_vector(7 downto 0));
end ACCUM;

architecture A_ACCUM of ACCUM is
    signal A         : std_logic;
    signal B         : std_logic;
    signal D         : std_logic;
    signal Q         : std_logic;
begin
  REG_A: process(Clk, Reset)
  begin
    if Reset = '0' then
      A <= (others => '0');
    elsif Clk'event and Clk = '1' then
      if En = '1' then
        A <= AD;
      end if;
    end if;
  end process REG_A;

  REG_R: process(Clk, Reset)
  begin
    if Reset = '0' then
      B <= (others => '0');
    elsif Clk'event and Clk = '1' then
      if En = '1' then
        B <= BD;
      end if;
    end if;
  end process REG_B;

  D <= A when Sel = '1' else B;

  ACCUM_P: process(Clk, Reset)
  begin
    if Reset = '0' then
      Q <= (others => '0');
    elsif Clk'event and Clk = '1' then
      if En = '1' then
```

```
          Q <= Q + D;
        end if;
      end if;
    end process ACCUM_P;
  end A_ACCUM;
```

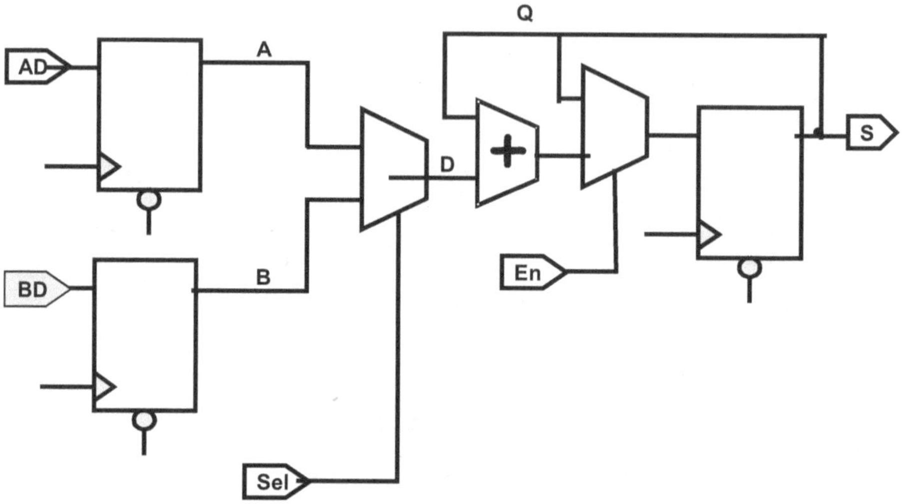

Figure 6. Elaborated model of ACCUM example.

Code 6. Example_2: ACCUM after parity encoding of the whole block and decoding output S

```
library IEEE;
use IEEE.std_logic_1164.all;
use IEEE.std_logic_unsigned.all;
use WORK.parity.all;
use WORK.fti_attr.all;

entity ACCUM is
  port(
    signal Clk      : in  std_logic;
    signal Reset    : in  std_logic;
    signal En       : in  std_logic;
    signal Sel      : in  std_logic;
    signal AD       : in  std_logic_vector(7 downto 0);
    signal BD       : in  std_logic_vector(7 downto 0);
    signal S        : out std_logic_vector(7 downto 0));
end ACCUM;

architecture Code_6 of ACCUM is
    signal A        : std_logic;
    signal B        : std_logic;
```

```
    signal D        : std_logic;
    signal Q        : std_logic;
    attribute fti of A : signal is "parity";
    attribute fti of B : signal is "parity";
    attribute fti of D : signal is "parity";
    attribute fti of Q : signal is "parity";
    signal AD_FT    : std_logic_vector(7 downto 0);
    attribute fti of AD_FT : signal is "parity";
    signal BD_FT    : std_logic_vector(7 downto 0);
    attribute fti of BD_FT : signal is "parity";
    signal s_FT     : std_logic_vector(7 downto 0);
    attribute fti of S_FT : signal is "parity";
    signal EOUT      : std_logic;
    attribute fti of EOUT : signal is "ERROR_SIGNAL";
begin
  REG_A: process(Clk, Reset)
  begin
    if Reset = '0' then
      A <= (others => '0');
    elsif Clk'event and Clk = '1' then
      if En = '1' then
        A <= AD_FT;
      end if;
    end if;
  end process REG_A;
  REG_B: process(Clk, Reset)
  begin
    if Reset = '0' then
      B <= (others => '0');
    elsif Clk'event and Clk = '1' then
      if En = '1' then
        B <= BD_FT;
      end if;
    end if;
  end process REG_B;

  MUX : D <= A when Sel = '1' else B;

  ACCUM_P: process(Clk, Reset)
  begin
    if Reset = '0' then
      Q <= (others => '0');
    elsif Clk'event and Clk = '1' then
      if En = '1' then
        Q <= WORK.parity.add(a => Q, b => D);
      end if;
    end if;
  end process ACCUM_P;
  S_FT  <= Q;
  AD_FT <= WORK.parity.encoder(bus_in => AD);
  BD_FT <= WORK.parity.encoder(bus_in => BD);
```

```
   S       <= WORK.parity.decoder(bus_in => s_FT);
   EOUT    <= WORK.parity.checker(bus_in => s_FT);
end Code_6;
```

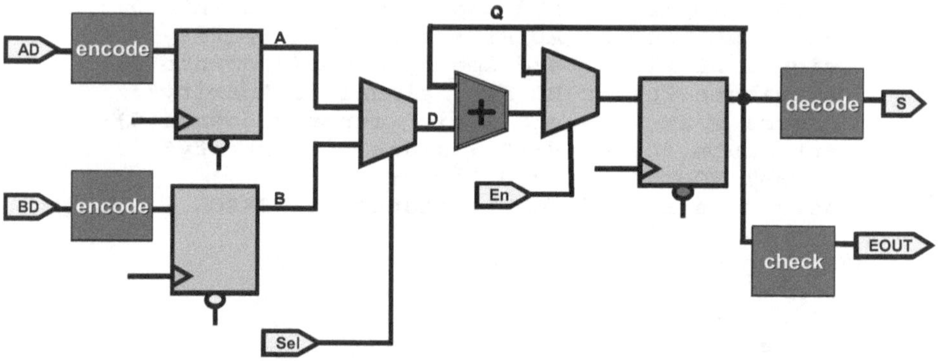

Figure 7. Elaborated model of Code 6, all block works with parity and parity checking is performed at the output S.

In the case of a Finite State Machine (FSM), the states are usually described in VHDL by means of a user-defined enumerated type. State coding is then performed automatically by the synthesis tool using a default code or a user-specified code. Rather than explicitly inserting an encoder and a decoder, the state coding can be easily changed to a redundant code in this case by setting a code to be used by the synthesis tool.

The VHDL Register Transfer Level Synthesis Standard IEEE 1076.6 [10] specifies the attribute ENUM_ENCODING that shall be supported for synthesis. The attribute ENUM_ENCODING provides the means to specify the encoding of enumeration type values. Redundant codes for enumerated types will be introduced by inserting a specification of the ENUM_ENCODING attribute.

5. Error recovery actions

Error recovery actions define the hardware to be inserted in order to process the error signals generated by techniques that do not have fault masking capabilities.

In a general case, the automatization of error recovery actions may be very difficult to implement because it requires specific knowledge of the functionality of the circuit. However, in most practical cases error recovery actions fall in one of the following categories:

- Disabling the load of a register
- Disabling the setting of a control signal (e.g., global reset)
- Controlling the multiplexing of spare units

These cases can be easily supported in an automatic way. The following example shows how a register model can be modified so that it is disabled in case an error occurs. A dual-rail register output can be produced (not shown in the example) by applying a hardware redundancy technique as described in section 3, if required.

Code 5. VHDL initial description of example_3: Register

```
process(clk, reset)
begin
        if reset = '0' then
          q <= '0';
        elsif rising_edge(clk) then
          if enable = '1' then
            q <= d;
          end if;
        end if;
end process;
```

Code 6. VHDL description of Example_3 with error disabling

```
process(clk, reset)
begin
        if reset = '0' then
          q <= '0';
        elsif rising_edge(clk) then
          if error = "01" then
            if enable = '1' then
              q <= d;
            end if;
          end if;
        end if;
end process;
```

6. Conclusions

In this work we present FTI tool, which automates the insertion of fault-tolerant structures in design descriptions at RT level. Techniques to automatically apply hardware redundancy and information redundancy to a design have been shown. The resulting fault-tolerant design is also obtained in the form of a RT level design description that can be further simulated and synthesized with commercial tools.

Proposed solution will imply higher design productivity thanks to the automation of hardening process. As well, reliability of circuits is enhanced due to the elimination of manual design mistakes and the possibility of early fault tolerance validation.

Developed tool has been validated with some benchmarks. Those are academic and industrial designs. First one is a PIC micro-controller and second is an aero-spatial FPGA (Rosetta SADE) developed by UC3M for Alcatel-Espacio. Obtained results are satisfactory, CPU time and memory used during application is negligible while VHDL descriptions obtained are 100% correct.

The proposed tool provides an automatic way to substitute current manual practices in the design of fault-tolerant circuits.

References

[1] "Minimizing Single Event Upset Using Synopsys". Application Note, Actel Corporation, July 1998.

[2] XTMR Tool. "http://www.xilinx.com/products/milaero/tmr/index.htm"

[3] R. O. Duarte, M. Nicolaidis, H. Bederr, Y. Zorian. "Efficient Fault-Secure Shifter Design". Journal of Electronic Testing, Theory and Applications (JETTA), vol. 12, p. 29-39, 1998

[4] M. Nicolaidis, R. O. Duarte, S. Manich, J. Figueras. "Achieving Fault Secureness in Parity Prediction Arithmetic Operators". IEEE Design and Test of Computers, April-June 1997.

[5] R. O Duarte, I. A. Noufal, M. Nicolaidis. "A CAD Framework for Efficient Self-Checking Data Path Design". IEEE International On-Line Testing Workshop, July 1997

[6] M. Nicolaidis, Y. Zorian. "On-Line Testing for VLSI- A Compendium of Approaches". Journal of Electronic Testing, Theory and Applications (JETTA), vol. 12, p. 7-20, 1998

[7] L.Berrojo, F.Corno, L.Entrena, I.González, C. López-Ongil, M.Sonza-Reorda, G.Squillero. "An Industrial Environment for High-Level Fault-Tolerant Structures Insertion and Validation". Proceeding of the 20th IEEE VLSI Test Symposium. 2002.

[8] Tauri™. FTL Systems Inc.

[9] AIRE CE "Advanced Intermediate Representation with Extensibility/Common Environment "John Willis, Technical Editor / Architect (FTL Systems, Inc.). Version. 7 Pre-Release (2000).

[10] "IEEE P1076.6/D1.12 Draft Standard For VHDL Register Transfer Level Synthesis". IEEE, 1998.

Test Facilities for SEE and Dose Testing

S. Duzellier[1], G. Berger[2]

[1] Office National d'Etudes et de Recherches en Aérospatiale (ONERA),
Space Environment Department (DESP),
2 avenue E. Belin, F-31055 Toulouse cedex, France
duzellier@onecert.fr
[2] Université Catholique de Louvain-La-Neuve, Centre de Recherches du cyclotron,
2 chemin du cyclotron,
B- 1348 Louvain-La-Neuve, Belgique
berger@cyc.ucl.ac.be

Abstract. This chapter addresses the field of testing and characterizing the response of electronic devices to radiation exposure. Firstly, a brief overview of the critical parameters that influences the devices degradation or malfunction will be given in order to show how the standards and guidelines deal with them. Then, the most widely used facilities will be described (particle accelerators, radioactive source…) and their domain of application defined.

1. Introduction

Modern electronic components are very sensitive to the space environment. The main effects of radiation can be sorted in three categories:

1. Cumulative effects induced by an homogeneous and continuous exposure to protons and electrons of the radiation belts (TiD : Total Ionising Dose),
2. Probabilistic functional anomalies induced by a single particle strike (heavy ions or protons from the GCR or solar flares). These anomalies are grouped under the name of Single Event Effects (SEE),
3. Gradual degradation of performance of devices sensitive to Displacement Damage (DD). This latter is also a cumulative effect and is induced by protons from solar flares.

All these effects are the result of dose deposition whose rate and mechanisms differ: 1) instantaneous and localized for SEE (localized ionized plasma), 2) cumulative and homogenous for TiD (ionization) and DD (displacements).

The reader is referred to others sections of this document and literature for details.

R. Velazco et al. (eds.), Radiation Effects on Embedded Systems, 201–232.

2. Response of devices to radiation

The disruptions induced by the space environment in electronics manifest themselves in many ways depending on the device type and effect. The following tables summarized the most frequent observed anomalies.

Table 1. Non destructive SEE phenomena.

Upset - SEU[1]	corruption of the information stored in a memory element	Memories, latches in logic devices
Multiple Bit Upset - MBU	several memory elements corrupted by a single strike	Memories, latches in logic devices
Functional Interrupt -SEFI	loss of normal operation	Complex dev. with built-in state/control section
Transient - SET	impulse response of certain amplitude and duration	Analog and Mixed Signal circuits, Photonics
Disturb - SED	momentary corruption of the information stored in a bit	combinational logic, latches in logic devices
Hard Error - SHE	unalterable change of state in a memory element	Memories, latches in logic devices

Table 2. Destructive SEE phenomena.

Latchup - SEL	high-current conditions	CMOS, BiCMOS devices
Snapback - SESB	high-current conditions	N- MOSFET, SOI devices
Burnout - SEB	destructive burnout	BJT, N-channel Power MOSFET
Gate Rupture - SEGR	rupture of gate dielectric	Power MOSFETs
Dielectric Rupture - SEDR	rupture of dielectric	Non-volatile NMOS struct., FPGA, linear devices...

Table 3. Typical TiD degradation modes in devices.[2]

device	*Degradation mode*
MOS	Shifting of threshold voltage
BJT	Current gain degradation
Digital circuits	Increased leakage current (Iccop, Icc-sb), reduced retention time (DRAMs and SDRAMs)
Linear devices	Increased Offset voltage and bias current

[1] SE… Single Event…

[2] An extensive presentation of total dose effects in MOS devices can be found in the Short Course of NSREC 2002 [8].

Table 4. Typical DD degradation modes in circuits.

device	*Degradation mode*
CCD/APS	Reduced responsitivity (Increase of dark current, RTS occurrence...°
Bipolar Linear devices	Gain degradation
Solar cells	Reduced efficiency (output power, short circuit current)
Laser diode	Threshold current increase

2.1 Critical parameters

Here below, only the irradiation conditions will be addressed. However, one should keep in mind that bias and testing conditions strongly affect the device response. All methods described later in the "standard and guidelines" section require applying "worst case bias" during all the experimental evaluation. This condition is not always easy to determine practically.

More detail can be found in the literature [1-7] and in other sections of this document.

2.2 Brief reminder of the different used parameters

2.2.1 Stopping power

When an ion enters into the target material, it looses its energy by 1) elastic collisions with the target nuclei, 2) inelastic collision with the electrons. The total stopping power is defined as the energy loss per path length:

$$\left(\frac{dE}{dx}\right)_{tot} = \left(\frac{dE}{dx}\right)_{nuclear} + \left(\frac{dE}{dx}\right)_{electronic}$$

Even if the two processes contribute to the total stopping power, the electronic stopping power dominates at energies above 100 keV for silicon (see Figure 1).

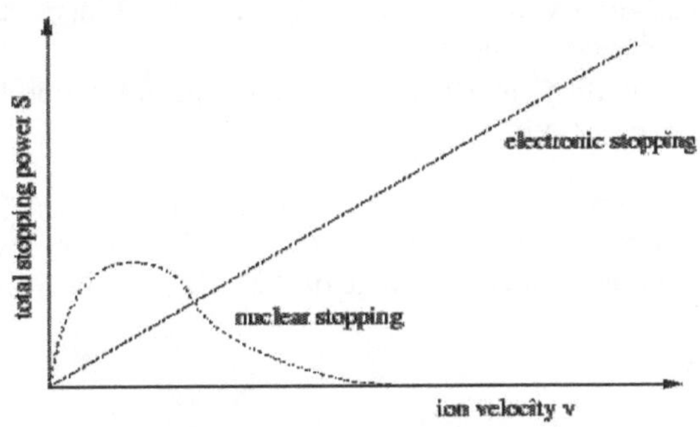

Figure 1. Evolution of the stopping power for ions.

2.2.2 Linear energy transfer (LET)

This is the energy loss per path length due to ionization:

$$LET = \left(\frac{dE}{dx}\right)_{ionization}$$

This value is generally expressed in MeV/mg/cm² or in MeV/μm

2.2.3 Range

This is the total distance traveled by a particle in the target material. It is correlated to the total stopping power by the equation:

$$R = \int_{E}^{0} \frac{1}{\left(\frac{dE}{dx}\right)_{total}} dE$$

2.2.4 Non-ionizing energy loss (NIEL)

The portion of the energy transferred to the material via process other than ionization; i.e. in total amount of energy that goes in displacements. It is measured in MeVcm²/g. The Displacement Damage Dose (DDD) is the product of the NIEL and the beam fluence (in Particle/cm²).

2.3 TiD

During irradiation, the oxide trapped charge density depends on:

- The incident particle type and energy, corresponding to a LET[3] value, which determine the ionised track structure (radial carriers densities profile) ,
- the dose rate (which corresponds to the average carrier generation rate),
- The electric field which modulates the carriers transport towards the oxide interfaces (bias conditions),
- The oxide "quality" that determines the carriers' mobility and the trapping factor (device parameter).

The following figure illustrates the influence of the radiation type (and bias condition) onto the fraction of non-recombined holes in a gate oxide. One can notice that using gamma rays ensure a conservative estimate of the TiD sensitivity of devices.

---

[3] LET: Linear Energy Transfer. This parameter corresponds to the electronic stopping power dE/dx, normalized to the material density, used by the physicists.

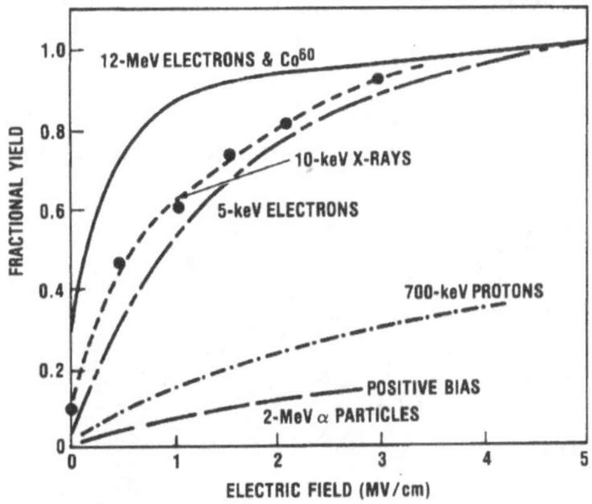

Figure 2. Fraction of holes escaping initial recombination versus electric field for different ionising radiation sources (after [9]).

The dose rate is by far the most important and critical parameter, especially, when dealing with bipolar technologies. The effect reported on the following figure is called ELDRS "Enhanced Low Dose Rate Sensitivity" and induces a much greater degradation of the performance at low dose rate. Using "standard" dose rates, in such case, would generate a strong under-estimation of the device sensitivity compared to the one in orbit.

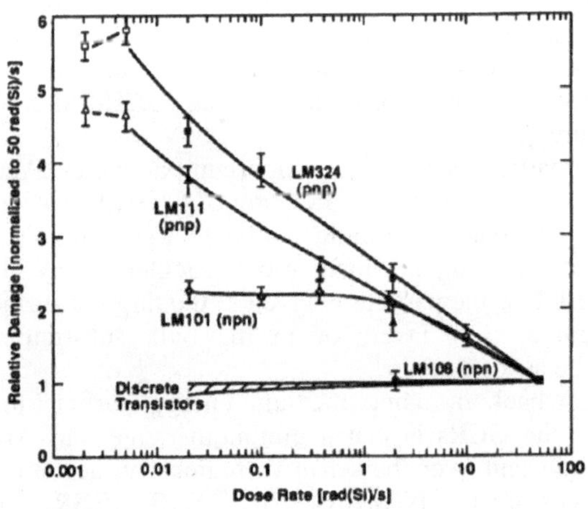

Figure 3. ELDRS effect in linear devices (after [10]).

2.4 SEE

The exact simulation of the heavy ion component of the space environment is not possible employing ground-based accelerators because of its wide distribution of ion species, high energies and omni-directional incidence. Galactic cosmic rays in space have a distribution of energies with a peak around a few 100 MeV/nucleon but ions of interest for the SEU issue cover the 0.1-100 MeV/mg.cm² LET domain. Few machines can deliver such high energy beams, thus "low" energy accelerators are used to reproduce an equivalent charge deposition by means of the LET parameter. The selection of ion species is usually limited to what the accelerator routinely produces (see Figure 4).

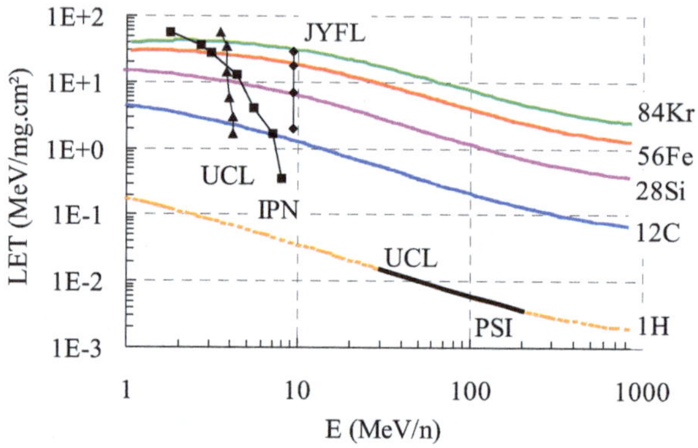

Figure 4. Space environment LET spectra compared with standard accelerator beams (described later in this section).

Up to now, no clear energy dependence has been observed on the majority of devices tested. Any energy dependence observed is usually associated with conservative values for low energy measurements [11, 12].

Contrary to heavy ions, protons with energies that correspond to those actually encountered in space are used for ground-testing (UCL-PSI available energy range is presented on figure 1).

Therefore, the particle energy is an important parameter. Practically, the measured cross section curves are plotted as a function of LET (heavy ion) or energy (protons) and the LET is calculated assuming a given target (atomic mass) and projectile (energy and species) using adequate tools (Ziegler tables [13], SRIM code). By convention, the LET of the particle is given at the target's surface. Energy loss in the package, lead-frame, over layers or in the bulk substrate in case of backside irradiation must be accounted for.

The other drawback of using medium energy particle to reproduce the LET characteristics of the GCRs is that a minimum range must specified to ensure the particle LET is constant over the sensitive region and account for cases where long collection distances are involved (SEL, SET, SEB, SEGR, MBU). This concern is even more justified by the use of the "effective LET concept" or "cosine law"[4].

[4] $LET(\theta)=LET(0°)/\cos\theta$

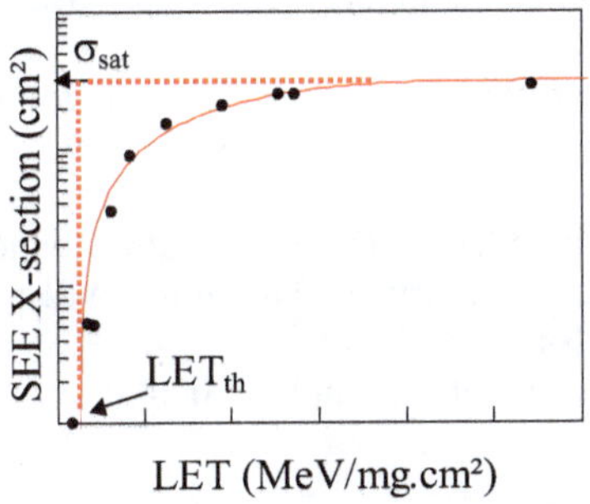

$$\sigma\ (LET) = \frac{Number\ \ of\ \ events}{Fluence}$$

(units in cm²)

$$LET\ (\theta) = LET\ (0°)\ /\ cos\ \ \theta$$

(units in MeV/mg.cm²)

Figure 5. Typical cross section curve for heavy ions (left) and expression for calculating the effective LET and σ (right).

In addition to the energy and range parameters, the absence of contamination (beam purity and the flux uniformity over the die surface are important parameters.

The beam purity impacts on the particle characteristics the same way the energy does (accuracy of the LET).

Accelerator testing is a broad-beam technique that means that the whole device surface is irradiated at one. Then the flux must be homogeneous over the entire die surface to ensure that each part of the device is equally exposed.

From the general standpoint, a proper dosimetry (particle counting and beam uniformity) will limit the statistical error on σ.

2.5 DDD

Three main parameters are critical when dealing with displacements: the species of the projectile, its energy and the received fluence.

The nature of the initial damage in the semiconductor depends on the species and energy of the projectile (see next table).

Table 5. Influence of species and energy on produced defects characteristics.

species	*energy*	*Interaction and produced defects*
electrons	all	Low mass => low energy transfer => point defects
protons	low energy (< 10MeV in Si)	Columbic scattering => point defects
protons	> 10MeV and higher	Nuclear scattering and spallation => cascade defects
neutrons	all	Nuclear scattering and spallation => cascade defects

However, in spite of the different nature of defects, the induced electrical effects may be the same [7].

As a matter of fact, for space applications, protons are usually chosen for DDD testing because representative of the space environment. Electrons (or low energy protons) are sometimes used for solar cell testing (no shielding).

It has to be mentioned here, that protons also induced TiD degradation which in turn is very dependent on biasing and temperature conditions. This point will not be discussed here and the reader is referred to [5-7] for more detail.

The received fluence is chosen to be representative of the final application.

3. Standards and guidelines

3.1 TiD

Standards for total dose evaluation of electronic components have been established in order to define methodologies adapted to laboratory evaluation of devices in view of applications in ionising environments. The aim of these standards is to guarantee a correct behaviour of the device under test in an ionising environment (conservative method) and to provide comparable results. As it is impossible to reproduce exactly all the device operating environment, choices has to be done to stay conservative but not too much. The main features of the ESA/SCC basic specification 22900 issue 4 (for lot acceptance testing) and MIL-STD-883E 1019.5 test method (in case of low dose rate applications) are presented in a simplified form and compared in table 6 for the irradiation phase and table 7 for the post irradiation effects.

Table 6. Main irradiation conditions in ESA and MIL strandards.

Method	Source	Dose rate	Part sample	Bias
22900.4	Co^{60} or electron accelerator	Standard: 36 to 360 Gy/h Low rate: 0.36 to 3.6 Gy/h	10 + 1 ref.	Worst case
1019.5	Co^{60}	50 to 300 rad(Si)/s (1800 to 10800 Gy(Si)/h)	Not specified	Worst case

The main difference concerns the dose rate ranges. In practice, the laboratory doses rates are always larger than the typical space dose rates and thus an accelerating factor always exists between space application and ground evaluation. This accelerating factor can lead to very large errors in device hardness [1].

The irradiation phase of device evaluation, for a total dose of interest D, must be completed by post-irradiation procedure.

Table 7. Main post-irradiation conditions for ESA and MIL standards.

Method	Room temperature	Over test	Annealing	Bias
22900.4	24 h	None	168h at 100°C	Worst case
1019.5	Time<D/R_{max} [5]	0.5xD (default)	168h at 100°C	Worst case

[5] R_{max} : maximum dose rate for the intended use.

In the 1019.5 test method, certain latitude is allowed on the room temperature annealing phase duration. This is justified by the fact that the large dose rate used enhances the induced leakage currents and a long room temperature annealing, within the limit of duration D/R_{max}, can help some devices satisfying the specifications in conservative conditions. In the 22900 method, the overestimation of degradation modes due to oxide-trapped charge is limited via the dose rate value which is lower than in the case of 1019.5.

To reveal possible degradations due to interface traps, it is necessary:

- To be sure that the interface traps production is conservative (room temperature annealing),
- To reduce the effect of oxide-trapped charge, that can compensate the interface trapped charge, at a level lower or equal to the level expected in the application (accelerated annealing at 100°C, rebound testing).

3.2. SEE

A SEE experiment aims at evaluating in real-time the device's response under consecutive exposures with several beam characteristics. The final objective is to obtain a good description of the device behaviour and an accurate measurement of its radiation response ($\sigma(E)$ or $\sigma(LET)$ for each error mode) to enable the calculation of reliable in-flight SEE rates.

The traditional approach for prediction consists in folding the adequate environment spectrum (energy or LET spectra) with the cross section curve of the device [14-16]. Indirect ionisation is considered to be predominant for proton SEE and direct ionisation to be prevalent for heavy ion events, and then both mechanisms are separately considered.

The purpose of standards and specifications is to provide a method and test procedure for helping designing and performing an experiment. These norms are tutorial in nature but provide useful recommendations ensuring that the data will be valid and meaningful.

Two main standards are widely used within the framework of SEE testing: 1) ESA/SCC 25100: Single Event effects test methods and guidelines, 2) JEDEC JESD57: Test procedures for the measurement of Single-Event Effects in semiconductor devices from heavy ion irradiation.

The SCC standard applies to proton and heavy ion testing whereas the JEDEC standard only addresses heavy ion testing. As previously mentioned, SEE affects all types of devices and technologies; therefore, no typical test approach can be adequately defined covering the whole range of devices and effects concerned. As a consequence, the device operating conditions are usually not specified but must be established prior to the experiment as they strongly impact the device's response.

The topics addressed in standard are listed in Table 8.

Table 8. Main requirements addressed in SEE standards and guidelines.

Requirements	ESA/SCC 25100	JEDEC JESD57
Scope	Single Event effects test method and guidelines	Test procedures for the measurement of Single-Event Effects in semiconductor devices from **heavy ion** irradiation
Radiation sources and characteristics	HI : range ≥ 30μm, $10^2 \geq$ Flux $\geq 10^5$ ions/cm².s p+ : 20-300 MeV $10^5 \geq$ Flux $\geq 10^8$ ions/cm².s	range large compared to the depth of the collection region, $10^2 \geq$ Flux $\geq 10^5$ ions/cm².s LET up to 120 MeV/mg.cm²
Dosimetry	Uniformity ±10% over the device area Flux ±10%	Energy ±10% Uniformity ±10% over the device area Flux ±10%
Testing requirements	Sample size ≥ 3 (same dtc) 5 measurements at different effective LETs (HI) or Energies (p+, normal incidence) Max fluence of resp. 10^7 and 10^{10} part./cm².s for HI and p+ or a meaningful number of events	Measurements at onset threshold, 10%, 25%, 50% and 75-80% of the saturated σ, Max fluence of 10^7 ions/cm².s for "hard" devices, 10^6 ions/cm².s or 100 events whichever comes first for "soft" devices Tilt angles limited to 60°

3.3 DDD

A standard method does not yet exist for displacement damage testing for the following reasons:

1) modes of degradation are very complex,
2) the induced electrical effects are very application dependent,
3) annealing mechanisms occur depending on the type of devices and application.

Then, DDD testing is always designed and performed according to specific requirements and applications.

However, some recommendations can be provided. They are summarized in the next table.

Table 9. Recommendations for irradiation conditions for DDD testing.

species	*energy*	*comment*
proton	50-60 MeV	Representative of "shielded" space environment Good penetration into package and device
proton	10 MeV	Detector array
electrons	1-3MeV	Solar cells

Testing at several energies the same device would allow for checking the correlation with the NIEL parameter. However, practical and funding constraints often limit this approach.

4. Test facilities and domain of application

The test methodology strongly differs depending on the phenomenon at concern. At first glance:

1. SEE characterization of devices requires real-time testing under exposure (functional testing mainly), and the use of particle accelerators,
2. TiD assessment implies the full parametrical characterization at different step of dose levels received (sequence of irradiation/testing phases). Mostly, 60Co sources are used for irradiating,
3. DD testing is quite similar to TiD characterization as parametrical measurements and functional checking occurs at different received fluence levels (leading to an equivalent displacement damage dose - DDD[6]). However, DD testing requires the use of particle accelerators.

Table 1 summarizes the basic requirements for irradiation facilities as defined in standards.

Table 10. Standard requirements for irradiation facilities.

Standards	Effect	Parameters of concern	Means
ESA-SCC 22900.4	TID	Total dose, dose rate	^{60}Co sources
MIL-STD 883E Method 1019.6	TID	Total dose, dose rate	^{60}Co sources
ESA-SCC 25100.1	SEE	LET/range (heavy ions), Energy (protons)	Accelerators of particles
JESD57	SEE	LET/range (heavy ions)	Accelerators of ions (Z>1)
ESA-SCC 22900.4	DD	Energy	Accelerators of particles (electrons, protons, neutrons)

4.1 TiD

Different ionising radiation sources are available for ground tests. They have complementary advantages and drawbacks summarised in table 11.

The main drawback in using particles such as electrons or protons is the associated cost. Furthermore, protons have a significant rate of displacements generation which can induce additional specific degradation. X rays generators are very convenient but,

[6] In this case the NIEL parameter (Non Ionising Energy Loss) is use to describe the particle matter interaction and subsequent dose deposition.

due to the low energy of the emitted photons, the deposited dose is not uniform versus depth near each interface between different materials. This effect, named "dose enhancement" effect [17], makes this irradiation facility very tricky to use.

Table 11. Main features of ground TiD radiation types.

Radiation type	Main advantages	Main drawbacks
Electrons (accelerator)	High dose rate available Representative of some orbits	Costly Not adequate for low dose rates
Protons (accelerator)	High dose rate available Representative of some orbits	DD contribution Costly
X rays (photons)	High dose rates available Low cost	Dose enhancement effect Not adequate for low dose rates
Cs^{137} & Co^{60} sources (gamma rays)	Very large dose rate range Dose uniformity	Heavy shielding necessary Non-dominant in orbit

Radioactive Cs^{137} and Co^{60} sources deliver gamma rays and, even if this type of radiation is minor in space environments, present two strong advantages which are the very wide range of dose rates available and the fact that the total dose is well controlled in the device thickness. As photons delivered by Co^{60} have a large energy, 1.17 and 1.33 MeV, dose uniformity is ensured. This advantage can be lost if the irradiation facility is not correctly filtered and delivers a sizeable ratio of low energy scattered photons inducing dose enhancements. Consequently, it is necessary to take care to correctly assess accurate dosimetry and use adequate filtering methods.

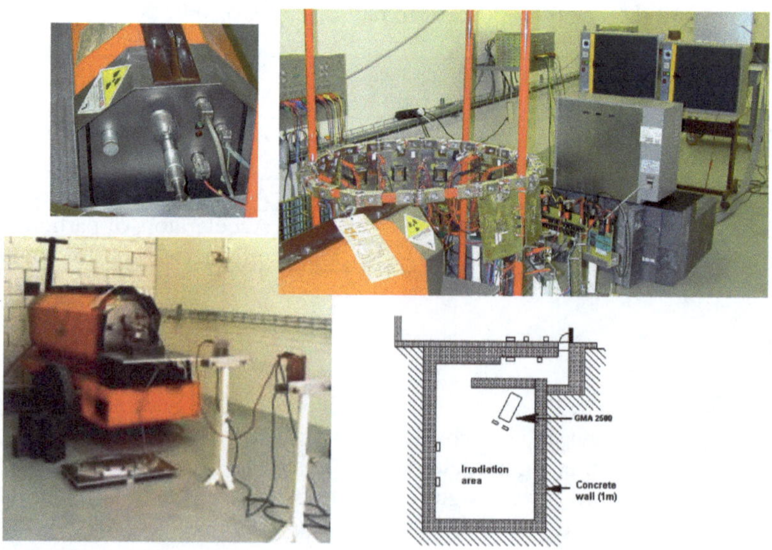

Figure 6. Panoramic GMA2500 Co60 source in its bunker (ONERA/DESP).

Gamma sources (mainly Co60) are the most widely used facilities for TiD testing and can be found in two features. The panoramic facilities allow for irradiating wide targets such as complex electronic boards and systems (Figure 6). Complex regulation systems such as temperature regulation can also be easily implemented. Because of the large exposure area, these facilities are commonly shared by industry and scientists. Fortunately, various experiments can be installed at the same time allowing easy scheduling of TID testing.

As the source activity is limited due to radiation protection purpose, this kind of facility is mainly devoted to low–to-medium dose rate experiments (typically in the 0.1-10 Gy/h range).

With a self-shielded facility, the source has a double positioning in one unique shielded apparatus: storage and use. The irradiation area is located inside the facility and limited to a few dm^3 (Figure 7).

Because these irradiators provide large and efficient shielding that limits problems of radiation protection for the experimenters, sources of higher activities can be used allowing medium-to-high dose rates (typically 10-5000Gy/h). In this kind of facility, low-energy scattered radiation[7] may lead to dose enhancement effects. It is therefore recommended to use filters (Aluminum/Lead foils for instance) to remove this low energy contribution and avoid dosimetry errors.

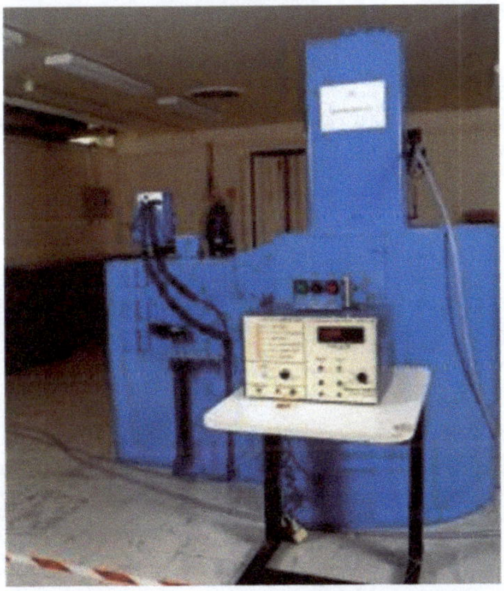

Figure 7. SHEPHERD 484 double Co60 source (ONERA/DESP).

4.2 SEE

Accelerator testing implies several restrictive requirements. A "component test line" is typically made of:

[7] Generated by the material surrounding the target (walls of the chamber)

1. A particle generator: tandem Van de Graff, cyclotron, synchrotron…,
2. A beam monitoring and dosimetry system: magnets, shutters, collimators, radiation monitors (counting, mapping of beam spot, energy measurement…),
3. With most heavy ion accelerators, irradiation exposure in air is impossible due to the limited energy and range of the particle. Consequently, experiments are performed in a vacuum chamber with cable feedthroughs, a test board/sample holder, collimators/shutters for the beam delivery ...

Particle accelerators were firstly built for the sole purpose to produce energetic particles to be used as projectiles to alter the structure of other nuclei or to act as probes in nuclear physics research. Latter, these machines were used in different fields: medicine (research, isotope production, cancer treatment …), solid state research, membrane production, nanotechnologies (nanopores, nanowires, nanotubes …), surface analysis and finally electronic devices testing.

Different particles accelerators classifications are possible, in the following section we will use their mode of operation for their descriptions and a summary of European accelerators used in our field will be presented.

4.3 DC accelerators

This type of machine consists of an isolated tube connected at one end to the particle source and to the target at the other end. The voltage is applied from end to end to produce the accelerating field. There are two types of these accelerators in function of the way to produce the DC voltage:

- "Cockroft-Walton" where the voltage is generated by a succession of voltage-multiplying rectified circuits. In 1932, this system was proposed at the Cavendish Laboratory in Cambridge with a DC potential of 600 kV.
- "Van de Graaff" (proposed around 1931) where voltage is created by transport and accumulation of electrical charges on the electrode containing the particle source.

The main withdraw of this type of accelerators is the discharge limit who set an upper limit to about 20 MV at the best. As the Van de Graaff is the most commonly used DC machine nowadays, the next section will describe it.

As seen on Figure 8 an isolated endless belt circulates between two pulleys, the first one is at the ground, the second one is inside an isolated metallic electrode. The belt is charged at the lower end of the machine and releases the charges by corona effect on the upper electrode.

The total current transported by the belt is $i_0 = \sigma v w$ where $\sigma = 2\varepsilon_0 E$

σ Is the superficial density

v, w Are the speed and width of the belt

ε_0 Is the vacuum permittivity = 9.10^{-12} C/m V

E Is the electric field on the belt surface

Main characteristic of this type of accelerator is its energy resolution. As the voltage is not generated by rectifiers in cascade, typical energy straggling is the same as the ions coming out of the source (about 100 eV).

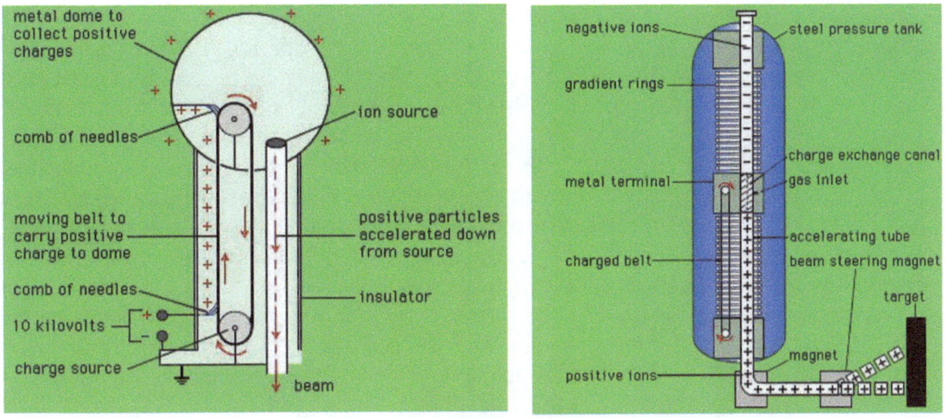

Figure 8. Van de Graaff accelerator (left) and TandemVan de Graaff (right).

On the technological point of view, several points are interesting to mention. First of all, the belt is rolling at high speed and need to hold the mechanical and electrical stresses. Secondly, the entire assembly is encased in a tank filled with several atmospheres (used gas: SF6 or 20% CO2 + 80 % N2 at 10-20 atm)

An approach to increase the final beam energy using the Van de Graaff method is tandem configuration as proposed by Alvarez in 1951. Two acceleration columns are placed on both side of a central electrode positively biased. The high voltage is produced by moving charges on the belt as before and the whole system in placed in the high pressure gas filled tank.

The ion source is placed outside; the produced positively charged ions are negatively converted by adjunction of electrons and are then injected in the first column. These negative ions are accelerated by the positive field created by the central electrode. Inside this one, the ions are stripped either by crossing a gaseous cell or a foil. Becoming positively charged, the ions see a repulsing field and are accelerated again in the second column.

Having an ion with the charge Z, and a high voltage V, the final energy is $E = (1+Z)V$.

4.4 Linear accelerators

In these accelerators (LINAC), the reference particle moves on a straight line along a high frequency electrical field. Particles passing through a series of drift tubes don't see the decelerating field and reach an accelerating gap at the proper phase.

The final energy of the particles is proportional to the sum of the voltages produces by the accelerating devices along the line.

The first used structure was the Widerö in 1928. Metallic cylindrical tubes are aligned along the axis, and as seen on are connected by pair to form the two electrodes of a capacity. This capacity and the inductance form a resonant circuit connected to the high frequency generator. During a half period, field is accelerating and the reference particle is between two tubes. Next half period, the field is decelerating and then the particle is inside the next tube and continues its path (the Widerö structure is also names mode). The lengths of the tubes are therefore almost as long as it takes the particles to travel for a half RF period. Due to the capacitive nature of this structure, only low energies were achievable. Above certain frequencies, this system is very loosy due to electromagnetic radiation.

A new approach was proposed by Alvarez in 1947, he proposed to create a chain of resonating cavities. For this, the drift tubes are enclosed in a long cylindrical. This LINAC configuration is based on the formation of standing waves inside the tank, the electric field is parallel to the structure. These accelerators operate at high frequencies (above 200 MHz). This structure is capable to accelerate protons and heavy ions from a few keV up to a few hundreds MeV total energy.

Last development in this category of accelerator is the Radio Frequency Quadrupole (RFQ). This machine is usable for low energies. The profile of the four electrodes is modulated along the axis. This shape gives an on axis contribution of the electrical field. The advantage of this structure is to be focusing (quadrupole effect) and accelerating (on axis field contribution).

4.5 Circular accelerators

In these machines, particle path is bent by the action of a magnetic field. The resulting trajectory may be a spiral or near circular.

4.5.1 Classical cyclotron

The classical cyclotron [20] is characterized by a constant frequency and a uniform magnetic field. The principle of these machines is the application of the Widerö linac in a coiled up version. The accelerating cavities have the shape of a pillbox split in two halves (DEE) placed between the poles of the magnet. The accelerating field is generated between those two dees, this implies that the particles must arrive at the proper instant to see an accelerating field (phase acceptance). The particle orbits occur mostly in the field free interior of the dees and traverse the accelerating gaps between the two dees twice per turn. As the particle energy increase, their trajectories have larger and larger radii. The magnetic field is just there to allow the repeated passage of the beam through the cavities.

An ion of mass m, charge Q and velocity v moving at right angle from a magnetic field B see a force $F_b = QvB$ normal to the magnetic field and in the direction of the ion. This force causes the ion to circulate in a circular path of radius r whose value is obtained by balancing the centrifugal force and the magnetic force:

$$\frac{m v^2}{r} = Q v B$$

The particle frequency of revolution is thus: $f = \dfrac{v}{2\pi r} = \dfrac{QB}{2\pi m}$

The problem of this type of machine is that the energy gained by a particle is limited by the relativistic increase of its mass. Above a certain limit, the orbital frequency of the particle decreases and the ion goes out of phase with the alternating voltage. This limit is around 20 MeV for a proton.

4.5.2 Synchrocyclotron

One way to get rid of this relativistic effect limitation is to keep the magnetic field uniform. Thus, to keep the synchronism lower the accelerating frequency while the particle energy increases. As only particles with a well defined energy are at the right phase with the accelerating voltage, the beam will be bunched. The particle flux has a pulsed macro structure equal to the cycling time of the RF modulation, and the average beam current is smaller than in a classical cyclotron.

4.5.3 Isochronous cyclotron

Another way to avoid the energy limitation is to keep the accelerating frequency constant and increase the magnetic field as the particle energy increases. In this way, the mean magnetic field is increasing with radius, keeping the particle orbital frequency constant.

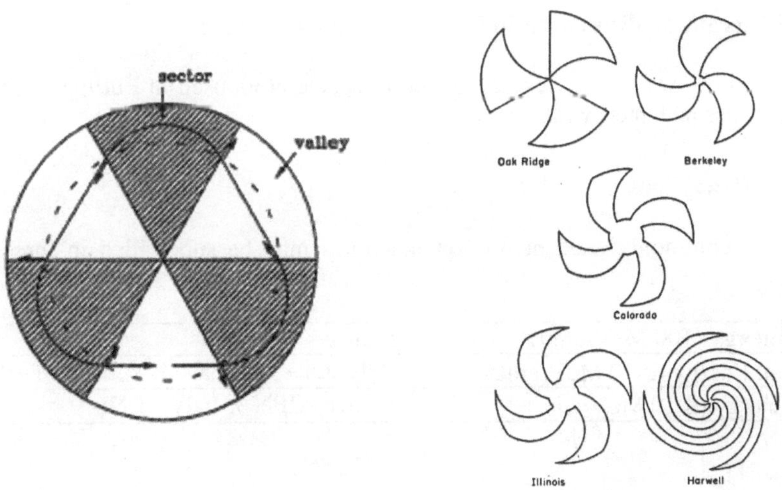

Figure 9. Isochronous Cyclotron (left), Spiral sectors (right).

Normally, this field modification would cause the beam to be axially defocused. But if the magnetic poles are constructed with sectors (Fig. 9 left), this induces azimuthal variation of the magnetic field inducing an axially restoring force on the particle at the edge of the sectors. Additional axial focusing may be added by twisting the sectors (Fig. 9 right).

There still exists an energy limit which is linked to the focusing capabilities of the magnet. The ultimate focusing limit is the separated sector cyclotron. Here, the magnet only consist of sectors, in between (in the valleys) there is no iron and practically no magnetic field. All additional equipment (RF system, extraction ...) may be placed between sectors (Figure 10).

Figure 10. Separate sector cyclotron.

4.6. Different facilities used in Europe for device testing

In this section, we will review the different accelerators used in Europe classified by particle types and energy range.

4.6.1 Heavy ions

The most commonly used heavy ion machines may be subdivided in three energy ranges

High Energy (100 MeV/amu)	France – GANIL
Medium Energy (≥ 10 MeV/amu)	Belgium – CYCLONE / Finland – JYFL
Low Energy (≤ 10 MeV/amu)	France – IPN / Italy – LNL

<u>GANIL</u>

The "Grand Accelerateur National d'Ions Lourds" (GANIL) is located in Caen, Normandy France. This centre is equipped with five cyclotrons:

- two compact cyclotrons (C01 and CO2) K=30
- two separated sector cyclotrons (CSS1 and CSS2) K=380
- one variable frequency cyclotron (CIME) K=265

With the combination of these different machines, high energies are achievable. The main advantage of this configuration is that devices may be irradiated in the air; there is no need for vacuum chamber. The disadvantage is a long beam setup time. For device testing, the cave G41 is the most used; Table 12 gives the available beams:

Table 12. Available beams at Ganil.

Ion	Energy [MeV/amu]	LET [MeV/mg/cm^2]	Range [microns]
36 Ar	95	2	4220
36 Ar	27	5.4	445
40 Ca	95	2.5	3812
58 Ni	52	7.6	1013
84 Kr	35	16.4	484
86 Kr	60	11	1223
93 Nb	31	22.7	349
112 Sn	47	24.2	601
132 Xe	35	33.5	393

CYCLONE (Heavy Ion irradiation Facility)

The "CYClotron of LOvaine la NEuve" (CYCLONE) [22] is a multiparticle, variable energy, isochronous cyclotron capable of accelerating protons up to 75 MeV, deuterons up to 55 MeV, alpha particles up to 110 MeV and heavy ions up to an energy of 110 Q^2/M, where Q is the ion charge state and M its mass in Atomic Mass Units. The energy range for the heavy ions extends from 0.6 to 27.5 MeV/AMU depending, among other things, on the ion's charge state. As can be seen from the above formula, heavier elements such as argon, krypton or xenon require high charge states to reach the energies needed for SEE works. Ions with such high charge states are produced in an external Electron Cyclotron Resonance source (ECR) type. CYCLONE is equipped with two such ion sources. The ions produced in the sources are accelerated to a low energy (± 10 * Q keV), analyzed in Q/M, transported to the cyclotron and injected axially for subsequent acceleration.

One of the advantages of this type of ion source is the possibility to produce "ion cocktails" (Table 13).

Table 13. M/Q = 3.33 cocktail (left) and M/Q = 5 cocktail (right).

Ion	Energy [MeV]	LET [MeV/mg/cm^2]	Range [microns]
$^{13}C^{+4}$	131	1.2	266
$^{22}Ne^{+7}$	235	3.3	199
$^{28}Si^{+8}$	236	6.8	106
$^{40}Ar^{+12}$	372	10.1	119
$^{58}Ni^{+18}$	567	20.6	98
$^{83}Kr^{+25}$	756	32.4	92

Ion	Energy [MeV]	LET [MeV/mg/cm^2]	Range [microns]
$^{15}N^{+3}$	62	2.97	64
$^{20}Ne^{+4}$	78	5.85	45
$^{40}Ar^{+8}$	150	14.1	42
$^{84}Kr^{+17}$	316	34	43
$^{132}Xe^{+26}$	459	55.9	43

These are elements with almost identical Q/M ratios will be accelerated at the same time in the cyclotron. By fine tuning either the magnetic field or the RF frequency each of them can be extracted separately from the cyclotron and transported to the user's target. All these elements have identical velocities but since their atomic numbers are different, they have different ranges and different LET. Its use allows a fast ion changing and subsequently a fast LET modification. Any one of the single beams from one cocktail can be adjusted within a few minutes on the target.

RADEF

University of Jyväskylä, Finland (JYFL) has recently opened their RADiation Effect Facility (RADEF). The used accelerator is a K130 cyclotron comparable the the Louvain la Neuve one and they use two separate ECR sources for ion production. The "cocktail" principle is also used here. The following table gives the used beams:

Table 14. 9.3 MeV/amu cocktail (left) and 3.6 MeV/amu cocktail (right).

Ion	Energy [MeV]	LET [MeV/mg/cm^2]	Range [microns]
$^{15}N^{+4}$	139	1.8	202
$^{20}Ne^{+6}$	186	3.6	146
$^{30}Si^{+8}$	278	6.4	130
$^{40}Ar^{+12}$	372	10.1	118
$^{56}Fe^{+15}$	523	18.5	97
$^{82}Kr^{+22}$	768	32.1	94
$^{131}Xe^{+35}$	1217	60.0	89

Ion	Energy [MeV]	LET [MeV/mg/cm^2]	Range [microns]
$^{12}C^{+2}$	43	3	51
$^{30}Si^{+5}$	108	11	38
$^{54}Fe^{+9}$	194	27	33
$^{84}Kr^{+14}$	302	40	39
$^{132}Xe^{+22}$	475	69	40

IPN

This accelerator belongs to the CNRS/IN2P3 in Paris, France. This is a 14 MV tandem Van de Graaff accelerator using a double stripper (foil and gas).

Table 15. Available beams at IPN.

Ion	Energy [MeV]	LET [MeV/mg/cm^2]	Range [microns]
C	84	1.63	143
F	120	4	93
Cl	199	11.8	60
Ti	160	21	32
Ni	182	29.9	29
Br	236	40	31
I	325	62	31

SIRAD

This facility is in Legnaro laboratory, (Padova - Italy). It is a 15 MV tandem Van de Graaf, double stripper machine using two different ion sources.

Table 16. Available beams at SIRAD.

Ion Species	Energy (MeV)	q_1	q_2	Range in Si (μm)	Surface LET in Si (MeV×cm^2/mg)
^{1}H	28	1	1	4390	0.02
^{7}Li	56	3	3	378	0.37
^{11}B	80	4	5	195	1.01
^{12}C	94	5	6	171	1.49
^{16}O	108	6	7	109	2.85
^{19}F	122	7	8	99.3	3.67
^{28}Si	157	8	11	61.5	8.59
^{32}S	171	9	12	54.4	10.1
^{35}Cl	171	9	12	49.1	12.5
^{48}Ti	196	10	14	39.3	19.8
^{51}V	196	10	14	37.1	21.4
^{58}Ni	220	11	16	33.7	28.4
^{63}Cu	220	11	16	33.0	30.5
^{74}Ge	231	11	17	31.8	35.1
^{79}Br	241	11	18	31.3	38.6
^{107}Ag	266	12	20	27.6	54.7
^{127}I	276	12	21	27.9	61.8
^{197}Au	275	13	26	23.4	81.7

4.6.2 Protons

Several facilities across Europe are available for device testing using protons. As energy losses for proton in the air are low, all irradiation are performed in the air avoiding the complexity of the vacuum system.

Table 17. Available European proton accelerators.

CYCLONE – Belgium	Up to 70 MeV
JYFL – Finland	Up to 45 MeV
CPO – France	Up to 200 MeV
IPN – France	Up to 20 MeV
SIRAD – Italy	Up to 28 MeV
PSI/OPTIS - Switzerland	Up to 63 MeV
PSI/PIF - Switzerland	Up to 300 MeV

Using the primary beam at the highest energy, it is possible to change very quickly the energy on the device by adding degraders.

4.6.3 Neutrons

Neutron tests mainly concern avionic people but start to be a major concern for all electronic equipment. The JEDEC JESD89 [21] specification states that the preferred facility to perform the test is WNL (Los Alamos – USA) because its energy spectra match the neutron spectra at sea level (Figure 11). However, the same norm explains how to reach data using quasi-monoenergetic neutron beams, which is available in Europe. Two accelerators are equipped with neutron beam lines.

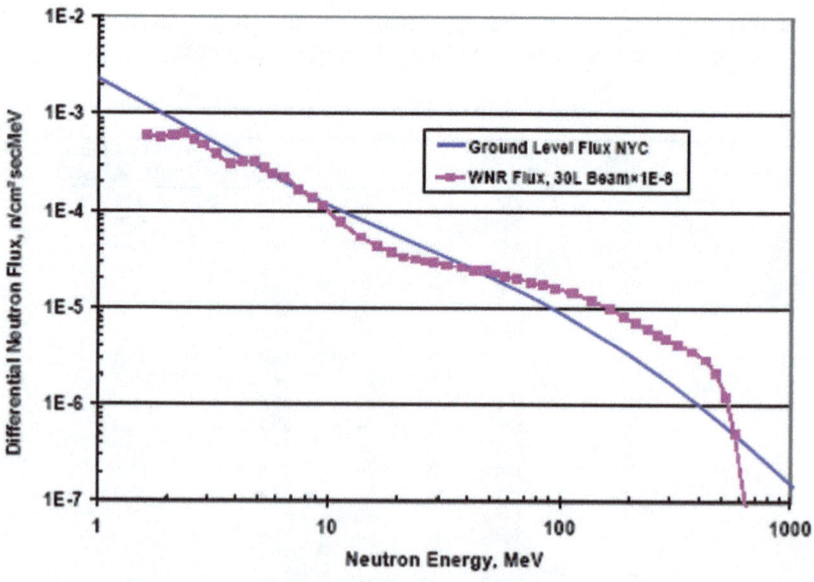

Figure 11. Differential neutron fluxes in NY City and at LANL (reduced with a factor 1E8).

CYCLONE

Cyclone is equipped with three different neutron beam lines:

- Quasi-monoenergetic line
- High flux line

4.7 Quasi-monoenergetic line

A collimated neutron beam in the energy range of 25 to 70 MeV can be produced using the CYCLONE proton beam [23-24-25]. The ^{7}Li(p,n)^{7}Be reaction (Q = -1.644 MeV) produces a quasi monoenergetic neutron beam.

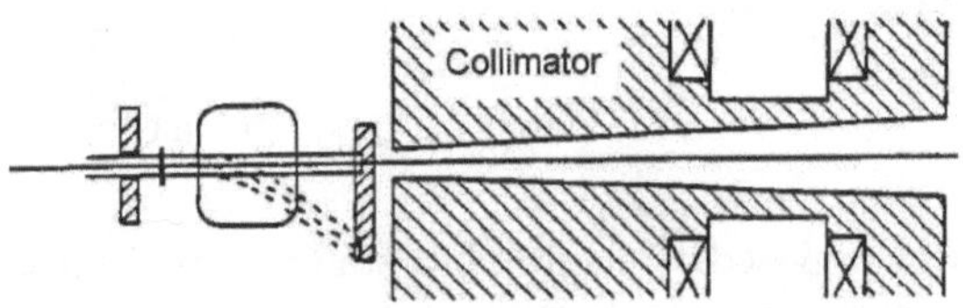

Figure 12. Neutron beam line, cave Q.

The collimated proton beam irradiates a thin lithium target and is then magnetically deflected on a graphite beam stop. This solution has been chosen to lower the background at the device position. It has the advantage to use the collimator and all its shielding to prevent neutrons and gamma generated in the beam dump to reach the irradiation position.

A second magnet, placed behind, is used to sweep out the charged particle contamination from the neutron beam. The whole set-up is surrounded by a concrete, a borax paraffin mixture and iron shielding. The neutron beam is collimated by several brass cylinders of increasing internal diameter. The collimator design took the following assumptions into account: the diameter of the collimator placed in front of the production target is 7 mm and the diameter of the last section is fixed to 20 mm.

The Figure 13 presents typical energy spectra for neutron emitted at 0° for three different proton energies (36, 48 and 63 MeV). On these graphs, we see that about 50 % of the neutrons are at a well defined energy. They are coming from the ^{7}Be produced in the ground state or in the first excited state (Q = 0,429 MeV). The mean value of this peak is about 2 MeV lower than the incident proton energy. The remaining 50 % are in a broad, rather uniform, low energy tail.

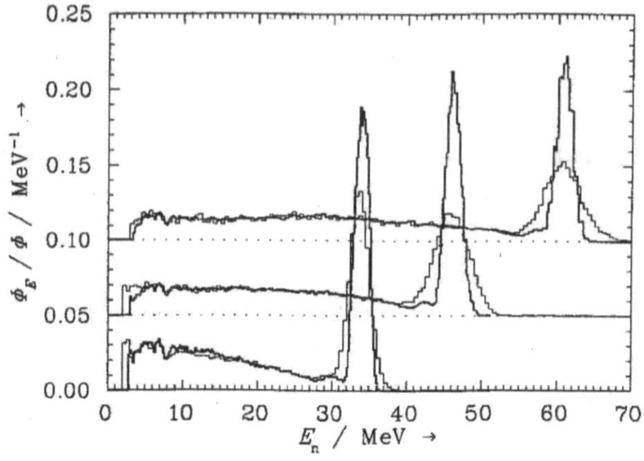

Figure 13. Neutron energy spectra in Q cave.

The used target is composed of natural lithium (6 % of ^{6}Li and 94 % of ^{7}Li), as the cross sections ratio at 0° between these two isotopes (^{6}Li / ^{7}Li) is 27 %, the produced neutrons from the reaction ^{6}Li(p,n)^{6}Be represent a very low contaminant in the spectra (2% of the total neutrons) [26-27].

As the reaction cross section above 30 MeV saturates at 35 mBarn, the produced neutron flux is not linked to the incident proton energy. At 3 m from the target, for a 3 mm thick natural lithium target and a 10 μA proton beam, the flux reaches 10^6 n°/s over an area of 30 mm in diameter. The peak widening limits the production target usable thickness (maximum 10 mm), typical FWHM value is 2 MeV.

4.7 High flux line

One of the CYCLONE beam line has been equipped to test miscellaneous equipments dedicated for CERN. The LHC environment constraint will be very large (1E14 n/cm²) for 10 years) it was thus necessary to reach such a high fluence in a reasonable time. A fast neutron beam based on the reaction $^9Be + d \rightarrow n + X$ using 50 MeV deuteron beam was developed; typical reached flux is 6.6E12 N/s sr. To avoid secondary beam contamination, a filter system was added behind the production target (1 cm thick polystyrene, 1 mm of cadmium and 1 mm of lead); the resulting contamination is 2.4% in gamma and 0.03% in light particles.

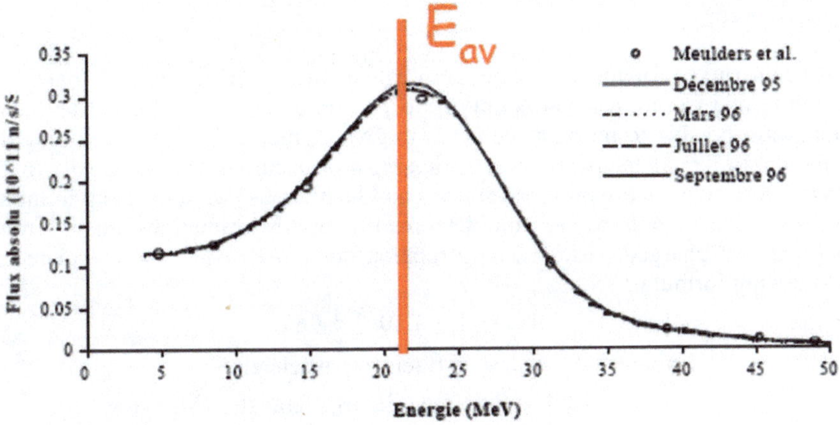

Figure 14. Neutron energy spectra in T2 cave.

Svedberg Laboratory (Uppsala, Sweden)

This Swedish laboratory proposes a large range of quasi-monoenergetic neutron using the thin target approach to produce quasi-monoenergetic neutrons in the energy range 20 – 180 MeV.

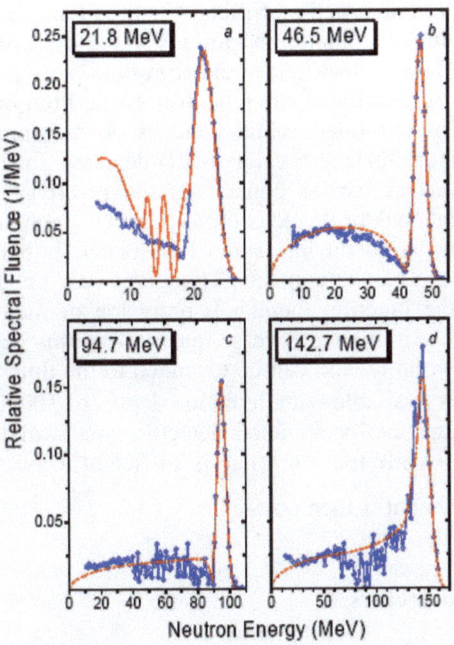

Figure 15. Neutron energy spectra in Uppsala.

4.8 Dosimetry

Used beam fluxes for testing range from a few ions/s cm² up to 1E5 ions/s cm² in heavy ions and from a few part/s cm² up to 1E9 part/s cm² in protons. The dosimetry system must be able to monitor the beam to give a precise fluence at any time during irradiation. Referring to specifications, flux must be accurate within a ± 10% margin.

While testing device with ions (protons and heavy ions) we must keep in mind that total dose is accumulated. A complete record of the irradiations history must be generated. For charged particles, a correlation dose in Rad may be calculated using the following formula:

$$D = 1.6\,X\,10^{-5}\ \phi\ LET$$

where ϕ is the ion fluence [particle/cm²]
LET is the Linear Energy Transfer [MeV/mg/cm²]

Different approaches are possible to monitor the beam flux; the final choice of the detection system is based on the type of particle to measure, the expected flux and the energy of the incident beam [35].

Most commonly detection systems around SEE facilities are scintillators, semiconductor detectors and ionization chambers.

4.8.1 Semiconductor detectors

These are based on a simple junction. Silicon Charged Particle detectors have a P-I-N structure [33-34] in which a depletion region is formed by applying reverse bias, with the resultant electric field collecting the electron-hole pairs produced by an incident charged particle. The resistivity of the silicon must be high enough to allow a large enough depletion region at moderate bias voltages. A traditional example of this type of detector is the Silicon Surface Barrier (SSB) detector. In this detector, the n-type silicon has a gold surface-barrier contact as the positive contact, and deposited aluminum is used at the back of the detector as the ohmic contact.

The new version is based on implanted (Passivated Implanted Planar Silicon or PIPS) rather than surface barrier contacts. This gives very good energy resolution and rugged detector. At the junction there is a repulsion of majority carriers so that a depleted region exists. An applied reverse bias widens this depleted region which is the sensitive detector volume, and can be extended to the limit of breakdown voltage. Detectors are generally available with depletion depths of 100 to 700 μm.

Using this technique, totally depleted detectors are available. These are used as transmission detectors while the ion range is sufficient to pass through the wafer. A direct dE/dx measurement is then possible.

4.8.2 Scintillator detectors

It is based on the property of certain materials to emit a flash of light when struck by a particle. This scintillating material is optically coupled with a photomultiplier tube (to convert the light signal in electric pulse usable by the acquisition chain) either directly or via a light guide.

The ideal scintillation material should have the following properties:

- High conversion efficiency of the lost beam energy in detectable light.
- Linearity of this conversion on a large energy range.
- Transparency of the medium to its own light for a good light collection.
- Short decay time of the luminescence to allow fast pulses.
- Index of refraction close to the glass for an efficient optical coupling.

There are two categories of scintillating materials: the organic and inorganic.

Inorganic presents the advantage to have a good linearity and light output, withdraw is a slow response time. Organic scintillators on their side present fast answer but have a lower light yield.

4.8.3 Ionization detectors

These are very often used for proton and heavy ion detection. These detectors are based on the collection of the ion pairs produced in a gaseous. The electrodes may be in flat, cylindrical, or other geometry, and the chamber can be filled with almost any gas but argon is generally used. Once particles enter in the detector, electron-ion pairs are generated. The mean number of pairs is proportional to the beam energy loss. Under the action of the electric field, electrons and ions travel towards electrodes where they are collected.

This type of detector may be used in different ways in function of the bias conditions as seen of Fig. 16.

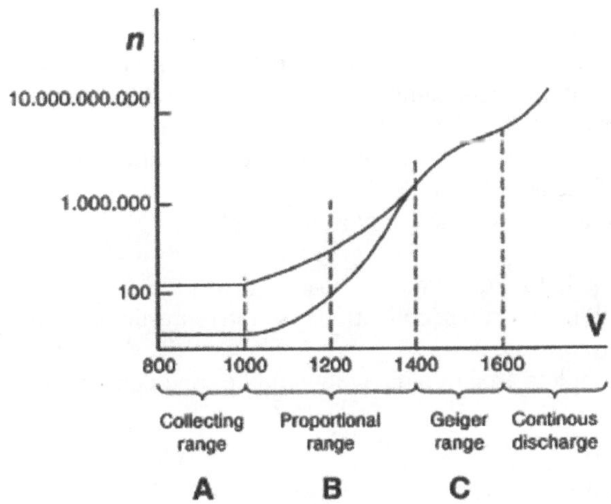

Figure 16. Ionization detector mode of operation.

- Ionization chamber: if bias is in the lower portion of the curve (A), the current starts to increases more and more the electron – ion pairs are collected before they can recombine. At a certain level, all created pairs are collected and an increase in voltage shows no effect.
- Proportional counter: if bias is raised above the first plateau, current increases again with voltage. Electric field is capable of accelerating electrons to energies where they can ionize gas molecule in the detector. The electrons created in this secondary process are then also accelerated to produce more and more ionization. This is the cascade or avalanche mode where the number of electron – ion pairs is directly proportional to the number of primary electrons. This results in a proportional amplification of the current which is linked to the applied bias.
- Geiger-Muller: for bias set above this second region, the total amount of ionization created through multiplication becomes sufficiently large that the space charge distorts the electric field about the anode. There is no proportionality anymore. Above this threshold, discharge occurs in the gas.

The Parallel Plate Avalanche Counter (PPAC) are gaseous detectors, with planar electrodes, working in avalanche mode. A single chamber consists of two planar metal (or metallized) electrodes kept at a fixed distance by a spacer. The gap between the electrodes is very small (1-2 mm) and is filled with gas which needs to be refreshed to avoid quenching effect. PPAC are routinely used for heavy ion monitoring.

4.9 Complementary tools (laser, ^{252}Cf, ^{241}Am, µbeams…)

Aside from the standard facilities used for routine SEE testing, some complementary tools bring some additional help when dealing with SEE.

Contrary to the broad-beam approach which consists in irradiating the whole surface of a device, lasers or micro-beams allows for performing localized exposure (spot diameter of the order of 1 µm) and then correlating a structure to an observed failure mode. A laser can also be triggered by the test application permitting the temporal characterization of the anomaly [18, 19].

However, these techniques have major drawbacks:

1. A laser is not able to cross metal or opaque layers,
2. Charge injection (and then collection) is not representative in case of the laser (track structure effect),
3. Mapping a complete area of a device would be time consuming.

4.10 Microbeam

SEE characterization using heavy ion beams is a global approach. It tells us if an event occurs, but no information on the location of its origin can be found. Testing with a laser overcomes this problem. A small laser spot and a precise localization of the laser light impact allow sensitive device nodes to be pinpointed with submicron accuracy [29-30-31-32]. The major problems of this technique is that the laser light is reflected by metallization layers and the penetration depth is only a few micron (a 815 nm laser penetrates 12μm in Si).

The test process generally starts using a defocused beam to define sensitive regions. A tightly focused laser spot at higher magnification is then used to pinpoint sensitive nodes within the regions identified previously.

During classical SEE characterization using heavy ion beams, the cross section is plotted in function of the LET which is not direct with lasers. An energy calibration needs to be performed in order to obtain a correlation between the laser energy and LET. It is to note that this calibration needs to be performed for each laser wavelength. Using a heavy ion microbeam overcomes this problem.

Table 18.

	Heavy Ion Beams	Laser	Heavy Ion Microbeam
Beam diameter on DUT	A few cm	Down to 1 μm	Down to 1 μm
Limitations	Range several tenths of μm	Small penetration depth	Range several tenths of μm
Localization of sensitive areas	NO	YES	YES
Study of rare phenomena	NO	YES	YES
Cross section determination	YES	NO	YES

In Europe, only one facility is equipped with a microbeam: GSI (Darmstadt, Germany). This facility uses its linear accelerator to produce ions from carbon to uranium energies between 1.4 MeV/amu and 11.4 MeV/amu.

In collaboration with CNES, CYCLONE (Louvain-la-Neuve, Belgium) team is studying a microbeam beam line [28]. The idea is to use one of the two available cocktails.

Then using this approach to characterize a device (measure of the σ(LET) sensitivity curve) is not realistic but these tools are very useful to help understanding failure modes.

Californium-252 is an artificial radioactive element that spontaneously fissions into several fragments (figure 17), α particles and neutrons. Only fission fragments that represent only 3% of the total amount of fission products are useful for SEE purpose.

Because of their low energy, the emitted ions are easily stopped (range of about 15μm in Silicon) and then are not representative of space particles. However, the mean LET value (around 43MeV/mg.cm^2) allows for inducing errors in devices and debugging test set-up before moving to the accelerator site.

Americium-241 is used as a α particles emitter in order to simulate the radioactivity of packages. For this purpose, it is recommended in the recent JEDEC JESD89 standard that addresses the avionics and terrestrial SEE issues.

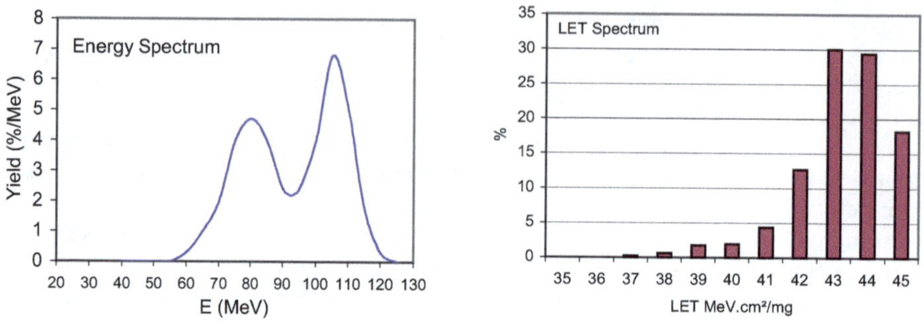

Figure 17. Fragments energy spectrum (left) and LET spectrum (right).

Acknowledgment

The authors would like to thank JP David from ONERA and F. Bezerra from CNES for their contribution in the TiD sections of this document.

References

[1] David J.P.: Total Dose Effects on Devices and Circuits, Space Technology Course (SREC04, June 2004), p.199.

[2] Duzellier S.: Single Event Effects: analysis and testing, Space Technology Course (SREC04, June 2004), p.221.

[3] Stapor W.J.: Single Event Effects Qualification, IEEE nuclear and space radiation effects conference, short course, section II, 1995.

[4] Titus J.L. et al: Experimental study of Single Event Gate Rupture and Burnout in vertical Power MOSFETs, IEEE Trans. Nuc. Sci., NS-43, n°2, p. 533, 1996.

[5] Marshall S.: Proton effects and test issues for satellite designers: part B. displacement effects, IEEE nuclear and space radiation effects conference, short course, section III, 1999.

[6] Johnston A.: Photonics Devices with Complex and Multiple Failure Modes, IEEE nuclear and space radiation effects conference, short course, section III, 2000.

[7] Hopkinson G.: Radiation Engineering Methods for Space Applications: Displacement Damage - Component Characterisation and Testing, Radiation Effects and Analysis, Radecs conference, short course, 2003.

[8] Schwank J. R.: Total-Dose Effects in MOS Devices, IEEE NSREC Short Cource, Section III, 2002.

[9] Ma T.P., Dressendorfer P.V.: Ionizing Radiation Effects in MOS Devices and Circuits, Wiley-Interscience, 1983.

[10] Johnston A.H., Swift G.M., and Rax B.G.: Total Dose Effects in Conventional Bipolar Transistors and Linear Integrated Circuits, IEEE Trans. Nucl. Sci. vol 41, n° 6, pp 2427-2436, Dec. 1994.

[11] Duzellier S. et al.: SEE Results using High Energy Ions, IEEE Trans. Nuc. Sci., NS-42, n°6, p. 1797, Dec. 1995.

[12] Dodd P.E. et al.: Impact of Ion Energy on Single-Event Upset, IEEE Trans. Nuc. Sci., NS-45, n°6, p. 2483, Dec. 1998.

[13] Ziegler J.F., Biersack J.P., Littmark U.: The stopping and range of Ions in solids. Volume 1 of "stopping and range of ions in matter", Pergamon Press (New-York).
[14] Petersen E.: Single Event analysis and prediction, IEEE nuclear and space radiation effects conference, short course, section III, 1997.
[15] Adams Jr. J. H.: Cosmic ray effects on microelectronics, Part IV, NRL memorandum report 5901, 1986.
[16] Pickel J. C., Blandford J. T.: Cosmic ray induced errors in MOS devices, IEEE Trans. Nuc. Sci., NS-27, n°2, 1006, 1980.
[17] Garth J.C., Burke E.A., Woolf S.: The Role of Scattered Radiation in the Dosimetry of Small Device Structures, IEEE Trans. Nucl. Sci. vol 27, n° 6, pp 1459-1464, Dec. 1980.
[18] Duzellier S. et al: Application of Laser Testing In Study of SEE Mechanisms In 16-Mbit Drams, IEEE Trans. Nucl. Sci., NS-47, n°6, December 2000.
[19] Makihara A. et al: Analysis of Single-Ion Multiple-Bit Upset in High-Density DRAMs, IEEE Trans. Nucl. Sci., NS-47, n°6, December 2000.
[20] John J. Livingood, "Principle of Cyclic Particle Accelerators", D. Van Nostrand Company Inc.
[21] JEDEC standard "Measurement and Reporting of Alpha Particles and Terrestrial Cosmic Ray-Induced Soft Errors in Semiconductor Devices", JESD89 August 2001
[22] G. Berger, G. Ryckewaert, R. Harboe-Sorensen, L. Adams « The Heavy Ion Irradiation Facility at CYCLONE - a dedicated SEE beam line » 1996 IEEE NSREC Workshop.
[23] A. Bol, P. Leleux, P. Lipnik, P. Macq, A. Ninane « A Novel Design for a fast intense neutron beam » NIM 214 (1983) 169.
[24] I. Slypen, V. Corcalciuc, A. Ninane, J.P. Meulders « Charged particles produced in fast neutron induced reactions on ^{12}C in the 48-80 MeV energy range » NIM A 337(1994)431-440.
[25] C. Dupont, « Mesures de Sections Efficaces Différentielles du Bremsstrahlung Neutron-Proton à 76.5 MeV », UCL PhD Thesis 1987.
[26] H. Shuhmacher, H.J. Brede, V. Dangendorf, M. Kuhfuss, J.-P. Meulders, W.D. Newhauser, R. Nolte and U.J. Schrewe, « Quasi-Monoenergetic Référence Neutron Beams With Energies From 25 MeV To 70 MeV », International Conference on Nuclear Data for Science and Technology, Trieste, May 1997.
[27] M. Lambert, S. Benck, I. Slypen, J.-P. Meulders and V. Corcalciuc, « Comparison of Fast Neutron Induced Light Charged Particle Production Cross Sections for Si and Al », International Conference on Nuclear Data for Science and Technology, Trieste May 1997
[28] F. Bezerra and G. Berger, "Heavy Ion Micro-Beam Study", RADECS Thematic Workshop on European Accelerators, Jyvaskyla May 2005.
[29] D. McMorrow et al., "Application of a Pulsed Laser for Evaluation and Optimization of SEU-Hard Designs," IEEE Transactions on Nuclear Science, Vol. 47, pp. 559–563 (2000).
[30] J. S. Melinger, et al., "Pulsed Laser-Induced Single Event Upset and Charge Collection Measurements as a Function of Optical Penetration Depth," Journal of Applied Physics, Vol. 84, pp. 690–703 (1998).
[31] G. C. Messenger and M. S. Ash, Single Event Phenomena (Chapman-Hall, New York, 1997).
[32] T. F. Miyahira, A. H. Johnston, H. N. Becker, S. D. LaLumondiere, and S. C. Moss, "Catastrophic Latchup in CMOS Analog-to-Digital Converters," IEEE Transactions on Nuclear Science, Vol. 48, pp. 1833–1840 (2001).
[33] H. A. Rijken, S. S. Klein, W. Jacobs, L. J. H. G. W. Teeuwen and M. J. A. de VoigtP. Burger " Subnanosecond timing with ion implanted detectors. " NIM B64 (1992) 272-276.

[34] E. Steinbauer, P. Bauer, M. Geretschläger, G. Bortels, J. P. Biersack and P. Burger "Energy resolution of Silicon detectors: approaching the physical limit." NIM B85 (1994) 642-649.
[35] Glenn Knoll, Radiation Detection and Measurement, 3rd Edition, 2000.

http://www.cyc.ucl.ac.be/
http://www.gnail.fr/public/presentation/index.html
http://pif.web.psi.ch/
http://www.phys.jvu.fi/research/accelerators/index.html
http://inpweb.in2p3.fr/%7edivac/divis/tandem.tandem.html

Error Rate Prediction of Digital Architectures: Test Methodology and Tools

Raoul Velazco, Fabien Faure

TIMA Labs, 46 Av. Félix Viallet,
38031 Grenoble (France)
http://www.tima.imag.fr
raoul.velazco@imag.fr
fabien.faure@imag.fr

Abstract. Evaluating the sensitivity to Single Event Effects of programmable digital integrated circuits (i.e. microprocessors, digital signal processors and field programmable gate arrays) requires specific methodologies and dedicated tools. Indeed, such an evaluation is based on data gathered from tests performed on-line during which the target circuit is exposed to a flux of particles having features (energy, range in Silicon) somewhat representative of the ones the circuit will encounter in its final environment. These experiments, usually called accelerated radiation ground testing, are performed by means of appropriate radiation facilities entailing thus significant development efforts and cost impact. In this chapter an approach will be presented, describing the corresponding hardware as well as software tools developed to deal with such experiments at a reasonable cost versus effort trade-off.

1. Introduction

Energetic particles hitting Silicon loose energy along their tracks. This energy is absorbed by the matter creating electron-holes pairs. This charge may be collected by diffusion and drift mechanisms resulting in a spurious current pulse at the struck node. The current pulse has a duration and amplitude which depend on both the incident particle features (energy, mass) and the circuit's technology. The phenomena caused by this current pulse are gathered under the appellation Single Event Effects (SEE).

Main SEEs are the so-called Single Event Upset (SEU) and the Single Event Latch-up (SEL). The former is not destructive and occurs when the current pulse resulting from ionization changes the content of one of the memory cells of the considered circuit. Its consequences depend on both the nature of the corrupted information and the occurrence instant. In case of SEL, the current pulse provokes a short circuit between ground and power by triggering a parasitic thyristor present in all CMOS circuits. Single Event Latch-ups can be easily detected by on-line measuring the circuit (or system) power consumption, shutting down the power supply if the consumption is higher than a predetermined limit. At the opposite, consequences of Single

R. Velazco et al. (eds.), Radiation Effects on Embedded Systems, 233–258.

Event Upsets can be difficult to detect or predict as they depend on the future use of the modified information after the SEU occurrence.

In a processor for instance, if a particular register is considered, it is straightforward to identify those periods during the execution of a given program that will be error prone if the SEU occurs, and those that will not. Indeed, if the considered SEU target is loaded with a value at time t and the value is used in an operation performed at instant $t + \Delta t$, the sensitivity to SEU will be limited (supposing that the register is not anymore used in the program) to the interval $[t, t + \Delta t]$ all SEUs occurring outside this interval being thus without effect at the program execution level. On the other hand, if the content of a register such as the program counter (PC), which is permanently used to point to the next instruction to be executed, is modified by an SEU, it is easy to forecast that in most of the cases the program execution will lead to results (or program execution durations) different than the expected ones. SEUs in PC leading to wrong program results are very difficult to be forecast; sequence-loss situations can in certain cases result in a program crash as the consequence of the execution of illegal instructions or infinite loops.

This rough analysis of the consequences of spurious modification of complex circuit's memory cells contents aims at showing that the consequences of SEU phenomenon are potentially critical for any application devoted to operate in radioactive environments and must thus be considered with care while selecting the parts and developing the architecture.

2. Radiation ground testing requirements and objectives

Predicting SEE error rates requires knowledge about the targeted circuit's response to a single particle. Terms, basic mechanisms, concepts and approximations used to successfully characterize programmable digital devices are described in the "Single Event Effects: analysis and testing" module of this document.

As stated earlier, accelerated radiation ground testing is necessary to study the response of the device under test (DUT) to a single particle. However, a radiation facility is not able to deliver such a granularity: particles are always delivered as a broad beam. Moreover, the DUT has many sensitive volumes. One can never know which particle of the beam will interact with a given sensitive volume of the device. As a consequence, radiation ground testing data have a statistical nature and are a function of the beam features (mainly the flux) and the device features: sensitive volume (SV) size, number of SV per surface unit. The experimenter thus needs a concept to remove these geometrical effects, and to measure the dynamic of the particle/device interaction. This is the concept of cross-section.

Let's imagine a virtual interaction between an incoming energetic particle and a device with a regular array of sensitive volumes. Moreover, let's assume that the particle strikes the device with a normal incidence. Finally, we require that the particle has a LET[1] high enough to trigger a single event if it passes through the sensitive volume ($Q_{col} \geq Q_{crit}$). The end-result of this experiment is a measure of the single event rate by unit of time (R_b). This virtual set-up is given in Figure 1.

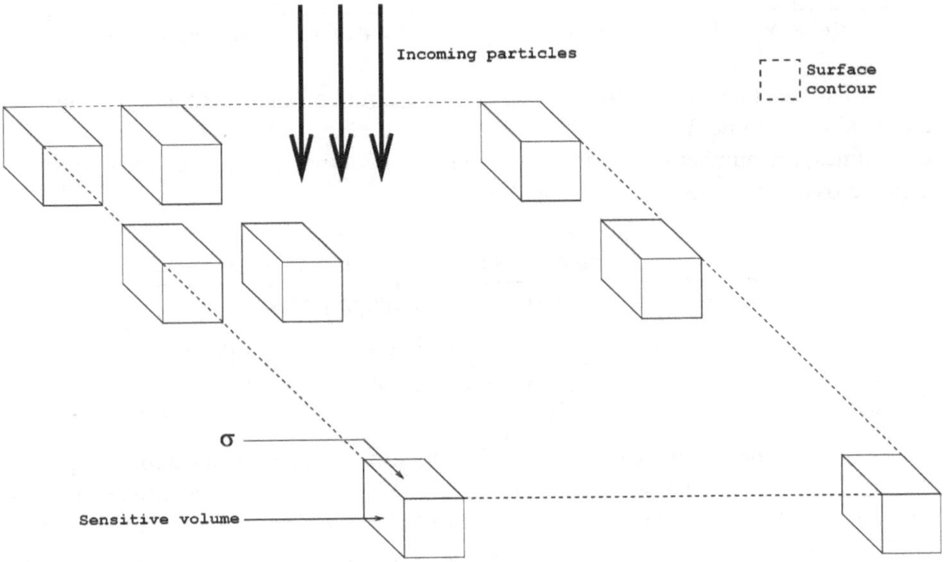

Figure 1. Virtual experiment set-up.

We can forecast that we will be able to observe a single event each time an incoming particle enters the top face of sensitive volume (its surface is noted σ). If the surface of the device seen by the beam is S and N_b is the number of sensitive volumes in S, the total sensitive surface of S is $\sigma \times N_b$ and we have:

$$\frac{\sigma \times N_b}{S} = \frac{R_b}{R_a}$$

where R_a is the number of incoming particles crossing the surface S by unit of time. So,

$$\sigma = \frac{R_b}{\frac{R_a}{S} \times N_b} = \frac{R_b}{\Phi_a \times N_b}$$

[1] The LET (Linear Energy Transfer) is a measurement of the ionization caused as the incoming particle looses energy along its path. For instance, in Silicon, a particle having a LET of 97 MeV.cm²/mg is depositing about 1 pC per micron.

where $\Phi_a = R_a / S$ is the beam flux (number of particles per unit of surface and time). If we introduce the experiment duration T,

$$\sigma = \frac{R_b \times T}{\Phi_a \times N_b \times T} = \frac{\text{\# of single event recorded during the experiment}}{\text{fluence of the experiment} \times \text{\# of sensitive volumes exposed}}$$

σ is called the interaction cross-section and is a direct measure of the top surface of the sensitive volume. Its unit is the cm² or it is expressed in barn (1 barn = 10^{-24} cm²). Sometimes, the number of sensitive volumes is not known, and σ is given as an interaction cross-section per device (or per bit).

$$\sigma_{dev} = \frac{\text{\# of single event recorded during the experiment}}{\text{fluence of the experiment}}$$

$$\sigma_{bit} = \frac{\text{\# of single event recorded during the experiment}}{\text{fluence of the experiment} \times \text{\# of bits}}$$

As a consequence, the end-product of a radiation ground testing will be a plot of the interaction cross-section versus particle LET. Two important magnitudes must be noticed: the LET threshold, which is the first LET provoking an event and the saturation cross-section which corresponds to whole sensitive area. From the LET threshold and the final environment nature, can be decided the assessment needed to use the tested circuit in a reliable way. Table 1 summarizes typical values of LET threshold and corresponding assessment. The obtained measures of the LET threshold allow determining a first diagnosis about the suitability of the DUT to fit with the requirements of the final radiation environment.

Table 1. LET threshold and environments to be assessed.

DEVICE THRESHOLD	*ENVIRONMENT TO BE ASSESSED*
LET_{th} < 10 MeV.cm²/mg	Cosmic Rays, Trapped Protons, Solar Flares
LET_{th} = 10 – 100 MeV.cm²/mg	Cosmic Rays
LET_{th} > 100 MeV.cm²/mg	No analysis required

In the following we will deal only with Single Event Upsets. To derive the SEU cross-section the circuit must be active during its exposure the selected particle beam. One important methodological aspect is the choice of the activity the DUT will achieve during the radiation experiment. This strongly depends on the type of circuit, being quite straightforward for those circuits having a single function, such as memories and quite difficult for circuits such as processors capable of executing a set of instructions or commands. In the next section some trends about main strategies usually adopted for representative circuits will be briefly exposed.

2.1 Static and dynamic SEU test strategies

A key aspect in radiation ground testing of microelectronic integrated circuits is the activity to be accomplished by the circuit during the exposure to the beam effects. This activity must be as representative as possible of the one the circuit will accomplish when installed in the final application and environment.

For chips such as memories (SRAMs, DRAMs, ...), the estimation of the sensitivity to SEUs can easily be done by loading them with a predetermined pattern, whose corruption by SEUs is identified by periodically reading out the exercised memory area and comparing it with the expected values.

Such a strategy is generally called "static test" because all accessible targets are exposed to SEUs at the same time, situation far from the real circuit activity. When used during radiation ground testing, a static test strategy will provide a worst-case estimation of the memory sensitivity to SEUs. Indeed, when included in an equipment the memory will be used to store information, some of them (a program code or a data area for instance) stay unmodified during the whole system operation but some other zones are used more dynamically. A way to determine a sensitivity close to the one of the final equipment consists in evaluating the average number of used bits and estimating the sensitivity as the product of this number by the per-bit sensitivity (derived of the memory cross-section issued form radiation testing divided by the number of bits).

When the DUT is a processor, attempts of adopting the static strategy were made by considering the target memory area comprising all accessible DUT registers and internal memory zones accessible by the execution of appropriate instruction sequences. A program implementing a static test strategy can be easily written using mainly LOAD and STORE instructions targeting all those processor's memory elements accessible to the instruction set (registers, internal SRAM ...). Running this program when the processor is exposed to radiation will provide a cross-section that corresponds to a very worst-case estimation of the circuit SEU sensitivity, not corresponding to the sensitivity of the DUT when running a real application. Indeed, registers and various memory zones are used in a dynamic and not controllable manner when running a given program: data stored in a register for instance will be sensitive to radiation only in the period during which the data will have a meaning for the program.

Extrapolating memory testing strategy to processors was shown to significantly overestimate the error rate. Indeed, in [1] [2] the authors performed ground testing with both a static test strategy (in which the program executed by the processor under test is restricted to loading and observing the register set and internal memory) and a set of programs intended to be somewhat "representative". Results of the application of such an approach, called dynamic strategy, to different processors including 16-bit processors (the Motorola 68000 and 68882) and a reduced instruction set processor with parallel capabilities (the Inmos Transputer T400) showed that

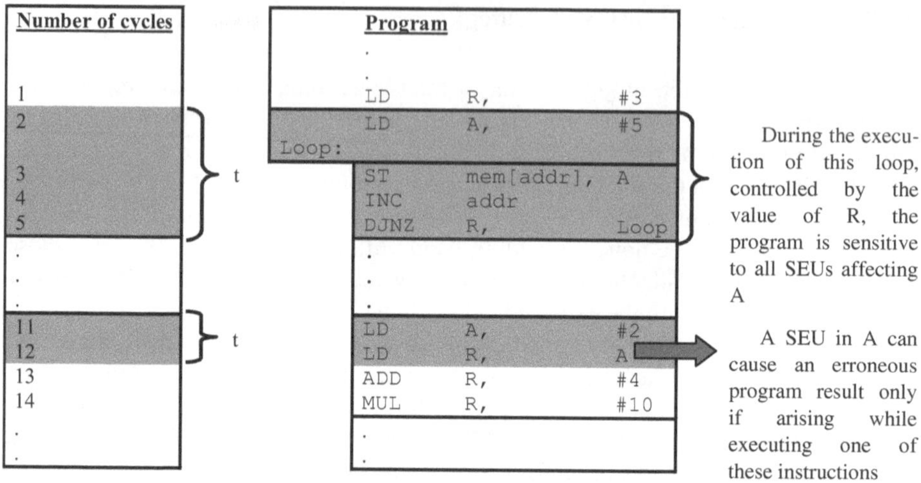

Figure 2. Example of duty cycles.

error rates were at least one order of magnitude less than those derived from ground tests performed with static test strategy.

These results are not surprising. Indeed, to induce an error at a program output, an upset must occur during the sensitive period of the target. Moreover, many registers and memory elements are not used by the program; upsets affecting these latches will have no effect on the program outputs, and consequently will not contribute to the final error rate. Unfortunately, there is little published work comparing error rates derived from ground testing using benchmark programs with error rates obtained during ground testing or observed in flight for an actual application.

Ideally, the final application program should be used during radiation experiments to get realistic cross-section measures, but often it is not available at the time radiation tests are prepared to select among candidate chips for space equipment. An interesting approach to deal with this difficulty was proposed in [3], where the authors stressed the importance for evaluating the application's cross-section, of calculating the "duty factors" of various processor elements used by the software. The time between loading a given register or memory location and writing it to memory is the period during which the register is sensitive to upsets. When this period is expressed as a percentage of the total execution time, it is called the duty factor of that particular register or memory element. The sensitivity to radiation of a program can be calculated as the sum, for all the sensitive elements, of individual sensitivities (SEU cross-sections issued from a radiation ground testing with a static strategy) weighted by the corresponding duty factors.

As an example, let us consider the piece of program depicted in Figure 2 in which only one register, register A, is considered for the purpose of this analysis. To ease the explanations we suppose that every instruction is executed in one clock cycle and that the maximum number of cycles needed to execute the program is 1200.

The first highlighted program portion is a loop controlled by the content of register R, which is automatically decremented while executing conditional branching instruction DJNZ whose execution will allow to enter the second highlighted program portion when R will contain 0. In the second program portion the content of A is first initialized to 2, to be then transferred to register R and used as an operand for different operations. It is clear that, as far as register A is concerned, its sensitivity to SEUs will be during this program portions (assuming that it is not anymore used outside them) and the contribution to the global sensitivity can therefore be calculated as the addition of the corresponding cycles: 10 cycles (1 for the initialization of R and 9 for the three times it will enter the loop) for the first portion and 2 cycles for the second one. The contribution of A to the error rate due to SEU in this program will therefore be 0.01, thus 100 times less than the underlying sensitivity of A derived from a static strategy.

This way of calculating the error rate of a program will provide good estimations needing only limited ground test experiments to calculate the underlying cross section of any of the potential SEU targets of the studied DUT. The duty cycles being calculated from an analysis performed on the assembly language source code can be used to follow any future evolution of the application program. Nevertheless, the main difficulty comes from the diversity of processors and instruction sets making very difficult in the practice to develop a tool automating the calculation of duty periods. Indeed, to our knowledge, from the first presentation in 1988 of this approach, no automated tool allowing the derivation of the duty factors for a given application has been presented. The development of such a tool is seriously impeded by the combinatorial explosion of execution paths (induced by software constructs such as loops and conditional statements) making analysis intractable. Moreover, variations in program inputs can result in different instruction paths, different duty factors and thus different SEU vulnerability. Another limitation comes from the fact that not all the SEU targets are accessible through the instruction set. Many temporary registers and flip-flops existing in a processor are not known at the architectural level and their contribution to the global error rate cannot be included in the calculations. Despite this intrinsic limitation, the error rates calculated taking into account duty periods of SEU accessible targets is generally considered as being quite close to the expected error rates when the circuit will operate in the final application.

2.2. Error types and guidelines

2.2.1 Error types

Several types of error may be encountered during SEE testing of a programmable device. Following are some situations used in this section:

- Corrupted output: The program output is not the one expected,
- Delay: The program execution duration is shorter or longer than expected. This is considered an error, especially in real-time systems,
- Loss of sequence: The program execution flow is broken and the processor needs a soft reset to restart,
- Functional interrupt: The processor does not respond to a soft reset. Exiting this situation requires a hard reset (power off/on).

Since they are programmable, most of these devices embed a high number of memory elements (registers, cache memories). All these memories can potentially be perturbed by SEUs. Nowadays, it is not uncommon to have processors with more than 500 KB of cache memory, making it a preferred target for SEUs. As a reference throughout the discussion, Figure 3 shows a napkin sketch of a virtual microprocessor.

Bus Unit
This component usually embeds a few registers to latch data as well as addresses. The main concern here is SEU. Upsets in these latches may lead to incorrect data/address read or write.

Integer Unit, Floating Point Unit
Most of the time, IUs and FPUs are pipelined. Upsets on these registers may lead to incorrect computations.

Register File
With recent compiler technology, register file usage is fairly well optimized, meaning that these memories will have a very high duty-factor. As a consequence, SEUs in these registers have an instantaneous impact on the program output, going from corrupted output to a complete loss of sequence.

Instruction Cache
These units are split in two areas. The largest one is an SRAM array that store fetched instructions. The second one is a tag array, whose purpose is to validate/invalidate fetched code. Upsets in the tag array have mainly two consequences:

- If an upset invalidates an instruction that will be executed, the direct consequence is a delay in the program execution since this instruction will have to be fetched again,

- On the other hand, if an incorrect instruction is validated, the program flow will be broken.

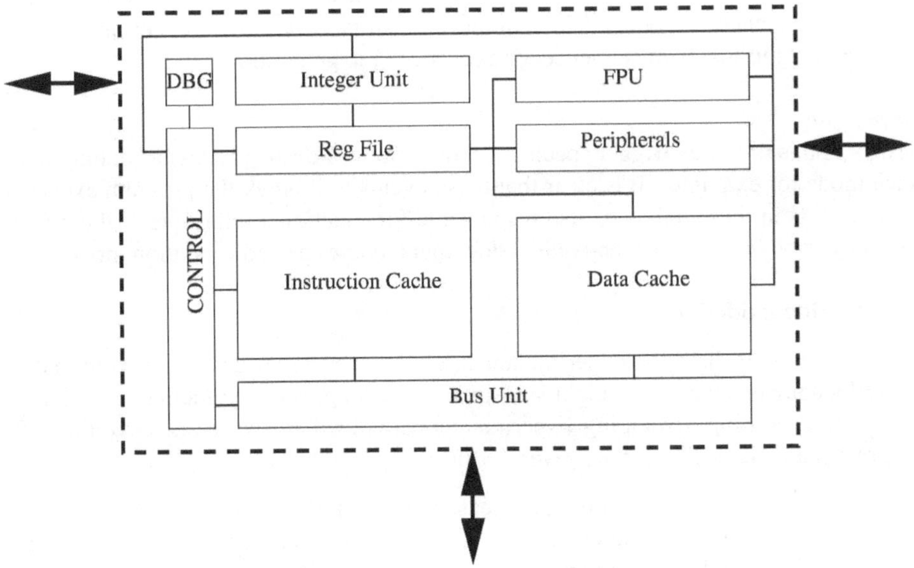

Figure 3. A virtual complex microprocessor.

Consequences of upsets in the instruction array are hard to predict. If the corrupted instruction is not validated by the tag array, no incorrect behaviour will be observed. If the code is validated by the tag array, three situations are possible:

- The corrupted instruction is not anymore in the processor instruction set. When the control unit tries to decode it, an exception/trap will be generated,
- The bit-flip changes the instruction (for example in Reduced Instruction Set processors, RISC, opcodes of load and store instructions differ only by one bit),
- The upset changes the operands of the instruction.

Again, the final consequence varies from corrupted outputs to loss of sequence.

Data Cache

Data caches are built like instruction caches i.e. with a tag array and a data array. Upsets in the tag array may invalidate data (introducing delay) or validate out-of-date data (leading to corrupted output). As expected, upsets in the data array may lead either to a corrupted output or to no observable effects if the data were out-of-date. Upsets in data cache memories are very unlikely to cause a loss of sequence. This situation can occur if the code is manipulated as data by the processor (auto-morphing code for example).

Control Unit
The control unit may implement complex algorithms depending on the processor (code pre-fetching, out-of-order execution…) by means of state machines or micro-code and sequencer/decoders. This complex circuitry may easily be perturbed by an upset, leading to losses of sequence, or exception/trap generation.

Debug unit
Upsets in this unit may trigger special execution modes of the processor (debug mode, trace mode for example). It is clear that these events will break the program execution flow. Upsets in these units are also responsible for functional interrupts of the processor (they may, for example, activate undocumented/unexpected execution mode).

2.2.2 Testing guidelines

Standards exist to help the experimenter in qualifying digital devices. Unfortunately, these documents concentrate mainly on memory testing. Some guidelines are given in this section, aiming at helping experimenters involved in the characterization of a programmable device. Hints are given for three types of tests:

- Static Test: A type of test aiming at measuring the static sensitivity (i.e. static cross-section) of a device's memory elements. The DUT should be inactive during irradiation. This condition is required in order to avoid Single Event Transients to be latched during read/write accesses to memories. This type of test should be conduced as following. At the beginning, the processor memory elements must be filled with a known pattern. Then, the processor is deactivated, for example by gating its clock. Beam is then turned on. At the end of the experiment, the beam shall be turned off, and then the clock be freed. The processor then dumps data hold in its memory elements,
- Semi-static Test: This type of test still aims at determining the sensitivity of memory elements, so they should adopt the same strategy as static tests. However, the DUT is active while the beam is on. By subtracting the cross-section obtained here with the static one, the experimenter is able to measure the consequence of Single Event Transients. Semi-static Tests shall be carried out at various frequencies,
- Dynamic Test: This kind of experiment is useful to obtain figures about the error-rate of a given application. It also helps determining exotic failure modes of a processor. The processor shall be active during irradiation, and performing some computations (for example a sort program, or a matrix multiplication). Dynamic tests should be performed at various frequencies and with cache memory on and off.

Whether the test is static, semi-static or dynamic, the experimenter should add the following features to his set-up:

- Watchdog: This feature is required to measure the execution time of a program. The DUT asserts an output pin if and only if it has reached the end of its program. If the recorded time is shorter or longer than expected, something happened during the test,

- Trap Tables: These tables must all be filled with codes identifying the nature of the trap. Ways of identifying may be writing a signature in the main memory, or asserting a dedicated output pin per trap considered. This helps the experimenter understanding what went wrong if an unexpected trap is triggered,
- NOP: After the program, the memory must be filled with the NOP instruction (No OPeration). Just before the end of the code memory, a specific program shall be written. This program asserts a dedicated output pin, and then loops forever. This feature aims at capturing unexpected jumps outside of the program address ranges.

2.3 Hardware set-up

Evaluating the error rate of equipments designed to operate during radiation exposure is a major concern. The general approach for such estimations is based on both the features of the final environment and the sensitivity to radiation of the parts included in the system. Among these parts, memories and processors, present in almost all complex digital systems, can be identified as the major contribution to the global error-rate. Indeed, the large number of memory elements they comprise make them potentially sensitive to Single Event Upsets which is one of the critical effects owing to the facts that as a consequence of a single particle hit either data having crucial importance for the application can be modified, or the program sequencing of a processor at the heart of the equipment may be loss provoking system unexpected and possibly critical behaviours.

Estimating the sensitivity of a circuit to SEE phenomena needs, as stated before, to expose the target circuit to a suitable flux of particles using radiation facilities such as particles accelerators (cyclotrons, linear accelerators, etc). To provide precise measures of the energy of incident particles, these tests are usually performed in vacuum chambers. Photo depicted in Figure 4 shows the vacuum chamber used for SEE testing in the cyclotron available at Lawrence Berkeley Labs. (LBL, Berkeley, CA, USA).

The main difficulty entailed by such tests, usually called radiation ground tests, is the fact that they need to be performed "on-line" i.e. the circuit under test must be active during the experiment duration. This means that the circuit under study must operate within a typical electronic/digital environment, including in the case of the tested circuit is a processor, external memory space and all required circuits allowing both to make the circuit capable to execute the selected program and to on-line detect the effects of SEEs.

From a hardware point of view, this means that a test platform must be developed. Such a platform should provide the above mentioned capabilities as well as the communication with the external world to store, for further analysis, information related with detected errors. A typical hardware set-up of an SEE experiment is provided in Figure 5. The photo was taken during a radiation ground test campaign realized at the cyclotron of LBL. The hardware platform is mounted in a mobile support, which will allow aligning the target circuit with the beam and, if needed/possible, to "tilt" the device under test in order to simulate with the same beam the effect of particles having higher effective LET. This experiment was performed using a tester called THESIC [4] which allows through its motherboard/daughterboard architecture, to reuse some of the functionalities for different DUTs. Indeed, the motherboard was designed to cope with the following requirements:

- Control the power consumption of the DUT to detect Single Event Latch-ups and switch off the power supply in order to avoid DUT damages,
- Starting and stopping daughterboard activities, detecting and recovering from SEUs provoking program sequence-loss in processors (by means of a programmable timer),
- On-line detecting SEUs and recovering related information. This last aspect is achieved, for those SEUs resulting in an erroneous DUT output by reading the content of a memory space shared by the motherboard and the DUT board in which the DUT is supposed to write results of the program execution. SEUs resulting in program sequence-loss are detected by means of a timer, which will indicate that the program duration is different than the expected one and stop the program execution by asserting an asynchronous interrupt signal.

Figure 4. The vacuum chamber for SEE testing with the 88" cyclotron facility of Lawrence Berkeley Labs.

The goal of these experiments, during which the circuit is exposed to the chosen particle beams while performing a given activity, is to derive the so-called SEE cross-sections which are a measure of the sensitive area. Typically used particle beams include heavy ions, protons and neutrons (atmospheric environment).

Figure 5. A typical hardware set-up for SEE testing of processor architecture.

3. Estimating error rates for a digital architecture

As seen in previous section, the estimation of a processor sensitivity to SEU can be far from the real sensitivity if a static strategy is adopted during radiation ground testing. Indeed, such a strategy considers that every bit flip affecting the content of a register will result in an error at the program execution level. This is obviously not the situation when a real application program is executed and the solution should be the use of the final application during the exposure to radiation beams. Unfortunately, the final applications are generally not available when the radiation ground testing devoted to select the processors to be included in the equipment are performed. Moreover, the final application may entail, to be exercised, the use of architectures including sensors communicating with the external world. This dependency of data coming from the environment makes difficult to establish realistic test scenarios which are accessible only to the designers or final users. Summarizing, the main problem when assessing the reliability of a processor with respect to SEEs is the difficulty to exercise the DUT in a way as close as possible to the one of the final application. In the practice, this difficulty was attempted to be tackled by defining program benchmarks supposed to provide more realistic Figures corresponding to the dynamic sensitivity with respect to SEUs. Such programs typically include a sort algorithm, an FFT and a matrix multiplication program.

The main goal of next sections is to describe an approach allowing predicting the SEU cross-section for any program from both the static SEU cross-section derived from radiation ground testing and error rates issued from fault injection sessions.

3.1 General methodology

Let's suppose that a method allowing injecting bit-flips randomly in both the occurrence instant and target can be implemented for the considered programmable DUT (processor, microcontroller, digital signal processor). The cross-section derived from the static test strategy provides the average number of particles of a given type which is necessary to provoke a bit flip of one of the memory cells included in the DUT.

If bit-blips are injected concurrently with the execution of a given program by software and/or hardware means, it can be derived an error rate, which will be called τ_{inj} in the following, as the number of detected errors divided by the number of injected bit flips:

$$\tau_{inj} = \frac{\#\,Errors}{\#\,injected\ bit\ flips}$$

As τ_{inj} can be interpreted as the number of bit-flips needed in the average to provoke an error in the program, the sensitivity to SEUs of the exercised program can be calculated as the product between the underlying cross-section (which provides the average number of particles needed to provoke a bit flip) and the error rate issued from fault injection:

$$\tau_{SEU} = \sigma_{SEU} * \tau_{inj}$$

This approach to estimate the error rate can be in principle applied to any processor. The accuracy of derived error rates when compared to the ones measured under radiation will strongly depend on the number of memory elements included in the studied circuit which are not accessible by means of the instruction set.

Main difficulty of this approach resides in the implementation of the fault injection strategy which must emulate as close as possible real bit-flips affecting the processor as the consequence of an SEU. In the next sub-section will be described a possible implementation of such an approach using the application hardware with minor modifications. Alternatives to that strategy based on the use of DUT software models can be also adopted but are out of the scope of this presentation.

3.2. A hardware-based strategy for SEU-like faults injection

We consider architectures organized around a device (a processor) capable of executing instruction sequences and with the possibility of taking into account asynchronous signals (i.e. interruptions, exceptions). In principle, this processor can be programmed to directly or indirectly perform read and write operations of any of the external SRAM locations, as well as its internal registers and memory area. On the other hand, as stated before, there is no way to access (to control and observe) a subset of the internal processor's memory elements (particularly the flip-flops and latches embedded in processor's control part), because they are either unknown or not directly accessible through the instruction set.

For most existing processors, bit flips can be injected by software means concurrently with the execution of a program, as the result of the execution, at a desired instant, of dedicated sequences of instructions [5,6,8]. In the following, such software simulated bit-flips will be called CEU (Code Emulated Upsets). The piece of code able to provoke the CEU occurrence will be called CEU code. The memory location in which the upset is injected will be called CEU target.

Typically, injecting a bit flip at a general purpose register or at directly addressable internal or external memory location needs only a few instructions to perform the following tasks:

- reading the existing content of the CEU target,
- XOR-ing it (perform the exclusive-or logic operation) to an appropriate mask value (having "1" for those bits that are to be flipped and "0" elsewhere),
- writing the corrupted value back to the CEU target location.

The only remaining step required to emulate an upset is to trigger the execution of the CEU code at the desired instant. If the CEU code is located in a suitable memory space (external SRAM for instance) at a predefined address, pointed to by the interruption vector (or an equivalent mechanism), this step can be achieved by asserting an interruption. Indeed, in response to an interrupt signal assertion, the processor will perform the following tasks:

- stop running the program after completion of the currently executed instruction,
- save the context (at least the program counter content) in the stack,
- jump to the CEU code and execute it, provoking the upset,
- restore the context from the stack and resume the program execution.

As a result to this four-step activity, depicted in Fig. 6, the program will be executed under very close conditions to those appearing when a bit flip occurs as the result of a particle having enough energy to provoke an SEU, hitting the circuit at the same considered instant and target bit. Except for the CEU target modified by the upset, the only processor registers used to perform the simulation are the stack pointer (which saves and restores the critical registers in the stack) and the program counter (used to point to the next instruction to be executed). These two registers will require a particular attention when targeted by the CEU injection.

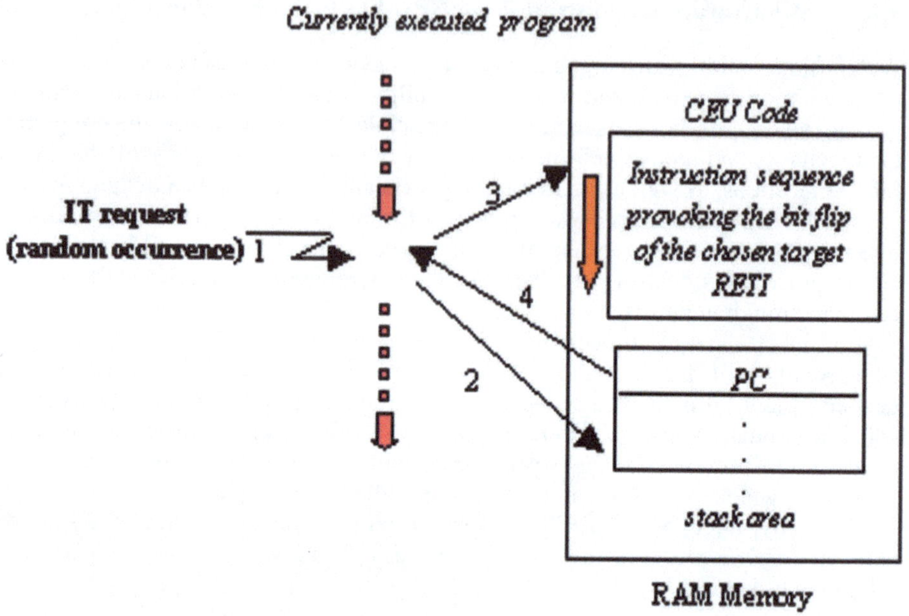

Figure 6. Single Event Upset injection by means of interruption.

In the following will be illustrated the determination of significant CEU sequences for typical targets: internal registers, internal or external SRAM memory and special function registers. First step consists in the analysis of the access nature and functionality of the various memory elements included in the processor architecture, this to determine families of CEU targets to which will be associated generic CEU codes. These generic codes will contain variables whose affectation with appropriate values prior to perform the CEU injection will allow to determine the target bits. Let us suppose a processor comprising:

- an accumulator ACC, implicitly addressable using LOAD ACC instruction,
- Registers (SFRs) which are mapped in the upper part of an internal SRAM whose content is addressable through the accumulator, using instruction LOAD ACC, addr,
- 128 bytes of internal SRAM, accessible by direct addressing mode through the accumulator,
- The stack pointer SP (1 byte) which points to the internal SRAM (where the stack is supposed to be implemented),
- The 2 bytes of program counter (PC).

Once the CEU families are identified, a CEU code must be associated to each of them. We recall that the CEU code is a sequence of processor instructions whose execution provokes the modification of one bit of a selected target. The simplest example of CEU code is the one associated to the accumulator. Indeed, to modify one bit of ACC only two instructions are needed: the exclusive OR (also called X-OR in the following) between the value stored in ACC and a byte indicating by the position of a 1 among 0s the bit of ACC to be inverted (noted BitPos), followed by the return from interrupt (RETI) instruction. The goal of CEU injection being the modification of a single bit of the SEU sensitive area, a special care must be taken to save the content of all memory elements different to the CEU target which must be used within the CEU code. For instance when the CEU target is a byte of an internal memory area, the use of accumulator ACC can be mandatory to get the value to be modified and stored it in the final targeted address. This needs to save the content of ACC, using for instance the stack and related PUSH and POP instructions, thus avoiding the injection (by mistake) of multiple bit-flips. Table 2 gives examples in pseudo-assembly language of these two CEU code types.

Table 2. CEU codes associated to the Accumulator and the internal SRAM. "addr" is the address of the SRAM byte to be perturbed, "BitPos" is a byte having a 1 among 0s (corresponding to the position to be inverted).

CEU TARGET	***CEU CODE (ASSEMBLY LANGUAGE OF 8051)***	***COMMENTS***
Accumulator	`XOR ACC, BitPos`	Modification of one ACC register bit
	`RETI`	Return to main program
Internal or SRAM	`PUSH ACC`	Save the content of accumulator ACC
	`LOAD ACC, addr`	Read the content of the target byte
	`XOR ACC, BitPos`	Modify the target bit
	`STORE ACC, addr`	Store the modified byte in target SRAM
	`POP ACC`	Restore ACC
	`RETI`	Return to main program

It is important to note that the CEU code may include instruction sequences to read, modify and overwrite values stored in the stack (this last operation being only available when the DUT uses a software implemented stack). This makes it possible to inject CEUs on PC and other control registers, often not directly accessible by the instruction set. In all processor having an architecture in which the stack is accessible through the instruction set, the following method can be adopted to inject faults in PC: perform an X-OR between the values of PC stored in the stack and the one of a mask corresponding to the target bit. In the case SP points to the top of the stack where is loaded the less significant byte of PC, the most significant byte (MSB) being in the following address, the CEU code associated with PC will begin by decrementing SP, to point to the MSB (see Table 3).

Finally, the injection of bit-flips in the stack pointer must be analysed with care. Indeed, the faults arising in SP have chances of having critical consequences to all program including operations with the stack (subroutine calls, interrupt processing ...). The CEU approach being based on the use of interrupts, the modification of SP, which is used to restore the program context (return address) at the end of the interrupt processing, must be accomplished with particular care. A general solution to this problem, avoiding the program sequence-loss that will result from loading PC with a wrong address as the consequence of a modification of SP was found in [5] and will not be presented here. It mainly consists in composing, at a particular address, an unconditional jump instruction branching to the address at which the program was interrupted (recovered from the stack) before perturbing SP. In that way; before changing the content of SP, the return address to the main program is saved at a predetermined address in the external SRAM and instead of using the return from interrupt (RETI) instruction to resume the program execution, a jump to this address is used provoking the jump to the real return address without using SP.

Table 3. CEU codes of PC.

CEU TARGET	***CEU CODE (PSEUDO ASSEMBLY LANG.)***			***COMMENTS***
Program Counter Low	`PUSH`	`R0`		Save the content of R0 in the stack
	`PUSH`	`ACC`		Save the content of ACC
	`LOAD`	`R0,`	`SP`	Use R0 to point where PCL is stored
	`LOAD`	`ACC,`	`@R0`	Load PCL in the ACC register
	`XOR`	`ACC,`	**`BitPos`**	Flip the content of the target bit in ACC
	`STORE`	`@R0,`	`ACC`	Store the modified value in PCL
	`POP`	`ACC`		Restore ACC
	`POP`	`R0`		Restore R0
	`RETI`			Return to main program
Program Counter High	`PUSH`	`R0`		Save R0
	`PUSH`	`ACC`		Save the accumulator content
	`DEC`	`SP`		Decrement the content of SP
	`LOAD`	`R0,`	`@SP`	Point with R0 to the second half of PC
	`LOAD`	`ACC,`	`@R0`	Load the content of PCH in ACC
	`XOR`	`ACC,`	**`BitPos`**	Change the target bit content at ACC
	`STORE`	`@R0,`	`ACC`	Transfer ACC content to PCH
	`INC`	`SP`		Increment SP
	`POP`	`ACC`		Restore ACC
	`POP`	`R0`		Restore R0
	`RETI`			Return to main program

The size of CEU codes, for an 8 bit microprocessor for instance, may go from 3 bytes (in the case of the accumulator) to a few tens of bytes (in case of PC). Considering, for the purpose of a rough evaluation, a clock cycle of 1 μs, the execution of the CEU code by the DUT will take no more a few tens of μs. To this time must be added the times needed to prepare the CEU code, operation that can be done by the DUT itself or by an external device and to execute the program under test itself. To get an order of magnitude of the duration of fault injection sessions realized with the CEU approach, the ones realized with THESIC tester on an 8-bit microcontroller, presented in the following sections, took about 1s per injected bit flip when running the microcontroller at 1 MHz. This duration is drastically reduced using faster clocks. As a comparison, if the fault injection is performed by software means (using a model of the studied processor), it can take 10 s per injected bit flip.

Once the CEU is injected, a mandatory step is the observation of its effects at the program outputs. This is strongly related to the type of program. For the purpose of this work we will consider that the correctness of a program can be determined by comparing to the expected values (obtained from an execution of the considered program):

- The content of a particular memory area where program outputs were stored,
- The execution time of the whole program.

In summary, the small impact on circuit operation of the CEU approach is one of the major features making this strategy suitable for comprehensive upset simulations. Indeed, the time needed to inject the upset is relatively short (maximum a few tens of clock cycles) and independent of the application complexity. This makes it affordable to iterate as many times as necessary, to derive statistics about the effects of upsets on complex programs. Indeed, by simply reloading (or modifying) the associated memory area with the CEU code corresponding to another target, the system is ready to a new error injection. Nevertheless, two limitations of the CEU approach must be mentioned: (a) upsets occurring during the instruction execution period cannot be simulated, as interruptions are always taken into account at predetermined instants, (b) not all possible upset sensitive targets can be corrupted. The quantification of the CEU target memory area size with respect to the actual total area sensitive to upsets is not straightforward because it needs information available only to the circuit designers or manufacturers. In spite of this limitation, we assume that the complexity of modern processors and their huge internal memory space make the accessible area represent a significant percent of the total sensitive area, thus giving some validity to the results of the proposed fault injection approach.

4. Combining radiation ground testing with fault injection sessions: an example

4.1 Target processor: the 8051 microcontroller

To illustrate the concepts and methods presented in the previous section, we have chosen the 8051 from Intel, which is a simple but powerful 8-bits microcontroller for which radiation ground testing results are available, because it was used in many space equipments.

The 80C51 architecture includes an internal SRAM accessible through LOAD and STORE instructions with standard addressing modes. Half of this memory can be used to store the data, the variables, the stack and, if required, the code of a program. In the other half are mapped the different microcontroller registers. For the purposes of the description that will follow, in Fig. 7 is given a schematic of the occupation of this internal SRAM for a particular case of a matrix multiplication program in which the operands and result matrixes are stored in the first SRAM half.

CEU targets for this processor are: the 128 bytes of internal RAM, the Special Function Registers, the Program Counter and the Instruction registers. As stated before some memory elements potentially sensitive to upsets (i.e. ALU registers, control part flip-flops) cannot be considered by the CEU upset injection approach. In the case of the 8051 microcontroller, available information allowed to estimate the non-accessible targets as representing around 7% of the upset sensitive area. The development of CEU codes for the 8051 sensitive areas is quite straightforward from a deep analysis of the instruction set. Except details about the assembly language syntax, the resulting codes are very close from the ones presented in Tables 2 and 3.

After determination of CEU codes, remaining feature needed to implement the fault injection CEU strategy is the availability of an asynchronous interrupt signal. In the case of the 8051, both maskable and non-maskable interrupt signals exist and can be used to trigger the execution of the CEU code at a desired instant. From a hardware point of view, this entails the implementation of a programmable timer. Before starting the execution of the selected program this timer is initialized to a randomly chosen value by an external controller (test platform) being periodically decremented. When the timer goes to 0, the interrupt signal is asserted, simulating thus the random occurrence of SEUs.

A program was developed to provide worst case conditions in terms of exposing the circuit accessible sensitive area to the effects of SEUs. Indeed, the selected program, a 6x6 matrix multiplication, with both the operand and result matrixes stored within the 128 bytes of the internal SRAM, occupies most of this memory.

As shown in Table 4, the first ten internal memory positions, from 00H to 09H, are used to store the variables of the matrixes multiplication program (basically the matrix dimensions, indexes and addresses). The next positions from 0AH to 051H are used to store the elements of the two operand matrixes, while the results of their multiplication occupy the memory area between addresses 052H and 075H. In addition, when implemented in THESIC daughterboard this application uses directly or indirectly 10 of the 21 special function registers (SFRs), and, obviously, the program counter PC and instruction register (IR). Thus, only 20 bytes (10 SFRs and 10 bytes of internal memory) are not used. Faults injected in them are without any effect on the final program execution result.

The CEU codes corresponding to the different targets were developed, validated and used to explore the effects of transient faults. A particular effort was devoted to extending the strategy to critical registers, such as the program counter (PC) and the stack pointer (SP), thereby simulating most of the possible upsets arising when the 80C51 operates in the final harsh environment.

4.2. Fault injection results and error rate prediction

12245 single CEUs were injected randomly in occurrence and location, while executing the matrix multiplication program. According to the consequences at the program level execution, the injected bit-flips can be classified in the following three types: effect-less, result errors, loss of sequence.

The first class, effect-less CEUs, corresponds, for instance, to those bit flips injected in memory elements whose content is not relevant for the rest of the program execution when the CEU occurs (for instance a register or memory byte not used or which will be written after the bit flip occurrence, thus "erasing" the fault). Injected CEUs leading to result errors are those for which the obtained program results differ from those expected for the program (the result matrix in our case) in at least a single bit. Finally, cases where after fault injection the program duration is different from the expected one are classified in the loss of sequence group. The consequences of CEUs belonging to this last dysfunction type may range from effect-less to unrecoverable, needing in this last case a hardware reset to restart program execution.

Table 4 summarizes the experiment results, giving for each error class the corresponding percentage, with respect to the total number of CEUs provoking it. It can be noted that practically half of the total number of injected faults caused errors on the results of the multiplication program, while only 2.8% of them caused loss-sequence faults. Furthermore, despite the reduced number of bits of SFRs used in this application (88 bits) compared to those used in the internal memory (944 bits), 44.8% of them resulted in lost of sequence. On the other hand, the result errors are caused mainly by CEUs on the internal memory with only 1.5% of them coming from the SFRs.

Finally, if we perform a new CEU injection experiment while running the matrix multiplication program, this time with the operand and results matrixes stored in an external SRAM, the effect-less CEUs rise to 94 %. This shows the potentially small impact of real SEUs and the corresponding overestimation obtained if error rate is derived from ground tests with static test programs. Table 5 summarizes the results of these two CEU injection sessions, putting clearly in evidence the need for both evaluating the SEU sensitivity with programs as close as possible to the final application and taking into account the actual organization (memory occupancy). It is important to recall that in the case of the 80C51 microcontroller, the CEU targets represent approximately 93% of the whole SEU sensitive area, giving to error rates predicted according to the studied approach potential good perspectives to be in good agreement with real error rates.

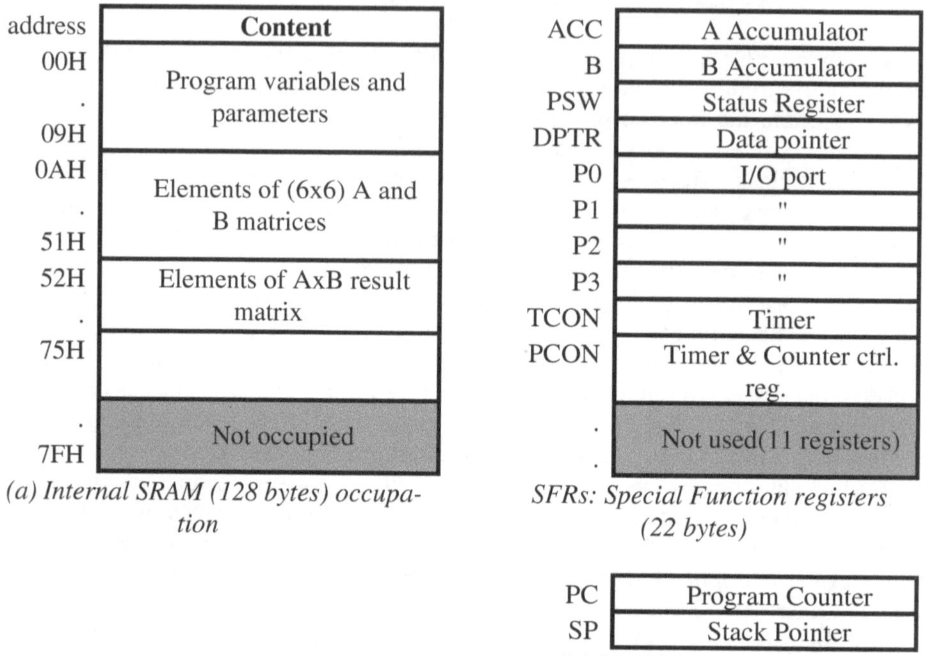

(a) Internal SRAM (128 bytes) occupation

SFRs: Special Function registers (22 bytes)

(b) 80C51 internal registers

Figure 7. Targets of CEU injections for the 8051: the internal memory area and registers.

Table 4. CEU injection experiment results for the 8051.

# INJECTED ERRORS	EFFECT-LESS CEUS	RESULT ERRORS	SEQUENCE LOSS
Internal memory 10780	4890	5700	190
SFRs 1465	1227	84	154
Total 12245	6117 (49.96 %)	5784 (47.24%)	344 (2.8 %)

Table 5. Fault injection results for two different memory occupancy strategies of the matrix multiplication program

TYPE OF ERROR	MATRICES STORED IN INTERNAL SRAM	MATRICES STORED IN EXTERNAL SRAM
No Error	50%	94%
Result Error	47%	4%
Sequence Loss	3%	2%

4.3. Radiation ground testing result

The 8051 included in a THESIC daughterboard was exposed to heavy ion beams by means of the "Cyclone" cyclotron of HIF (Heavy Ion Facility) facility available at Louvain-la-Neuve (Belgium). The hardware set-up is shown in the photo depicted in Fig. 8. The THESIC daughterboard developed for the 8051 included, in addition to the tested microcontroller, memory areas and logic circuits needed to execute the selected programs while exposing the DUT to radiation. Particularly, a common memory area, so called MMI (memory mapped interface) is included to allow the DUT to store the program results and the motherboard to compare them to the expected ones.

Main goals of this radiation ground testing were to measure both the static and dynamic SEU cross-sections for the 8051 microcontroller while executing the matrix multiplication program and to assess the methodology of error rate prediction. To perform these measurements; the 8051 was exposed to different heavy ion beams, while executing the considered programs. First column of Table 6 summarizes the features of used beams. To cover a large spectre of LET for some of the experiments the daughterboard was tilted to simulate beams of higher LETs with a given particle.

The SEU cross-section curves given in Fig. 9 summarize the very good agreement between predicted and measured dynamic cross-section for the 8051 microcontroller. Two important types of information extracted from these curves are the LET threshold and the saturation cross-section. The former, here close to 3 Mev/mg/cm^2, clearly shows the need for carefully assessing for all space environments the sensitivity to SEU of this processor.

As stated before, static and dynamic SEU cross-sections are calculated as the rate of observed errors and number of particles hitting the DUT. This number, the particle fluency, is provided by the cyclotron facility control. A radiation test experiment is stopped only when the error rate is stabilized, i.e. when the observed number of errors divided by the particle fluency is less than a predetermined accuracy. Results of the radiation test campaign performed for the 8051 while executing the matrix multiplication program are summarized in Table 6. As shown in §3.1 predicted SEU error-rates are calculated as the static cross-section multiplied by the error rate derived from CEU fault injection session. The comparison of last two columns of Table 6 clearly shows the accuracy of error-rate predictions: a difference less than 3% between the prediction and the measure was observed for all used heavy ion beams.

Such a good fit between estimated and measured error-rates proves the suitability of the proposed two steps SEU error-rate prediction presented in preceding sections. For more complex processors, the ratio between accessible memory cells, the CEU targets, and total number of memory cells (real SEU targets) will determine the accuracy of error rate predictions. However, it is important to mention that the estimations being based on sessions during which very huge numbers of bit-flip injections are performed, have chances to be closer of the real error-rates than those derived from ground testing results in which a few tens of errors are detected. The beam time cost together with the complexity of architectures are the two reasons making radiation ground testing experiments not often performed in the right way.

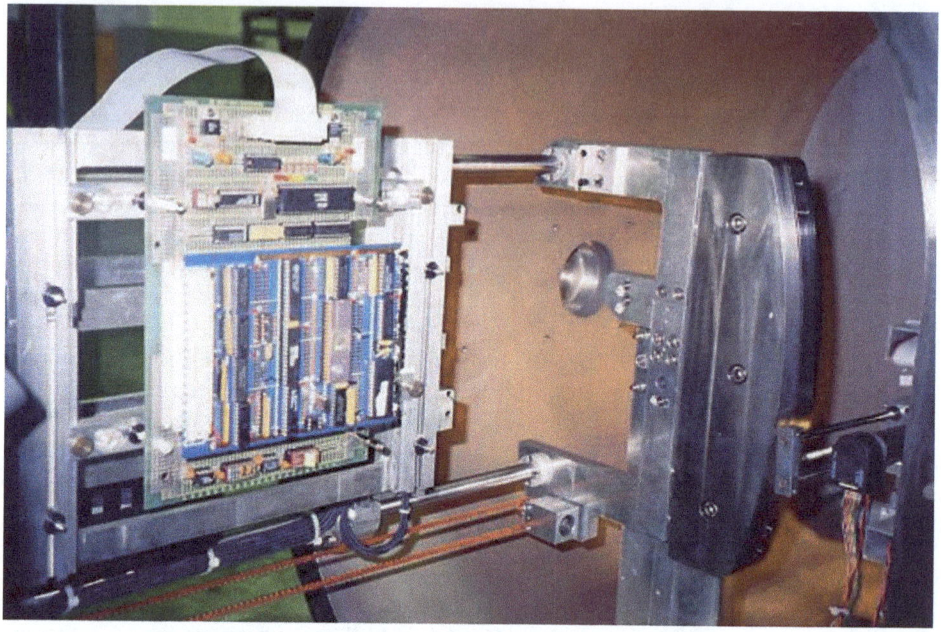

Figure 8. 8051 THESIC daughterboard installed in the vacuum chamber of the cyclotron available at Heavy Ion Facility (HIF, Louvain-La-Neuve, Belgium).

Table 6. Predicted vs. Measured SEU dynamic cross-sections.

PARTICLE BEAM	*EFFECTIVE LET (MeV.cm²/mg)*	*ERROR RATE (cm²/composant)*	
		MEASURED	*PREDICTED*
Nitrogen	2.97	$2.00\ 10^{-6}$	$2.00\ 10^{-6}$
Neon	5.85	$1.02\ 10^{-4}$	$1.55\ 10^{-4}$
Chlorine	12.7	$3.96\ 10^{-4}$	$3.78\ 10^{-4}$
Argon	14.1	$4.50\ 10^{-4}$	$4.33\ 10^{-4}$
Chlorine tilted at 48°	19.5	$6.63\ 10^{-4}$	$6.00\ 10^{-4}$
Chlorine titled at 60°	25.4	$7.13\ 10^{-4}$	$7.55\ 10^{-4}$
Krypton	34	$9.12\ 10^{-4}$	$8.86\ 10^{-4}$
Bromine	40.7	$8.85\ 10^{-4}$	$9.00\ 10^{-4}$

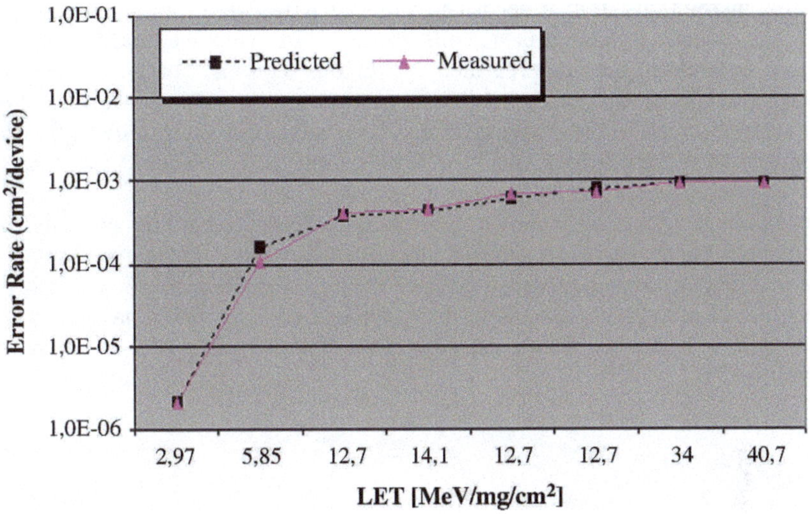

Figure 9. Predicted vs. Measured error-rates for the 8051 while executing the matrix multiplication program.

To get confidence on the studied error-rate prediction strategy accuracy, the two-step strategy has been applied to different processors in the frame of cooperation with space agencies (CNES, INTA, and NASA/JPL). In references [5][6] can be find results obtained for complex processors, 16 bit microcontrollers and digital signal processors using heavy ion beams available in different facilities in Europe and the US. In all these experiments, predictions and measures had the same order of magnitude which is considered as acceptable by the radiation test community. It must be noted that for some of these experiments, the fault injection sessions were performed by means of an instruction set simulator [7] or using a HDL (Hardware Description Language) model of the target processor. Particular care must be taken when developing the fault injection strategy for processors having cache memories. Indeed, for those architectures the significant contribution of the cache memory to the sensitive area makes mandatory to include the cache in the target area. However, injecting faults in the cache using the CEU approach is not an obvious task and requires a very deep knowledge of both the processor architecture and the instruction set capabilities. An example of such an experiment can be found in [8].

5. Dealing with more complex architectures

Up to now, this text deals mainly with the characterization of microprocessors. However, the family of programmable digital devices includes also circuits such as Field

Programmable Gate Arrays. With these devices, the "application program" is not a sequence of instructions, but rather a hardware design. The desired functionality is obtained by connecting together basic cells of the FPGA. Thus the qualification strategy is quite different to the one presented in previous sections. The experimenter has to find simple designs, each of them allowing measuring the upset sensitivity of each and every different cell of the device. Among these cells, one can usually find look-up tables, flip-flops and embedded block of RAM, some new devices embedding also digitally programmable phase locked loops.

But the picture is not complete yet. The space industry is thinking about using reprogrammable FPGAs. For these devices, the connection list is stored in small SRAM cells. As a direct consequence, basic cells of those circuits are not only sensitive to radiations, but also connections among them. Even if some attempts have been made to qualify reprogrammable FPGA, the correct methodology is up to know far to be well established.

References

[1] Bezerra F. et al, "Commercial Processor Single Event Tests", 1997 RADECS Conference Data Workshop Record, pp. 41-46.

[2] R. Velazco, S. Karoui, T. Chapuis, D. Benezech, L. H. Rosier, Heavy ion tests for the 68020 Microprocessor and the 68882 Coprocessor, IEEE Trans. on Nuclear Science, Vol. 39, N° 3, Dec. 1992.

[3] J. H. Elder, J. Osborn, W.A. Kolasinsky, R. Koga, A method for characterizing microprocessor's vulnerability to SEU, IEEE Trans. on Nuclear Science, Vol. 35, N° 6, pp. 1679-1681, Dec. 1988.

[4] Velazco R., Cheynet Ph., Bofill A., Ecoffet R., "THESIC: A Testbed Suitable for the Qualification of Integrated Circuits Devoted to Operate in Harsh Environment", IEEE European Test Workshop (ETW'98), Sitges, Spain, pp. 89-90, May 1998.

[5] Velazco R., Rezgui S., Ecoffet R., "Predicting Error Rate for Microprocessor-Based Digital Architectures through C.E.U. (Code Emulating Upsets) Injection", IEEE Transaction of Nuclear Science, Vol. 47, No. 6, Dec. 2000, pp. 2405-2411

[6] Rezgui S., Velazco R., Ecoffet R., Rodriguez S., Mingo J.R., "Estimating Error Rates in Processor –Based Architectures", IEEE Transaction of Nuclear Science, Vol. 48, No. 5, Oct. 2001, pp. 1680-1687.

[7] R. Velazco, A. Corominas, P. Ferreyra, "Injecting bit flips faults by means of a purely software approach : a case studied", Proc. of 17th IEEE International Symposium Defect and Fault Tolerance on VLSI systems (DFT 2002), Vancouver (Canada), 6-8 Nov. 2002, pp.108-116.

[8] F. Faure, R. Velazco, M. Violante, M. Rebaudengo, and M. Sonza Reorda, "Impact of Data Cache Memory on the Single Event Upset-induced Error Rate of Microprocessors", IEEE Transactions on Nuclear Science, Vol. 50, No. 6, pp. 2101-2106, Dec. 2003.

Using the SEEM Software for Laser SET Testing and Analysis

Vincent Pouget, Pascal Fouillat, Dean Lewis

IXL – Université Bordeaux 1 – UMR CNRS 5818
351 Cours de la Libération – 33405 – Talence - France
pouget@ixl.fr, fouillat@ixl.fr, lewis@ixl.fr

Abstract. This chapter describes the possibilities of a pulsed laser system for studying radiation-induced single-event transients in integrated circuits. Three case studies are presented to illustrate the benefits of the spatial and temporal resolution of the technique. We use a dedicated software tool for analysing the transient responses obtained during laser testing. This software can extract all the information required for sensitivity evaluation or design hardening.

1. Introduction

Interaction of charged particles like heavy ions with the semiconductor material of space embedded integrated circuits leads to the generation of electron-hole pairs that may be separated and collected by the electrodes of a device, inducing a perturbation in the electronic function. The different effects that can result from a single interaction are called single-event effects (SEE). Among these, the single-event transients (SET) correspond to a transient perturbation of a voltage or current. In linear devices, the SET phenomenon is the analog counterpart of the single-event upset (SEU) in digital devices. More generally, the term transient implies that the perturbation is locally self-reparable and that, contrarily to an SEU, an SET does not require a reprogramming or a clock front for the corrupted information to come back to its normal value. However, the transient perturbation may propagate to a sequential logic element where it can be latched into a persistent perturbation if it satisfies certain threshold conditions. Thus, SET is a concern for almost any kind of integrated circuits that include analog functions or combinatorial logic.

In digital devices, SET will mainly appear as a clock frequency dependence of the error rate. In linear devices, the analog nature of an SET makes it more difficult to characterize than a binary SEU. If we consider a simple analog function composed of a small number of transistors, the temporal shape and the amplitude of the transient generated at the output of the function by the impact of a particle can vary considerably with respect to the impacted transistor and the electrical environment of the function. This has several implications in terms of SET sensitivity evaluation and design hardening. In the context of part qualification, one has to make sure that the part is tested in a configuration that is as close as possible from the final application or, in a conservative approach, using a worst case setup. In the context of design hardening, it is important to identify the circuit nodes at the origin of the most critical transients. The laser testing method is useful in both contexts because it allows easy

R. Velazco et al. (eds.), Radiation Effects on Embedded Systems, 259–268.

test setup variations and its spatial resolution provides the information required for analysing the sensitivity of a design.

In this paper, we present several case studies of SET analysis in linear and mixed-signal devices using the pulsed laser technique in conjunction with a dedicated software tool. The first section presents a basic example on the LM124. The next section presents an example of time-resolved analysis on an analog-to-digital converter (ADC). The last section presents a detailed study of the SET signatures generated in a portion of the LM6142 using a dedicated software tool.

2. Laser induced SET

2.1 Laser testing of analog devices

As a first example, we discuss here the characterisation of the SET sensitivity of the LM124, a quadruple operational amplifier widely used in linear systems. A typical application of the LM124 is as a voltage amplifier. The input voltage is maintained constant during the test, which consists of monitoring the transients induced in the output by the laser pulses.

Figures 1-a and 1-b present experimental results obtained respectively with a particle accelerator at the Nuclear Physics Institute in Orsay (France) and with the pulsed laser set-up at the IXL laboratory in Bordeaux [1] with having a wavelength of 800 nm and pulse duration of 1 ps. Comparison between the transient responses obtained with boron ions and with laser light pulses shows that the signals are almost identical. The transient is just one example among the many different waveforms that can be observed when striking this device. The fidelity of the laser technique was verified for the different observed waveforms. This demonstrates the validity of using pulsed laser light to simulate SETs for this circuit.

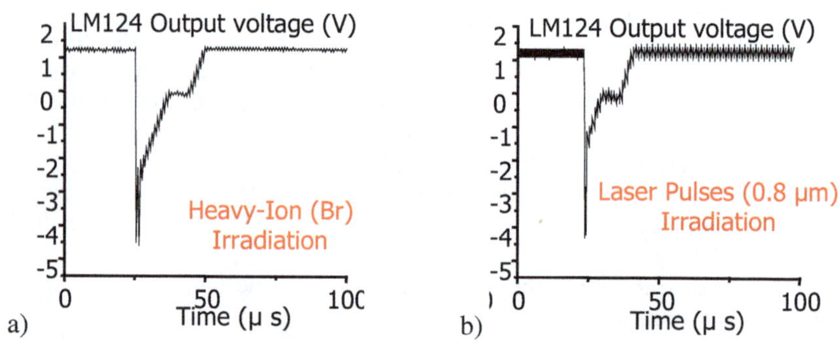

Figure 1. LM124 output transient response induced by a Br ion (a) and a front side laser pulse (b).

One of the main advantages of the laser method is its capacity to provide SET sensitivity mappings. Among the different parameters that can be extracted, we present in Figure 2 the peak-to-peak amplitude of the transient response versus impact position [2]. These figures were obtained by using front-side (Figure 2-a) and backside

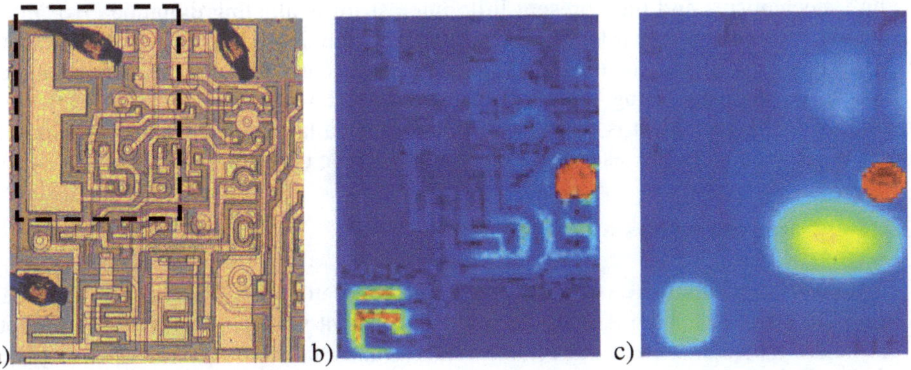

Figure 2. Microphotography of one quadrant of the LM124 (a), corresponding SET amplitude mappings obtained by front side (b) and backside (c) laser testing of the zone delimited by the dashed rectangle.

(Figure 2-a) irradiation. The laser pulse energy was adjusted in the backside mode to induce SETs with the same amplitudes as in the front-side approach. A color map is used to represent the amplitudes, with red points corresponding to the highest values. One can observe the good similitude between both figures, and the occultation by metal interconnections in the front-side mapping. These mappings allow a detailed analysis of the electrical origin of the sensitivity. In particular, after several experiments, the area most sensitive to SETs was clearly identified as a floating base npn transistor in the vicinity of two transistors of the Darlington amplification stage [3].

2.2 Laser testing of mixed signal devices

2.2.1 Laser characterisation of analog-to-digital converters

Since ADCs output codes are composed of several bits coding an analog value, single-event upsets (SEU) in ADCs can be regarded as a digital conversion of single-event transients (SET) in linear devices. Indeed, in most cases, an event can be characterized by its amplitude (in term of digital code) and its duration (in term of conversion cycles). The SEU susceptibility of ADCs is not totally characterized by a single cross-section curve as it is the case for memory devices. The characterization of SEE in ADCs is a complex issue partly because of the variety of electronic functions implied in their design. From the sample-and-hold function to the voltage comparators and the coding logic, there are many different ways for the data to get lost or corrupted in a flash ADC, with different impacts on the system operation. This is also true for other ADC architectures. Although the result is a digital error, an event can be classified either like an SEU or an SET depending on where it takes place in the architecture.

Among the limitations from classical testing approach in particle accelerators, errors histograms, obtained for static input voltages are useful when comparing different devices architectures but they do not always provide a clear insight of the failure mechanisms and they present little interest for evaluating dynamic error rates. To help the designers in hardening their circuits to ionizing particles, there is clearly a need to control both the location and the temporal occurrence of the ionizing impact. This can be done by using the spatial and temporal resolution of the laser testing method [4], [5]. Indeed, it is possible to synchronise a laser pulse with a test clock in order to set accurately the laser pulse arrival time in the conversion cycle.

2.2.2 Laser test results with AD7821

The following results were obtained at the IXL laboratory while testing an 8-bit half flash ADC (AD7821 from Analog Device). A constant voltage is applied to the input of the ADC [6]. After each conversion and laser pulse, the output bits are acquired and the digital error is calculated by comparing them to the expected conversion result. A conversion error map is built by associating each calculated error to the corresponding location of the laser beam in the scan window. The delay between the beginning of a conversion cycle and the triggering of the laser pulse is controlled by a delay generator.

In a flash (or half-flash) ADC, some of the most interesting structures are the parallel comparators that represent the limit between the analog and digital signals. Figure 3 presents the scan window we defined around two comparators as well as the corresponding error maps for three different delays. Each map contains more than 100 000 test points (grid step of 1μm) and took around 30 minutes to complete. The levels of gray represent different error codes defined as the difference between the observed conversion result and the expected value. A top to bottom evolution is clearly observable from one map to another. These maps clearly indicate which parts of the block contain sensitive information at a given time. On the first map (delay of 30ns), an SET sensitivity is located in the input nodes of the comparator. For longer delays, the sensitivity logically appears as SEU in the output latches of the comparators. By integrating the sensitive areas for each delay, it is possible to plot the evolution of the cross-section during the conversion. This is interesting in a design hardening approach to identify the most critical phases of the timing diagram.

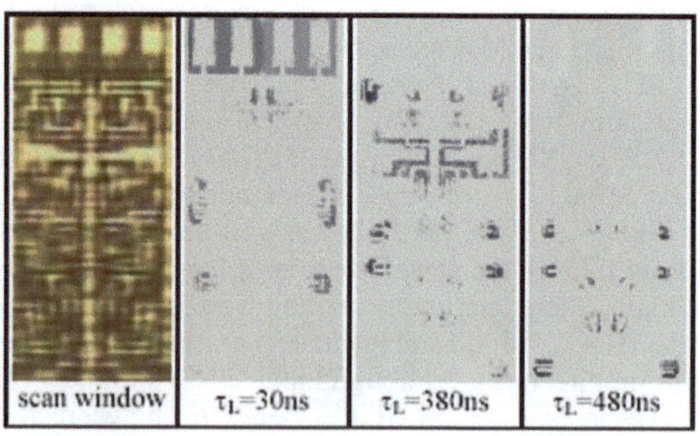

Figure 3. Scan window on two comparators of the AD7821 and corresponding error maps for three laser pulse delays with respect to the beginning of the conversion cycle.

The mappings of Figure 3 provide a unique view of information propagation in the device. Besides the SET analysis, this view may be used for internal signal inspection, reverse engineering, or defect localization. One advantage of this technique is that the resulting image is not filled with non-critical information, as it is the case for potential contrast images obtained with electron beams in which any single metal line is visible. The transient fault injection technique only reveals the zones containing critical information for the functionality of the DUT at a given instant.

3. SET analysis with laser testing and SEEM

In this section, we present an SET study of the LM6142 double operational amplifier using the laser test set-up of the IXL laboratory and a dedicated software tool called SEEM (Single-event effects mapper). This device has been studied in detail elsewhere because it exhibits unusually long SETs [7]. In this paper, we only consider a region of the device that presents a more typical SET response for a linear device. The device is configured as an inverting amplifier with gain 10, polarized with ±10V and tested with a constant input of -60mV.

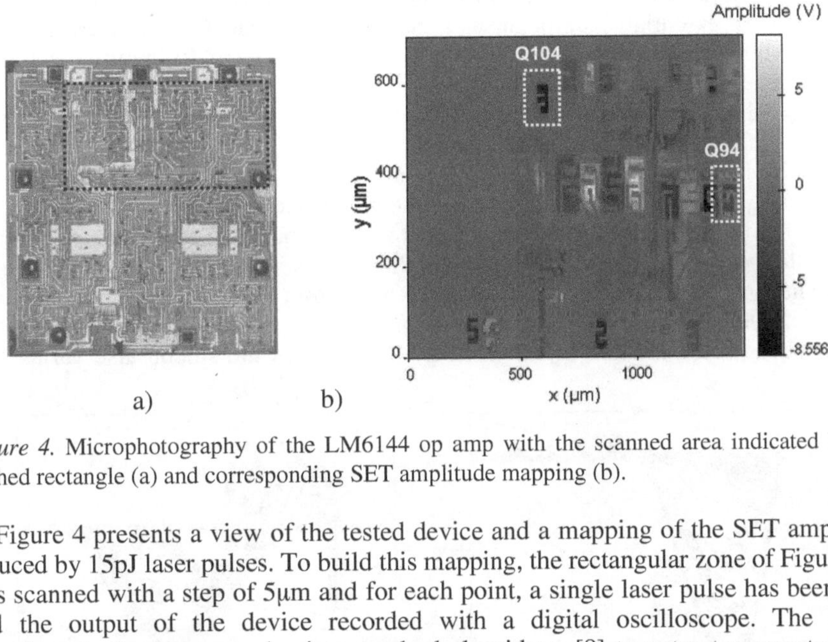

Figure 4. Microphotography of the LM6144 op amp with the scanned area indicated by the dashed rectangle (a) and corresponding SET amplitude mapping (b).

Figure 4 presents a view of the tested device and a mapping of the SET amplitude induced by 15pJ laser pulses. To build this mapping, the rectangular zone of Figure 4-a) was scanned with a step of 5µm and for each point, a single laser pulse has been fired and the output of the device recorded with a digital oscilloscope. The stored waveform is then processed using standard algorithms [8] to extract parameters like the amplitude, the pulse duration… The extracted parameters are finally represented as a function of the impact position. All these tasks are performed automatically by the SEEM software [5]. A light version of the software that includes only the data visualization and analysis portion of the code has been developed. It is named SEEM Reader (Figure 5) and can be provided with the experimental data to users of the IXL laser facility.

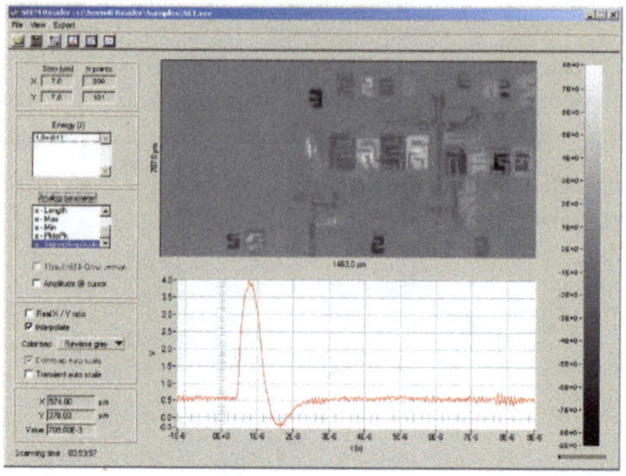

Figure 5. Graphical user interface of the SEEM Reader software.

On Figure 4-b), one can clearly observe different bipolar transistors that induce positive (clear points) or negative (dark points) transients. Two particular transistors, labelled Q104 and Q94, are indicated. Using SEEM Reader, we can plot all the transients generated by laser pulses impacting in the rectangles plotted around these transistors. The result is presented in Figure 6. The family of transients associated with Q104 is essentially unipolar and negative, whereas Q94 generates almost symmetric bipolar transients. These different shapes may have different impact at the system level, depending on the application.

The transients of Figure 6 have different amplitudes varying from zero to a maximum value although they were obtained for a unique value of the laser pulse energy. This is due to the fact that the different points do not have the same efficiency in generating the transient depending on their distance to a participating junction. Thus different amplitudes are observed for different impact locations. Another contribution to the different SET amplitudes is the occultation by metal interconnection layers that prevent the laser light from reaching the semiconductor. Since the level of occultation varies from one point to another in the vicinity of an interconnection, the amount of laser energy that reaches the silicon also varies with

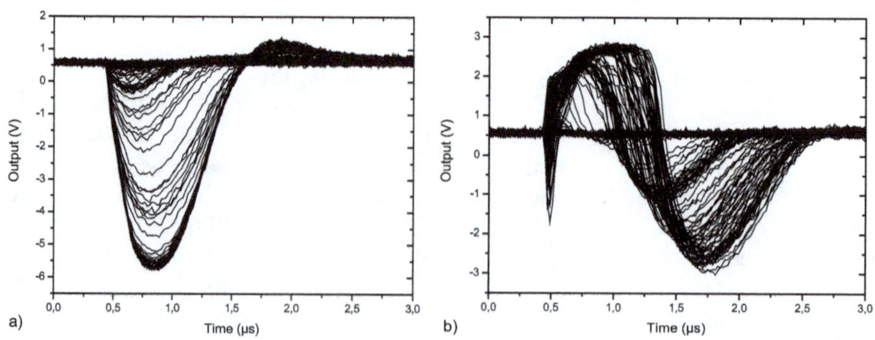

Figure 6. Collections of transients observed in the rectangles around transistors Q104 (a) and Q94(b) indicated in figure 4-b).

the position, leading to different SET amplitudes. For most linear devices that are not integrated with too aggressive technologies, and as long as one does not look for a excellent accuracy in the leading edge of the transient, it is a good approximation to consider that the family of transients observed in Figure 6 represents all the transients shapes that would be observed for laser pulse energies smaller than or equal to the energy actually used. This means that a high energy measurement is sufficient to determine all the possible transients shapes at lower energy. However, this certainly doesn't imply that the amplitude of the transients varies linearly with the laser pulse energy.

Since many transistors will lead to specific transient shapes, it is difficult to have a global view of the sensitivity of a device simply by plotting all the transients waveforms. Moreover, the main technique for preventing SET from propagating in an analog system is based on low-pass filtering. In this context, the detailed transient shape is not necessarily relevant and only the SET spectral power density is of major interest. In a first approximation, if we exclude exotic transient shapes, this can be estimated from the transient duration and amplitude. Plotting these two parameters also provides a convenient way of visualising the SET sensitivity of an analog circuit [9]. The so-called SET diagram generated by SEEM corresponding to the scan of Figure 4 is represented in Figure 7.

Figure 7 allows observing that the maximum SET durations for the scanned region of this device are in the microsecond range. This falls well inside the operating frequency range of this device, which has a bandwidth of 17MHz. This result implies some limitations concerning its utilization at full speed in space.

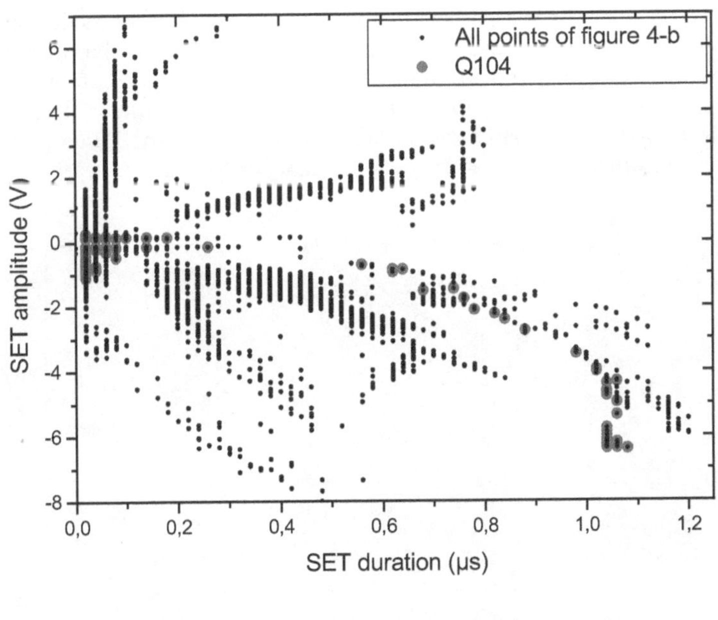

Figure 7. SET diagram representing the transient signed amplitude versus its duration. Each point corresponds to a point of the mapping of Figure 4. Grey circles represent the contribution of Q104.

It appears clearly that the points in Figure 7 are not regularly distributed but rather organized in different groups. Using SEEM, it is possible to make a direct connection between these groups and their origin in the mapping of Figure 4. A group is generally a signature of a limited set of transistors. Reciprocally, one can identify the contribution of a particular transistor to this diagram. The contribution of transistor Q104 is represented in Fig. 7. This transistor is one of the most critical since it is at the origin of the highest absolute amplitudes when considering pulses longer than 1μs. Such information is of particular interest in a design hardening approach.

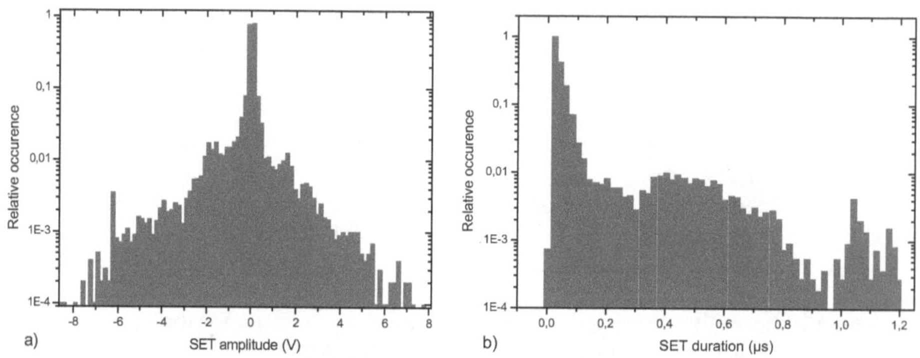

Figure 8. Normalized histograms of the SET amplitudes (a) and durations (b) corresponding to the scan of figure 4, and associated to Figure 7.

Figure 8 presents the normalized histograms or the SET amplitudes and durations. These plots provide an interesting view of the device response that is complementary to Figure 7. Indeed, Figure 7 does not inform on the probability of having a given set of amplitude and duration. On Figure 8, one can observe that many transients have an amplitude of -6V. Also, an SET duration of 1.05μs seems quite probable. These sample results simply illustrates the detailed analysis that can be realized using SEEM software.

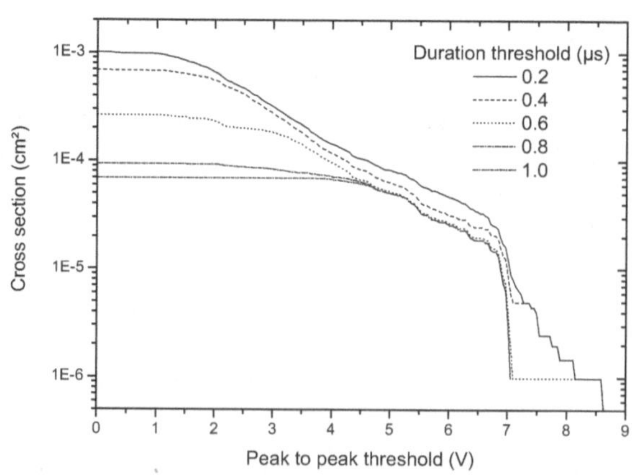

Figure 9. SET cross section as a function of the peak-to-peak amplitude threshold for different minimum duration criteria.

In the end, just as in the case of SEU, the main parameter that needs to be extracted for rate prediction is the SET cross section. Using SEEM, we can calculate the number of transients that satisfies particular threshold conditions in terms of amplitude and duration. In particle accelerators, SET tests are usually performed for one or a couple of fixed thresholds. With SEEM, the transient waveform obtained at every single point in the scan window is recorded whether or not it exceeds a threshold. This means that the data can be processed later and the amplitude and duration threshold used for extracting the cross section can be adjusted as a function of the application.

Figure 9 presents the SET cross section versus the peak-to-peak amplitude threshold for several duration thresholds. These curves correspond to the mapping of Figure 4, i.e. a single energy mapping. They can be obtained for different pulse energies in order to plot the cross section as a function of the laser energy or the equivalent laser LET [10], for a given couple of amplitude and duration thresholds.

4. Conclusions

We have presented the application of the pulsed laser testing technique for analysing the sensitivity of integrated circuits to SETs. Three case studies have been reviewed that illustrate the different resolutions of the laser tool. A dedicated software tool has been developed and can be provided with the experimental data to users of the IXL laser facility. It allows extracting from a laser experiment all the necessary information in order to evaluate quantitatively the SET response of a device and to understand the origins of its sensitivity in the context of a design hardening approach.

References

[1] P. Adell, R. D. Schrimpf, H. J. Barnaby, R. Marec, C. Chatry, P. Calvel, C. Barillot, and O. Mion, "Analysis of single-event transients in analog circuits," *Nuclear Science, IEEE Transactions on*, vol. 47, no. 6, pp. 2616-2623, 2000.

[2] D. Lewis, V. Pouget, F. Beaudoin, P. Perdu, H. Lapuyade, P. Fouillat, and A. Touboul, "Backside laser testing of ICs for SET sensitivity evaluation," *Nuclear Science, IEEE Transactions on*, vol. 48, no. 6, pp. 2193-2201, 2001.

[3] A. L. Sternberg, L. W. Massengill, R. D. Schrimpf, Y. Boulghassoul, H. J. Barnaby, S. Buchner, R. L. Pease, and J. W. Howard, "Effect of amplifier parameters on single-event transients in an inverting operational amplifier," *Nuclear Science, IEEE Transactions on*, vol. 49, no. 3, pp. 1496-1501, 2002.

[4] S. P. Buchner, T. J. Meehan, A. B. Campbell, K. A. Clark, and D. McMorrow, "Characterization of single-event upsets in a flash analog-to-digital converter (AD9058)," *Nuclear Science, IEEE Transactions on*, vol. 47, no. 6, pp. 2358-2364, 2000.

[5] V. Pouget, D. Lewis, and P. Fouillat, "Time-resolved scanning of integrated circuits with a pulsed laser: application to transient fault injection in an ADC," *Instrumentation and Measurement, IEEE Transactions on*, vol. 53, no. 4, pp. 1227-1231, 2004.

[6] S. Buchner, A. B. Campbell, A. Sternberg, L. Massengill, D. McMorrow, and C. Dyer, "Validity of using a fixed analog input for evaluating the SEU sensitivity of a flash analog-to-digital converter," *Nuclear Science, IEEE Transactions on*, vol. 52, no. 1, pp. 462-467, 2005.

[7] Y. Boulghassoul, S. Buchner, D. McMorrow, V. Pouget, L. W. Massengill, P. Fouillat, W. T. Holman, C. Poivey, J. W. Howard, M. Savage, and M. C. Maher, "Investigation of millisecond-long analog single-event transients in the LM6144 op amp," *Nuclear Science, IEEE Transactions on*, vol. 51, no. 6, pp. 3529-3536, 2004.

[8] "IEEE Standard 181-2003 on Transitions, Pulses and Related Waveforms," July1, 2003.

[9] S. Buchner, J. Howard, Jr., C. Poivey, D. McMorrow, and R. Pease, "Pulsed-laser testing methodology for single event transients in linear devices," *Nuclear Science, IEEE Transactions on*, vol. 51, no. 6, pp. 3716-3722, 2004.

[10] V. Pouget, H. Lapuyade, P. Fouillat, D. Lewis, and S. Buchner, "Theoretical Investigation of an Equivalent Laser LET," *Microelectronics Reliability*, vol. 41, no. 9-10, pp. 1513-1518, 2001.

INDEX

FSC
www.fsc.org
MIX
Papier | Fördert
gute Waldnutzung
FSC® C083411

Kolibri 360 GmbH
Ferdinand-Jühlke-Straße 7
99095 Erfurt, Deutschland
produktsicherheit@kolibri360.de